Tagore the Eternal Seeker: Footprints of a World Traveller

Tagore the Eternal Seeker: Footprints of a World Traveller

Editors
Suryakanthi Tripathi
Radha Chakravarty

Associate Editor
Nivedita Ray

Vij Books India Pvt Ltd
New Delhi (India)

Indian Council of World Affairs
Sapru House, New Delhi

Published by

Vij Books India Pvt Ltd
(Publishers, Distributors & Importers)
2/19, Ansari Road
Delhi – 110 002
Phones: 91-11-43596460, 91-11-47340674
Fax: 91-11-47340674
e-mail: vijbooks@rediffmail.com
web : www.vijbooks.com

First Published in India in 2015

ISBN: 978-93-82652-95-3

Price : ₹ 1650/-

Printed in India
at Aegean Offset Printers

Contents

Foreword

Gurudev Rabindranath Tagore is one of the greatest luminaries of modern times, who continues to evoke immense respect and admiration all over the world. He is not only a poet, writer, philosopher, painter, musician and educationist, but also perhaps one of the most able and astute goodwill ambassadors of modern India, who through his travels, acquainted the world about its vast and rich cultural ethos and outlined a vision of unity of mankind.

Tagore's fascination for travel was a significant aspect of his multifaceted personality. He travelled extensively, visiting about 34 countries on five continents between 1878 and 1932. These travels left a deep impact on him. It broadened his universal sensibility and global outlook. His interactions with other cultures allowed him to integrate the best aspects within his self and his institution *Visva Bharati*. Debates and discussions with the brightest intellects and creative personalities, on issues of philosophy, politics and aesthetics, helped Tagore in shaping his ideas of internationalism. His mission was to synthesize the East and the West. He was India's greatest cultural envoy. Very aptly Jawaharlal Nehru describes him in his book *The Discovery of India:* 'He (Tagore) has been India's internationalist par excellence, believing and working for international cooperation, taking India's message to other countries and bringing their message to his own people.'

Tagore had the ability to think through the horizon of the future, to imagine India and the world's future possibilities in terms at once universalist, as well as rooted in the particularities of myriad cultures. He was a great votary of universal brotherhood and an advocate of a global order where man is not restricted by the narrow boundaries of nationalism. He immensely contributed to the making of modern India and inspired leaders such as Jawaharlal Nehru and Sarojini Naidu and many others of that generation and of subsequent generations.

Nehru's pan-Asianism, articulated through the Asian Relations Conference and the subsequent creation of Non-aligned Movement, following the Bandung Conference, bears the mark of Tagore's thought. Nehru was greatly inspired by Tagore's idea of Asian Unity. Tagore's ideas also provided significant inputs to the policy of decolonisation, racial equality, civilisational commonalities, unity of the decolonised, re-engagement with Asia and the 'Panchsheel' or the five principles of co-existence. Similarly South-South cooperation, universal education and tolerance of dissent, draw sustenance from Tagore's philosophy.

Tagore's aspirations for free India and his vision of global and Asian unity contributed to shaping the tenets and ethos of the foreign policy of independent India. This is a facet of Tagore that has not been extensively explored yet. In this context, therefore, the initiation of a book project on the theme, 'Tagore - Envoy of India: His Vision of India and World', by the Indian Council of World Affairs (ICWA), assumes immense significance. As part of the book project, ICWA organised an international conference on 9 and 10 May 2013 to debate, discuss and reflect upon his thoughts and vision about India and the world and examine their contemporary relevance, especially in the global domain.

This volume, an outcome of this project, explores Tagore's travels and his influence on the ethos of Indian foreign policy. It examines how he conveyed the essential values of India to the world and assesses the extent to which his journeys influenced his own views of the Universal Man and his world view. With contributions from leading Tagore experts both from India and abroad, this compilation encompasses an array of perspectives on Tagore's thoughts and vision, which enables us to re-read Tagore and introspect their relevance to unfolding developments in the 21st century.

Ambassador Rajiv Bhatia
Director General
Indian Council of World Affairs

Preface

Rabindranath Tagore has been the theme of innumerable conferences. In India and abroad, people have got together to discuss his poetry, his painting and philosophy, and other facets of his remarkable personality and genius. Rarely, if ever, however, was a conference held on his extensive international travels, how they influenced his work and world view, and, in turn, how the world heard him passionately voice the concerns of the colonized and the oppressed.

The international conference organized by the Indian Council of World Affairs on 'Rabindranath Tagore - Envoy of India: His Vision of India and World' from 9-10 May, 2013 in New Delhi, therefore, was particularly significant. Through an exploration of Tagore's travels, it drew attention to his espousal of humanism and universalism as the path to a peaceful and harmonious relationship among countries and peoples.

According to Humayun Kabir, Rabindranath Tagore was the first great Indian in recent times, who went out on a cultural mission for restoring contacts and establishing friendship with peoples of other countries. Travel not only broadened his selfhood, it also contributed to his philosophy of internationalism and the development of his institution, Visva-Bharati. He believed that inclusivism and synergic interaction between cultures would propel the world towards harmony and global fellowship. His association with great minds such as William Butler Yeats, Romain Rolland and Albert Einstein, to name a few, helped him to forge personal and intellectual bonds abroad that reinforced the idea of international brotherhood. Within his internationalism, Tagore had a clear vision of how India should stand in the community of nations.

For this conference, the ICWA had invited scholars from the several countries that Tagore had travelled to in order to discuss his global legacy as an envoy of India, and to highlight his vision of independent India in a peaceful and unified world. How was he influenced by the culture and traditions of the countries he visited? How did his thoughts impact the overseas audiences? How did these voyages contribute to his philosophy of internationalism, universalism, humanism, spiritual unity and the religion of man? While evaluating Tagore's legacy to the world, the discussions also covered the influence of his ideas upon the evolution of Indian foreign policy, and Tagore's contemporary relevance. This is the commemorative volume of the conference, and it also includes the papers of a few authors who were unable to personally present them at the conference.

Some of Tagore's earliest journeys were to England and Europe. His reception in the British Press is described in Kalyan Kundu's article, while Supriya Roy offers a well-documented account of his relationship with France. The Italian angle is explored by Stefano Beggiora and Kalyan Kundu, while Philippe Falisse, Martin Kämpchen and Annemette Karpen offer accounts of his travels to Belgium, Germany and Denmark, respectively. The essays by Ana Jelnikar, Martin Hirbek, Imre Bangha, Alexander Shurbanov, Andreas L. Katonis and Demetrios Th. Vassiliadis offer a fascinating account of Tagore's reception in Eastern Europe, including Czechoslovakia, Hungary, Bulgaria, Slovenia and Greece. In his provocative essays, Sergei Serebriany tracks changing responses to Tagore in Russia, arguing for the need to re-introduce his works to an audience that has all but forgotten him.

Tagore made several journeys to America. Sugata Bose analyses Tagore's engagements with America with reference to poetry, painting, politics and philosophy. Gustavo Canzobre, in his illustrated essay, describes Tagore's relationship with Victoria Ocampo as a turning point in Argentina's attitude towards the East.

Tagore was convinced that Asia would lead the world in the years to come. He believed that India's ties with East Asia were crucial for world peace. He also had a deep interest in the civilizational past of countries like China, Japan, Korea, Indonesia and Thailand. Tagore's journeys to China, Japan, Indonesia and Thailand are discussed in the essays by Sreemati Chakravarty, Dong Youchen, Yukifumi Takeuchi, Gde Dwitya Arief Metera and Sawitree Charoenpong. Tagore's travels to the Middle East have not received enough critical attention. The essays by Ehsan Abbaslou and Mohammad Ajam present vital information about his journey to Iran. Tagore was also deeply aware of India's links with her immediate neighbours. Reba Som's essay dwells on the rich cultural exchanges that emerged from his visits to Sri Lanka.

While most of the essays in this collection focus on specific places visited by Tagore and the impact of these journeys, a few take a more general line, exploring Tagore's worldview and his contribution to international relations. In her perceptive essay, 'Tagore – an International Relations Perspective Finding a Home Everywhere in this World', Suryakanthi Tripathi dwells on the potential as well as the challenges implicit in Tagore's dual position as an international-national personality. Partha Ghosh's article traces Tagore's footprints on Indian foreign policy. Alok Bhalla describes Tagore's lifelong internal struggle between a dark vision of humanity and a conviction that human life has a divine purpose. Radha Chakravarty's article dwells on the implications of Tagore's role as cultural envoy. Two of the essays, by Sandra Chistolini and Kathleen O'Connell, elaborate on Tagore's vision as an educationist. Alim-ur-Rahman Khan's essay adds a spiritual dimension to our understanding of Tagore's vision of the Unity of Being, and his ability to cross the mental boundaries that divide people from people.

Taken together, the essays in this collection will, it is hoped, open new windows to our understanding of Tagore's legacy in the field of international relations, and add another dimension to his stature as a world personality.

Suryakanthi Tripathi

Radha Chakravarty

Contributors

1. **Alexander Shurbanov** has taught English literature, stylistics and translation at Sofia University for over forty years, as well as Bulgarian language and literature at British and the American Universities. He has also translated a large selection of Rabindranath Tagore's poetry.

2. **Alim ur-Rahman Khan** taught Islamic History (Art & Architecture and Philosophy & Mysticism) at the Central Women's University, Abhoy Das Lane, Tikatuly (Bangladesh). Currently he is involved actively as Chairperson of Sadhana, an institution to encourage and highlight the classical music and dance of the sub-continent. He is also engaged with lecturing on Islamic philosophy and mysticism in universities and on the TV.

3. **Alok Bhalla** was Visiting Professor at Jamia Millia Islamia and professor at the English and Foreign Languages University (former CIEFL). He is also on the Executive Council of the Sahitya Akademi. He has been a Lady Davis Visiting Professor, Hebrew University, Jerusalem (2007), a Fellow at the Rockefeller Centre, Bellagio, Italy and at the Indian Institute of Advanced Study, Shimla.

4. **Ana Jelnikar** was born in Slovenia and educated in Ljubljana and London. She holds a PhD from the School of Oriental and African Studies (SOAS), University of London. She is one of the foremost translators of contemporary Slovenian poetry into English. Jenlikar has translated the first Slovenian edition of C. G. Jung's *Man and His Symbols,* as well as seven collections of contemporary Slovenian poetry.

5. **Andreas L. Katonis** is an Associate Professor in the Department of Italian Philology of the Aristotle University of Thessaloniki (Greece). He is a member of ELINEPA, i.e. the Indo-Hellenic Society of Culture & Development. He graduated in Classical Philology (Ancient Greek and Latin) obtaining the title 'Lecturer of Greek and Latin'.

6. **Annemette Karpen** has been associated with teaching and directing dance and drama since 1977. She has been writing on South Asian dance and drama in *Den store danske encyklopædi* (The Big Danish Encyclopedia) and in *Gyldendal teaterleksikon* (Gyldendal's Theatre Lexicon). She has staged several of Tagore's plays and has been giving lectures on Tagore and his works.

7. **Demetrios Th. Vassiliades**, a Greek Indologist, teaches Hindi and Sanskrit at the School of Foreign Languages, University of Athens (2005 to present) and is Chairperson of the Indo-Hellenic Society for Culture and Development. He received his M.A. and Ph.D. from the Dept. of Philosophy & Religion at Banaras Hindu University.

8. **Dong Yochen** is a professor at the Center for South Asia Studies, Beijing Foreign Studies University and the Vice Director of the Literature - History College of the Party School of Central Committee CP of China. At present he is holding the position of Professor Emeritus at the Party School of Central Committee CP of China He is known for his Chinese translations of the works of Rabindranath Tagore.

9. **Ehsan Abbaslou** is head of English Department at Imam Khomeni Higher Educational Centre. He is also Head of the Translation Committee to Planning Research Institute and Chairman of judging Panel in field of Foreign Literature.

10. **Gde Dwitya Arief Metera** graduated from Center for Religious and Cross-cultural Studies, Gadjah Mada University (Indonesia). His research interests include cultural transformation, Bali studies, popular culture as well as religion and the state in Indonesia. Currently he is the editor for etnohistori.org, a website for publishing research draft on Indonesian ethnography and history by young Indonesian researchers.

11. **Ghada Ahmed Nabil Mahmoud Mostafa** has been Deputy editor-in-chief and columnist and head of the weekly cultural page in the daily newspaper *Al Joumhorya*. She was also founding member of 'Ghadab' - the movement of 'Anger Poets' founded after the 25th January 2011 Egyptian revolution.

12. **Gustavo Canzobre** is Director of Philosophy Teachers School of Fundación Hastinapura, Buenos Aires, Argentina, since 1991. He was co-founder of *Editorial Hastinapura* in 1983, and director up to 1991. Argentine Indologist since 1980, he has been studying Indian Art and Philosophy, lecturing and giving seminars on Indian Classical Culture.

13. **Imre Bangha** is a Lecturer in Hindi at the University of Oxford and the Head of the Alexander Csoma de Kőrös Centre at the Hungarian University of Transylvania, Romania. He holds a Ph.D. from Visva-Bharati and his publications include Hungarian translations from various South Asian languages.

14. **Kalyan Kundu** has been a specialist in Rabindranath Tagore for over thirty years. He has a Ph.D. in Botany from universities in both Kolkata and London. In London, whilst pursuing Biomedical Research on Muscular Dystrophy at the Institute of Neurology, he became interested in the work, philosophy and life of Rabindranath Tagore.

15. **Kathleen M. O'Connell** teaches courses on South Asia at New College, University of Toronto, Canada. Her interest in Rabindranath Tagore and the Bengali cultural and literary tradition began when she did her M.A. in Comparative Literature from Jadavpur University, Kolkata.

16. **Martin Hirbek** is a Senior Lecturer in Bengali and Indian Studies at the Institute of South and Central Asia, Faculty of Arts, Charles University in Prague. Trained in both ethnology and modern indology, his doctoral research was on Durgapuja festival in contemporary Calcutta.

17. **Martin Kämpchen** is a writer on India and is a revered Tagore scholar who has translated Tagore's biography from Bengali to German, and has done extensive research on Tagore's visits to Germany. He lives at Santiniketan.

18. **Mohammad Ajam** is a well-known scholar in Iran. Many of his research papers and books have been published. He is an ex-university professor and Minister Counselor for Legal Affairs and Education in the Embassy of Iran.

19. **Partha S. Ghosh** is a Senior Fellow at the Nehru Memorial Museum and Library, New Delhi. He was the Professor of South Asian Studies at the School of International Studies, JNU, and the ICCR Chair on Indian Studies at Victoria University of Wellington, New Zealand.

20. **Philippe Falisse** was the former Counsellor (culture) at the Embassy of Belgium, India. He studied the Bengali language and literature and passed the senior diploma in music at Rabindra Bharati University, Kolkata.

21. **Radha Chakravarty** is a critic and translator. She teaches English Literature in Gargi College, University of Delhi. She has co-edited *The Essential Tagore* (Harvard and Visva Bharati), a major anthology of Tagore's works, nominated Book of the Year 2011 by Martha Nussbaum.

22. **Reba Som** received her Ph.D from Calcutta University. She has been the recipient of several fellowships from the Indian Council for Historical Research (ICHR), Indian Council for Social Science Research (ICSSR) and was Jawaharlal Nehru Fellow in 2000-2002. Reba Som is also a trained singer of Rabindra sangeet. She has published English translations of Tagore's lyrics.

23. **Sandra Chistolini**, holds a full professorship in General and Social Pedagogy at the University of Rome Tre, Italy. She is responsible for General Pedagogy, Citizenship Education, Pedagogy of Communication and Vocational Training for student, teachers and social workers.

24. **Sawitree Charoenpong** is an Assistant Professor at Department of History, Faculty of Arts, Chulalongkorn University, Thailand. Earlier she was Director of Chulalongkron University Memorial Hall, and Head of the Department of History, Faculty of Arts in the same University. Her areas of interests are World Civilization, Indian Civilization, South-Asian History, and American History.

25. **Sergei Serebriany** is attached to the Russian State University for the Humanities (Moscow). From 1970-1992, he worked at Gorky Institute of World Literature (Moscow), doing research in the field of Indian and comparative literature.

26. **Sreemati Chakrabarti** is Professor of Chinese Studies at the East Asia Department of University of Delhi and Honorary Fellow of the Institute of Chinese Studies, Delhi. She has a Ph.D in Political Science from Columbia University and holds Master's degrees from Delhi University (Political Science) and Harvard University (Regional Studies – East Asia).

27. **Stefano Beggiora** is Researcher in History of India and lecturer in Anthropology (Indian Tribal Religions and Society) in the Department of Asian and North African Studies at the University Ca' Foscari Venice (Italy), where he received his Ph.D. in 2006. He conducted extensive fieldwork missions and specializes in South-Asian shamanism.

28. **Sugata Bose** is the Gardiner Professor of History at Harvard University. He has served as Director of Graduate Studies in History at Harvard and as the Founding Director of Harvard's South Asia Initiative (now Institute). He was educated at Presidency College, Calcutta, and the University of Cambridge where he obtained his Ph.D.

29. **Supriya Roy** has worked in Rabindra-Bhavana, the Tagore Memorial Museum, Archives and Research Centre of Visva-Bharati, Santiniketan, as a librarian and archivist for three decades. She has been the content writer of the Multimedia Encyclopedia on Tagore and series editor for the Tagore Travelogues for Visva-Bharati.

30. **Suryakanthi Tripathi** retired from the Indian Foreign Service, with her last assignment being as India's Ambassador to Spain. In Delhi, she was posted in the Ministry of Commerce, headed the Southern, Economic Relations, SAARC and Policy Planning Divisions of the Ministry of External Affairs, and as Director General of the Indian Council for Cultural Relations. She runs a small firm, Nasadiya Arts, which undertakes website and graphic design work.

31. **Yukifumi Takeuchi** is a Visiting Scholar at the Reischauer Center for East Asian Studies of the U.S. Johns Hopkins University. He worked at the Asahi Shimbun as the head of the New Delhi office and also served on the editorial board before leaving the company in March 2011. He is a visiting professor at the Center for South Asian Studies at Gifu Women's University.

Tagore – A Perspective of International Relations: Finding a Home Everywhere in the World

Suryakanthi Tripathi

Rabindranath Tagore was an intuitive philosopher. He was a seeker after truth, an idealist. He was not a power broker, a hard-headed negotiator of advantage, or engrossed in the minute analysis of the words and clauses of international agreements. Obviously, therefore, one should not expect conventional or convenient foreign policy prescriptions from him. However, given the innumerable and unending number of foreign policy disasters that the world can lay claim to, perhaps, in the perceptive wisdom and humanism of Tagore lies the answer to more productive, peaceful and harmonious relationship among countries and peoples. Of course, that again is too simplistic for a complex world.

For his time, Tagore was an Indian, who travelled abroad extensively. In his speeches and letters, during and after every trip, some fundamental beliefs of his become more evident, and, one may say, even to himself. And, they influenced Tagore's subsequent ideas in literature, as well as his work pertaining to education, social responsibility and international interactions.

Tagore derived much of his insight and inspiration from Bengal, his native land. To the external world, as an Indian, he represented the wisdom of an ancient civilisation. Then, as an individual, he identified himself with all of mankind. It becomes apparent that, as both an Indian and a humanist, he realized the imperative of feeling and being local and global at the same time. He was a Bengali, he was an Indian, but he was universal.

Rabindranath Tagore's vision for his own country was that her doors should be opened and none turned away from 'the shore of this vast ocean of humanity that is India' His song '*Jana Gana Mana Adhinayaka*', now India's national anthem, resonated with that same universality he considered sacred, and to which one had to awaken.

To understand any aspect of Tagore's life and work, one has to recognize the importance he attached to thinking and living in freedom. If everyone was free to exist in such freedom, then one would find a home anywhere in this world. Only through harmony, through an acceptance of diverse cultures, would the whole world become a single shelter. He himself did not find such a home in his lifetime. The world, of course, never considered such a quest as even remotely important.

Rabindranath Tagore died in 1941 at the age of eighty, a somewhat disappointed man. It was before India became independent and when Indians were still struggling for freedom, still longing for light. It was also the period that the world war had so bloodied with its unspeakable atrocities, negating every concept of freedom and universal humanity.

Tagore has been called an international nationalist. However, even for him, this was not always an easy or seamless reconciliation. Although an ardent protagonist of international peace and understanding, he did have to contend with the challenging international developments of the first half of the twentieth century. Further, Tagore's pan-Asian approach, his views on contact and conflict between the east and the west, particularly in the context of India's own freedom struggle, presented difficult philosophical contradictions.

To live as this dual international-national personality required, to use Tagore's own phrase in another context, a very high level 'of judgment and stamina'. It also required the flexibility to constantly contour one's responses to changing and conflicting situations within the country and abroad, while adhering firmly to one's principles.

Tagore's views were unambiguous and clearly expressed. Victories of war, according to him, were victories of armed might that would be unfruitful and only sow seeds of fresh dissensions. He was opposed to unbridled and militaristic nationalism, political aggressiveness, commercialism driven by an uncontrolled desire for profits that crushed the creative capabilities of the people, and the love of mechanisation for its own sake.

Nationalism and Patriotism

To Tagore, the freedom of the mind was of elemental import, irrespective of the situation an individual found himself in, owing to historical and geographic circumstance. To be free was also not merely to free oneself, but to be free to cast off the chains of all others. Another core affirmation of his was that no one should ever make the mistake of identifying the people with the state.

It was, however, a time that India was in the throes of its freedom struggle. As a Gandhian from Tagore's region said in 1928, it was the time of the 'sacred sacrificial rites' for freedom.

Tagore was one whose insistence on the need for Indian independence could never be doubted. As usual, however, Tagore was not articulating ideas in the manner of most leaders of the Indian independence struggle of the time. For him, it was the duty of people to struggle for their self-respect and for their freedom. However, if their nationalism meant being engrossed in their own claims, their own culture and heritage, to the exclusion of the heritage of others, then it could jeopardize the achieving of true freedom, a freedom that reached beyond all borders.

Nationalism implied working with power and that would not usher in either peace or cooperation. It would limit freedoms, lead to decreased harmony between the individual and the society, the society and the state, and between nations. The use of power led to misunderstanding, anger suppressed or hate expressed violently. Leading to the use of more power. Thus, by its very impulse, nationalism based on power was diminished in its legitimacy.

Tagore was an exemplar to whom a narrow definition of patriotism meant enhanced barriers to the world outside. In his extensive writings, it is evident that he was absolutely against internal divisions of caste, religion and class. His assertion, at all times, was universal and undivided. Only then would every individual be protected and participate in the great story of humanity, in the drama of life and liberty.

Much the same was said later in the Universal Declaration of Human rights, where the inalienable rights of all members of the human family is the foundation of freedom, justice and peace in the world. And that all human beings are born free and equal in dignity and rights. They are endowed with reason and conscience and should act towards one another in a spirit of brotherhood.

In his poem, *Sunset of the Century*, written on the last day of the nineteenth century, he said that the hungry self of a nation would burst in a violence of fury 'from its own shameless feeding, for it has made the world its food.'

He could never allow nationalism or patriotism to triumph over humanism, while he was ever committed to serving his country. In his novel, *The Home and the World*, Tagore's sentiment is unambiguous: 'I am willing to serve my country, but my worship I reserve for Right, which is far greater than my country.' He could worship only what was right for in that lay compassionate humanism.

Even in India, at that time, he did not have many takers. In August 1925, the *Ananda Bazaar Patrika*, a Bengali newspaper dismissed Tagore's stand and said that his views could appeal to those who lived in a dream world, but not 'those who are grounded in the soil of this country and know its realities'.

As a result, Indians were proud of Rabindranath Tagore as a great poet, a cultural ambassador of India, a literary genius that the world had recognized. However, in their view, India needed more practical prescriptions to obtain independence. They, hence, saw in the exchanges between Tagore and Gandhi the greater promise of freedom from colonialism in the latter's approach.

East and West

Tagore was initially prone to be limited by definitions – as he was in the case of the east-west balancing. The east was the east and the other was different, much as Rudyard Kipling, who won the Nobel Prize six years earlier than Tagore, had said with a tinge of colonial superiority.

But Tagore's definite intellectual east-west juxtaposition had to change. Tagore agreed that neither the east nor the west could be complete in itself. One was engrossed in the objective seeking of the smallest element, of prediction and control through science. The other in the deeply subjective appreciation of the indivisible whole through philosophy and spirituality, of becoming truly free, as the poet said, by merging our will with the world dharma.

But the east-west equilibrium, he found, was in reality vastly different. In the East, he had already been acknowledged as an Asian who had earned the admiration of the West. It was wartime and the aggression of one nation against another was at its height. Tagore decided in 1929 to set out on a voyage to Asia and during that trip, he also went to Japan, a country he had visited thrice before. On this visit, he was very disconcerted to find a rising imperialism, which he later referred to as 'essentially a case of borrowed pugnacity'. The Japanese, for their part, did not quite see eye-to-eye with him on Asian brotherhood, and viewed his anti-nationalism as passivity and not acceptable during wartime.

Writing to Rash Behari Bose, the Bengali revolutionary leader, Tagore had said that he 'did once admire and look up to Japan and did once fondly hope that in Japan, Asia had at last discovered its challenge to the West.' Japan, for him, had betrayed that hope. As a result, to his great disillusionment, Tagore's own view of the East also had to undergo a significant change.

He had admired the power of European intellectual traditions of social good, and their modern and rational organizational forms. He was greatly disappointed that, in its imperial ambitions, Britain had shed this aspect of its western civilisation. In a lecture on his last birthday, called *Crisis in Civilisation*, he says India had gained from 'the large-hearted liberalism of nineteenth-century English politics.' The tragedy was that this large–heartedness was not to be extended to the administration of colonised India. What 'was truly best in their own civilisation, the upholding of dignity of human relationships, has no place in the British administration of this country.' 'If in its place they have established, baton in hand, a reign of 'law and order,' or in other words a policeman's rule.' It was a mockery of civilisation that could claim no respect from him.

His travels and his contacts with writers, philosophers, artists, as well as the common people were extensive. To his delight, he met eminent men of science and men of mysticism in the west, and they were all relevant and true. Europe too had a soul, he realized, a spiritual longing, a centre that was not just materialistic, scientific, or purely of the intellect, but also came alive by joyous instinct.

These connections also provided Tagore with a deeper understanding, revealing to him that the east and the west were not essentially all that distinct and did have insights that both could share. As he said in the Gitanjali, the traveller has to knock at every door to come to his own, and one has to wander through the outer worlds to reach the true shrine. But, in that journey, every stop and path had something to offer.

He had done that, travelled to the heart of each continent. Both the scientific west and the philosophical east would together lead to unity of understanding. For in that was veritable freedom. It was in the synthesis of both that human advance lay. In this process, Tagore was also able to invite many to that arena of Indian contemplation and experience that had seemed either strange or daunting to the non-Indian mind. Though not explicitly stated, it could also have been Tagore's belief that India, with its millennial philosophical traditions, could provide the impetus for a new global coming together.

Particularly for the West, India in those days was only a colonised land and, therefore, one which did not merit much attention. As with Swami Vivekananda before him, it was also Tagore who showed the west that India did have a level of interlocutors in art, architecture, literature and philosophy that could add new energy to the quest of the intellect.

The earth is spherical and wisdom is universal, and Tagore showed us that in the search for knowledge, one can go east and west at the same time.

International Relations

In foreign relations, our immediate neighbourhood is a vital area, in every respect. Then there is the extended neighbourhood, one that defines the larger geographical, economic and strategic context. The country's relations with the major world powers, regions and regional groupings also have a determining say in foreign policy approaches. Then, there are the international organisations and the multilateral fora.

Each of these interactions, whether bilateral or multilateral, affects a country's present or long-term security, peace, development and growth. These exchanges could lead to not only great benefit, but also to tension if not conflict. Protecting one's self-interest takes priority in this landscape of dense and shifting vectors, and the voice of humanism is generally not heard. The voice of power is.

In international relations, power, not philanthropy, makes the moves. Power that is military or economic, or derived from size or resources. Hence, every country wishes to maximise its own power and safeguard itself from the muscular manoeuvring of another. Thus balance-of-power came into the common parlance of international politics.

It is not that the urge for the greater goal of humanism does not show itself. It does, mainly in the preambles to agreements that, sadly, few read. There are these lofty aims, constituting the salute that international protocol makes to philosophy, the bowing before the wrestling begins. The fact is that noble aims make for the underpinning to any cooperative arrangement. Unfortunately, the functional text could well be quite another document altogether, where every word is bitterly fought over and no party concedes a tenth of a centimetre. Nationalism does remain the sole basis for organizing international relations, an approach that Tagore, as a great patriot, had warned against.

Even as the real salutes the ideal, the ideal must accept the constraints of the real. Tagore was conscious of the imperative of the real even as he longed for the ideal. In 1926, in his first interview to the press in Greece, he did say, 'Every nation has its individualism and its mission, but I cannot imagine mankind divided in groups. For me, humanity constitutes a single unit.'

It was this spirit that actually did inspire independent India to a certain extent – to acknowledge differences but to seek a greater unity among the world's peoples. Once colonialism had ended, India decided to set aside past rancor, start afresh and have equitable relations with Britain. Pandit Jawaharlal Nehru, the first Prime Minister of independent India, had always acknowledged the influence of Rabindranath Tagore that he described was like the 'coming of the dawn in the mountains, it crept on us and permeated us'. In Nehru's non-alignment, where newly-independent nations could opt out of factions that were shearing the world, and speak unitedly for moderation, there was again inherent the wisdom of Tagore.

The world, or some part of it, is almost always in some degree of turbulence. Hence, in its international relations, each country does seek alliances, zones of security and cooperation arrangements that can assure and enhance its own well-being. Even in relatively hostile relations, maturity dictates that we try to set aside the issues of dissent to identify and pursue common interests that do exist, so that a conflict does not arise or escalate. For example, a long-standing and contentious border issue, while it is not abandoned, is not allowed to dominate or derail a larger bilateral agenda when, say, trade ties are of growing importance and need to be safeguarded. Further, each country does defend itself through a mesh of arrangements, resulting in bilateral cooperation agreements, agreements between countries that are contiguous, countries that share a sub-continent, a continent, a hemisphere or an ocean.

In his letter that Tagore wrote from Santiniketan to Pandit Nehru on 17 August 1939, talking of the Asian tradition of cultural exchange, he says that we have not fought with each other in the name of a hungry nationalism as the western countries have been doing in Europe. About India, he goes on to say that 'the pathway which led from India to her great neighbours is partly closed through centuries of neglect; we have to remove the weeds and also the recent barriers erected by fratricidal politics so that once more the traffic of human interchange can continue ... India's great awakening had crossed deserts and mountains, the overflow of her glorious epoch of culture touched far continents and left permanent deposits in distant shores of Asia.'

Half a century later, the world began talking of 'soft power', but again it was a concept energized by the idea of exercise of power and not necessarily by the humanism of Tagore. Half a century later, as well, India began talking of a Look East Policy in the hope of reviving a centuries' old tradition of the friendship of peoples and their unfettered contact.

Tagore went on to urge Nehru, calling him a messenger from India's youth, to give strength to the historic forces of Asiatic unity. To Tagore, India was herself passing through an eclipse when her own reality was lost to her in a haze of parochial politics, sectarianism and domestic contention. He was hopeful, that contact with the greater world of eastern culture would also help India enliven her national existence with a new humanity.

Whether it is the Ganga-Mekong Cooperation that allies a group of countries along two great civilisational rivers, the Indian Ocean Rim grouping bringing together countries with shared oceanic shores, IBSA that finds cooperation benefits among India, Brazil and South Africa, three countries of the southern hemisphere located in different continents – such examples that we can refer to as far as India is concerned, are innumerable at the global level. Not with staggering success perhaps, and slower than most of us would like, they constitute, nevertheless, the real working towards the ideal, despite constraints. They represent the sanguinity inherent in good foreign policy.

Seeking to find national good in the common good of the people in other lands, they aim towards the universal. The poet would have approved.

As Tagore did say, 'Not hammer-strokes, it is the dance of the water that sings the pebbles to perfection.'

All the Air is Vibrant

Tagore once said, 'With the tune of thee and me all the air is vibrant.'

When the tune of either, of thee or of me, is missing, the air is static and lethargic. There is no lyric and there is no song. We are always being asked to choose, to choose between nationalism and humanism, between our own well-being, the imperative of one's own country, and what is morally right, the global good. We are reluctant and apprehensive about moving beyond the needs and security of our own countries. We, therefore, find it difficult to reach out to the neighbourhood, the region and the larger world. Self-interest takes precedence over everything else. There are innumerable fears and forces that induce the building of fortifications, of walls of apprehension.

It is, after all, a world inhabited by practical people. Between the mine and the universal remains the chasm. Tagore wanted to reach a stage where there were no such divisions, where the whole world was one single nest, '*vishvam ekanidam*', both in mind and action. To Tagore, a feverish nationalism was certainly not the route.

Finding, achieving and abiding by the human and the true – therein lies the challenge in international relations. But miracles, they take long to happen. That 'heaven of freedom', the world is yet to awake to.

It is not that the urge for humanism is completely absent in those, who work in foreign ministries and in areas connected with international exchanges. Agreed, this humanism is not present in them in the same abundance that the great poet had. And, moreover, even the little that there is tends to get buried so far down that it does not guide the script in actual bilateral and multilateral proposals or negotiations. As a result, power and practical rationality get to lay down the agenda.

It is, therefore, that people in the field become foreign policy practitioners and not foreign policy philosophers. The world is not yet a single nest and humanity is not a single family. Foreign policy works to make friends. But it does not, cannot easily give up on borders, boundaries and barricades – both physical and mental – and only allows some gates to be opened selectively to facilitate transactions and contacts.

Humanism is mankind's great inspiration. Above all, this goal requires explicit and universal acceptance. It requires ceaseless and converging effort to ensure that there is nothing unequal, insensitive or forced in the relations among different peoples, countries and continents.

People across the globe can and should rightly expect the world's governments to set out on such a path and work towards achieving this common purpose. If we do succeed, as the poet has said, one is in possession of that which 'is broadly human and deeply true and can be safely shipped to distant times and remote countries'.

Nation, Nationalism and Internationalism: Footprints of Tagore on India's Foreign Policy

Partha S. Ghosh[1]

Introduction

The first Asian to be awarded the Nobel Prize in Literature, in 1913, Rabindranath Tagore was born in 1861. Till 1941, when he died, he wrote on almost every topic, although he is most remembered for his contribution to Bengali literature in the form of poetry, plays, novels, short stories and songs. He is probably the only poet and musician in the world who wrote and put to tune the national anthems of two countries, India and Bangladesh, and composed the tune of yet another, that of Sri Lanka.[2] He was also an essayist par excellence as well as a prolific letter writer. In those writings, one may discern his thinking about India as a nation and its place in the world. The same was also reflected in some of his political novels. His collected works run into 18,000 printed pages in Bengali (Guha 2010: 185). Although he did not write specifically on India's foreign policy, as he had no reason to do, since India was still a part of the British Empire when his end came, several major contours of India's foreign policy bear his imprint, as we will see in this paper. He was a nationalist with a difference, just like India's foreign policy, as it was conceived in the beginning, was also a policy with a difference. It was a departure from the conventional understanding of foreign policy, particularly for a new nation.

The Trinity

The trinity of India's freedom struggle consisted of Tagore (1861-1941), Mahatma Gandhi (1869-1948) and Jawaharlal Nehru (1889-1964). All three had considerable international exposure, Tagore having had the maximum. His education was primarily at home. He joined a school in England but soon left it to resume self-education. But, besides his home education, he educated himself through his various foreign trips. He visited thirty-four countries, which notably included Japan, United States, China, Russia and several countries of Europe, South East Asia, and Latin America. His Indianness, therefore, was homespun yet tempered by his experiences abroad. Compared to him, Gandhi's Indianness developed during his long exile that served as the 'nursery of nationality'. After living abroad for twenty years, Gandhi returned to India at the age of forty-six to soon lead the nationalist movement of India (Natarajan 2013:21). Nehru had his schooling and law education in England's elite institutions and was thoroughly familiar with the liberal educational system of the country.

All three profoundly contributed to India's foreign policy in the early phase. Nehru was indeed its chief architect but the influence of the remaining two, particularly that of Tagore, as well as other leaders, permeated through its essentials. The reason was that while Gandhi had his ideas fixed on independence with social justice closely integrated to it, Tagore's ideas were free-flowing as could be expected of a litterateur and social thinker. His relationship with Gandhi was, therefore, one of disagreement, for example over non-cooperation movement and the use of the spinning wheel (charkha), yet of deep respect for each other (Bhattacharya 1997). Tagore's imprints on India's foreign policy were noticeable as we will see here, although they can be deciphered only by careful scanning. His influence on the policies had to be indirect because he died six years before India's Independence. In this paper we will see, first, the broad contours of India's foreign policy as they were articulated by Nehru, and second, deal with two neighbouring nations, Pakistan and China, which consumed much of India's attention immediately after independence and have continued to do so.

Essentials of India's Foreign Policy

Preservation and promotion of national interest is the essence of all foreign policies. But where national experiences differ is the way those ideas are articulated and projected. The three pillars of India's foreign policy, as they were conceptualized in the early years of India's independence under the leadership of its first Prime Minister Jawaharlal Nehru, were non-alignment, peaceful coexistence of all nations, and solidarity with the nations newly emerging from colonial bondage. Since the process of decolonization was set in motion with India's independence, these theories were of immense relevance for all nations. This was the time when, on the one hand, colonization was coming to an end while, on the other, new grounds for conflict were being created in the shape of a Cold War between the United States-led capitalist world and the Soviet Union-led communist world

In India's reckoning, its policy of non-alignment and that of support for the Afro-Asian decolonization process were closely linked. Even prior to independence, India had hosted the Asian Relations Conference in New Delhi (April 1947) and, a few years later, had played a leading role at the Bandung Afro-Asian Conference (April 1955). In both these conferences, India expressed the hope to establish, and committed itself to strive for, a group of states which would stay away from the US-Soviet rivalry and work out its own cooperative development strategy, recognizing that the problems faced by these newly-liberated nations were largely identical. Though the phrase 'Third World' was yet to capture popular imagination, its building blocks were already in place. However, the systematic deterioration in the US-Soviet relations and the unabated spread of the Cold War seriously threatened these efforts. Leave alone Afro-Asian solidarity, even intra-regional cooperation within the South Asian region became a victim of the Cold War machinations, with Pakistan increasingly coming into the orbit of US global strategy of entangling alliances to counter Soviet influence.

Non-alignment, as a policy, is often misconstrued as something limited to being hands-off Cold War rivalries. It was meant to be a movement of all those states, a sort of Third World trade unionism, to bargain from the dominant bi-polar system without subscribing to either in particular. In essence, its constituents were asking for freedom of action in global affairs in pursuit of individual national goals. As far as India was concerned, it was inherent in its cultural background which was ever argumentative (to borrow Amartya Sen's phrase) and never static. The only un-changeability was its changeability itself. It must be underlined that though it was a long-drawn-out nationalist movement of about six decades against British imperialism, it was never against the British as a people or their intellectual attainments. Tagore belonged to this genre, just as Gandhi or Nehru or almost everybody at the helm of the movement did. In this regard, Tagore's transformation, from an uncritical admirer of the British heritage to an Indian nationalist questioning certain aspects of British imperial ways, is noteworthy.

Nationalism versus Patriotism

Nationalism and territoriality are the warp and woof of the fabric of foreign policy. Nationalism is often interchangeably used for patriotism. But both Tagore and Gandhi tended to disagree on this common connotation. In one of his writings, Ashis Nandy makes a fine distinction between the two through his understanding of the philosophies of these two modern thinkers of India. This strain in Tagore's political thought was discernible during his visit to Japan in 1916, the first country that he visited after winning the Nobel Prize. Tagore was a darling with the Japanese and he had met them in their country more than once. Japanese interest in his writings is evidenced by the fact that, in the 1980s, a twelve-volume *Works of Tagore* in Japanese was published. Compared to this, the translation of Mahatma Gandhi's writings into Japanese has been rather sporadic (Yamaguchi 2006: 6).

Ashis Nandy argues that both Gandhi and Tagore were patriots, not nationalists — patriotism is a non-specific sentiment while nationalism is an ideology. Since Gandhi never visited Japan, as Tagore did, the experience of the latter in Japan is interesting. After Tagore won the Nobel Prize, it catapulted him into a pan-Asian hero. His 1916 Japan visit evoked considerable enthusiasm amongst his hosts. In the midst of the First World War, an imperially-ambitious Japan was expecting a further boost to its concept of nationalism from this great poet hailing from an Asian country then under British tutelage. But Tagore's lectures proved to be a disappointment to his Japanese audience, if not an outright embarrassment. 'What is dangerous for Japan', said the poet, was 'not the imitation of the outer features of the west, but acceptance of the motive force of the western nationalism as her own.' It was a clear indictment of Japanese imperialism. No wonder the Japanese considered Tagore a pain in the neck for they were basking, at that time, in their newfound imperial glory. Tagore was given the most unceremonious sendoff — just one individual, his host, came to see him off, and because this could not have been avoided anyway. (Nandy 2006: 1, also see Mehta 2012, Engelmeier 2009: 175). Like Tagore, Gandhi also disapproved of Japan's West-inspired imperial adventurism and hence, his natural sympathy was for the Chinese, who were the victims of that adventurism. During the Quit India movement, Gandhi expressed his desire to be sent to Japan by the Government of India once India became independent in order to persuade the Japanese to free China (Yamaguchi 2006: 6).

Even in India, many did not approve of this Tagorian strand, more so because they were confused. While his patriotic writings and songs inspired scores of Indian nationalists, his novels, *Gora* (1909), *Ghare Baire* (1916) and *Char Adhyay* (1934) 'were seen as direct attacks on hard-edged, masculine nationalism' (Nandy 2006: 2). The following quotation from Nandy is instructive:

> This is not as inconsistent as it looks. There are two clues to how Tagore reconciled his two selves in his works themselves. First, in his Bengali writings, Tagore used something like twelve to fifteen expressions to denote one's love for one's country ranging from 'deshabhiman' and 'swadeshprem' to 'deshbhakti' and 'swadeshchetana'. But he used none of them as a synonym or translation of the word 'nationalism'. When he meant nationalism he used the English word 'nationalism' in Bengali script to distinguish it from the first set of words. Tagore was a patriot but not nationalist. He thought there was nothing in common between the territoriality associated with the various vernacular concepts of patriotism and the new idea of territoriality grounded in the idea of nation state and ideology of nationalism. I suspect that he thought the former to be tied to the idea of home and the latter to an artificial concoction that looked instrumentally at the former and, indeed, was often built on the ruins of the former. Indeed, nationalism was partly a reaction to a sense of being uprooted or rendered homeless. (Nandy 2006: 2).

It is not difficult to find connections between the fine distinction that Tagore made between patriotism and nationalism, on the one hand, and Nehru's concept of non-alignment and nation-building strategies, on the other. Nehru's emphasis on secularism as a state policy was a patriotic move which was forcefully articulated in the teeth of all odds and against the background of millions of deaths and displacements that the Partition had brought in its train. For Nehru, whosoever chose to remain in India, truncated though it had become, deserved state protection as full citizens of India. Thus, although he could not prevent Pakistan from being created on the basis of the two-nation theory, he steadfastly clung to the idea that Kashmir should not be allowed to secede from India, because on the existence of this only Muslim-majority state depended so critically the future of secularism in India, and hence the Union of India. While this was true as far as the nation-building outlook was concerned, he did not display any motive that even remotely could suggest a policy of territorial aggrandizement.

Nationalism to Internationalism

In foreign policy terms there is an umbilical connection between nationalism and internationalism. Dialectically if there is no nationalism, there cannot be any internationalism. All three, Tagore, Gandhi and Nehru, were diehard internationalists and it was possible because they were diehard patriots as well. While all three had many grievances and anxieties in respect of the manner in which the West looked at India and its quest for freedom, none of them nurtured any hatred for it. There was no difference between Tagore's idea of setting up a global knowledge centre called Visva-Bharati and the way Nehru conceived of the essential contours of India's foreign policy in which respect for the United Nations and its strengthening mattered a lot. It was probably this faith in the

UN system that drove Nehru to take the Kashmir problem to the United Nations, which was ideally not necessary.

Pragmatic Approach

Tagore had a love-hate approach towards the British. While on the one hand, like Gandhi, he made a distinction between the British as India's imperial rulers and the British as a people, on the other, he was impressed by their achievements in the fields of science and literature. Compared to other imperial rulers, Tagore found them less cruel. No wonder that none of his writings was proscribed by the British government, while many contemporary Bengali writings by other authors were (Kar 2009: 103). Of course, he was spied upon and all those who even indirectly associated themselves with him had to face the ire of the British secret police and were sometimes jailed. There could be two reasons for this. One, after he became a Nobel Laureate in 1913, any proscription of his writing could have embarrassed the British. Two, Tagore knew the limits to his patriotic expositions. Moreover, he was opposed to the terrorist fringe of the nationalist movement. This was revealed when Saratchandra Chatterjee's novel, *Pather Dabi* (The Claim of the Road), was banned in 1927 for its seditious content. Hoping that Tagore would come out in his support, Saratchandra approached him with a copy of his book for him to read. Tagore not only did not oblige him, he even politely admonished him to desist from such adventures, to Saratchandra's great dismay. Here is what Tagore wrote to Saratchandra in his letter:

> The book is stimulating. It attempts to provoke the readers against the British government. As a duty of a writer, it might not seem to be an offence. If the writer feels that it needs to be condemned, he cannot remain a silent spectator. However, he should be ready to face the consequence of voicing his opinion. There is no valour or dignity in thinking that we are free to criticize the British because they are tolerant and forgiving. The experience I have gathered after my visits to various countries prompts me to say that no ruling power except for the British accepts anti-government activities or criticism so patiently (Kar 2009: 256-59).

India's joining the British-sponsored Commonwealth of Nations after independence was in tune with this love-hate approach to the British. Even otherwise India's foreign policy was pragmatist. Indeed, no foreign policy can be pursued in black and white terms. India was non-aligned yet, whether it was in the Hungarian revolt of 1956 or the Chinese invasion of India in 1962, India could not strictly adhere to the tenets of its policy of nonalignment. Further, although India was committed to democracy, it did not mind seeking help from Stalin's Soviet government to suppress the Communist uprising in Telengana in the early years of India's independence (Mehra 2008: 2). India's non-alignment itself was rooted in India's politics, as this author has elsewhere explained (Ghosh 1978). The pragmatist strain in India's foreign policy was noticed even prior to India's independence when Gandhi maintained a considered silence at the news of the atomic bombing of Hiroshima and Nagasaki in August 1945, thinking that it might unnecessarily ruffle American feathers, something that was not desirable at that final stage of India's freedom struggle. Gandhi spoke on the subject only a year later when the dust had settled and it was almost certain that the British would leave (Rothermund 2009: 107).

Meeting the Pakistani Challenge

India's relations with Pakistan have been murky from the beginning. Three wars (1947, 1965, 1971), twenty years of low-intensity conflict in Kashmir (since 1989), the Kargil half-war (1999), innumerable violations of the LOC by both sides during the past decade, and over and above, the activities of Pakistan-sponsored terrorists on Indian soil (recall the attack on the J&K Assembly on 1 October 2001, the attack on the Indian Parliament on 13 December 2001, and the Mumbai massacre during 26-29 November 2008 carried out by sea-borne terrorists from Pakistan, just to mention a few), have made both the countries develop a militaristic mind-set about each other. Not that important civil society initiatives are not being taken to change the trajectory, such as dialogues at the Track II level, mutual trust-building initiatives by newspaper groups (TOI-Jung *Aman ki Asha*, meaning Hope of Peace), or the regular interactions between the two countries' intellectuals under the auspices of the Pakistan India People's Forum for Peace and Democracy (PIPFPD). Nevertheless, all these efforts come to naught at the slightest indication of a military face-off, which the electronic media takes no time to flare up. In their scramble to improve their ratings vis-à-vis other channels they indulge in mutual one-upmanship to project themselves as greater nationalists.

There is no way of knowing what Tagore would have thought of Pakistan because the country did not exist during his life time, although a year before his death the Pakistan Resolution had been passed in the Lahore Conference of the Muslim League. But the manner in which he had conceptualized Indian nationhood in civilizational terms, where all the races and religions were to co-exist, (see the Section on Approach to Refugees and Migrants) indicates that he would not have approved the Partition of India on the basis of a two-nation theory.

Relations with China

One of the important foreign policy challenges for India soon after its independence was how to deal with an emergent China where a long civil war had ended in the victory of the Communists just two years after India's freedom. In the first few years, the relations between the two countries were cordial. India even pleaded for the replacement of Nationalist China by the Communist China as the legitimate member of the United Nations Security Council. Intellectual and cultural exchanges were many and the Chinese were particularly keen to learn from the Indian planning processes and the use of statistics in those matters. Chinese and Indian statisticians exchanged several visits and the doyen of Indian statistical sciences, Prafulla Chandra Mahalanobis, was held in high esteem in China. But the respective countries' attitudes towards the borders that demarcated them and the occupation of Tibet by the Chinese in early 1950s came in the way of their bonhomie. Throughout the fifties, it was a relationship of tension which culminated in 1959 when the Dalai Lama, the spiritual and temporal head of Tibet, fled from Tibet to take shelter in India. China suspected that he did so under Indian prodding. The deteriorating relationship came to a head when China attacked India in 1962. India's total defeat in this short war, which resulted in the loss of 23,200 sq km of territory to China, bruised India's ego irreparably and is said to have caused the death of Nehru within two years of the war.

Relations with China have indeed improved over the years, though mutual suspicions persist. In the immediate aftermath of the war, 'Hindi-Chini Bhai Bhai' (India and China are brothers) was replaced by 'Hindi-Chini Bye Bye'. Soon China's military aid to Pakistan started and it was not just a coincidence that, in 1965, Pakistan inflicted a war on India. Meanwhile, in 1964, China became a nuclear power and, in 1968, it joined the NPT. From 1969 onwards, there were indications of a Sino-US rapprochement, which culminated in the visit of President Richard Nixon to China in 1972. There were simultaneously some improvements in Sino-Indian relations. Given the texture of the relationship, it was important that Chairman Mao had smiled at the Indian envoy at a Beijing party (the famous 'Mao smile' of 1970). It may be noted that it happened against the background of 1967 Indian decision to make the North East Frontier Agency (NEFA) a full Indian state called Arunachal Pradesh, which China had been claiming all these years. In 1976, full diplomatic relations between the two countries were restored. Under the visionary leadership of Deng Xiaoping, improvement in the relationship continued apace and soon it was evident that Chinese aid to the North East insurgency had come to an end. Between 1981 and 1986, the complex border negotiations took place and, through the Samdurong Chu military stand-off of 1986-87, it was clear that India's armed preparedness had come of age. It was not the India of 1962. In December 1988, Rajiv Gandhi visited China and soon several visits were exchanged. In the Kargil war, China's neutrality was noticed and, in 2006, the Nathu-la Pass was reopened to facilitate bilateral border trade.

In spite of some conflicting indications over the years, China is now India's biggest trading partner. In 2011, bilateral trade was to the tune of $73 billion and, by 2015, it is poised to grow to $100 billion. What is worrisome, however, is that, in commodity terms, the trade is lopsided and there is also a trade imbalance. Both the countries are aware of their difficulties and want to strike a balance. They have a Strategic and Cooperative Partnership for Peace and Prosperity and an agreed 'Shared Vision for the 21st Century'. In the Shanghai Cooperation Organization (SCO), India has an observer status just like China's status in the South Asian Association for Regional Cooperation (SAARC). Increasingly there is some coordination in multilateral fora on issues such as climate change and trade talks.

It is not expected of Tagore that he would have anticipated all that has happened in India-China relations since 1947. When he died, China was in the midst of a civil war as well as an international war in which it was on the side of the Allied Powers. Whatever official dealings Indians had was with the Chiang Kai-shek government. In 1938, Indians had sent Dr. Dwarkanath Shantaram Kotnis to China to provide medical assistance during the Second Sino-Japanese War. Besides his becoming famous for his service to the battle-wounded, this humanitarian gesture by the Indians was regarded as a fine example of Sino-Indian friendship and collaboration. Tagore had anticipated the importance that India should pay to knowing China for it was almost during this time that, in April 1937, he had established the famous *Cheena Bhavana* (meaning the China House, Institute of Chinese Language and Culture) at Santiniketan. Tagore invited Prof. Tan Yunshan to serve as the first chairperson of the institute. International scholars like Jan Yun-hua worked there on topics ranging from Sino-Indian studies, Buddhism and Chinese philosophy. Chiang Kai-shek and, after the Communist takeover, Zhou En-lai donated a large number of Chinese books to the institute to make it one of the most important libraries for classical Chinese studies in India. In one of his letters to Nehru, Tagore wrote on 28 March 1937:

> The huge library together with a fund of Rs. 50,000 is a gift from the Chinese people to India and it would be unfortunate if we do not see it in its proper perspective. The Sino-Indian Cultural Society, which has sponsored this move, has amongst its organizers all the leaders of Chinese life including Marshal Chiang Kai-shek, President Dr. Tsai Ti-tao and the Director of the Chinese National Academy of Researchers. It is incumbent on us to receive the gift in a proper spirit of friendship and cooperation and the formal inauguration of the work of the Society should be done in a manner which will readily impress our Chinese friends that India will worthily reciprocate this beautiful gesture. I cannot think of anybody better fitted than you for the opening ceremony (Nehru 1960: 226).

Approach to Refugees and Migrants

From the beginning, it has been India's avowed policy to welcome all migrants and refugees in distress seeking shelter in India. To keep the policy as flexible as possible, India has deliberately avoided signing the international Refugee Convention of 1951 and the Refugee Protocol of 1967 which smacked of Cold War trappings. Whether it was the 10 million Hindu and Sikh refugees from Pakistan after the Partition of the subcontinent in 1947, or the 10 million Bengali refugees from East Pakistan in the wake of the unprecedented repression unleashed by the Pakistani military junta to suppress the freedom struggle of East Bengalis in 1970-71, or the arrival of hundreds of thousands of Tibetan and Sri Lankan Tamil refugees in the 1950s and 1980s, respectively, or the arrival of Hindu, Sikh and even Muslim refugees from Afghanistan, India has always been welcoming.

In Tagore's thought, this hospitality dimension of India's foreign policy is clearly noticeable. Historically, people from all over the world, most notably from India's extended neighbourhood, have come to India and made it their home. Tagore had poetically expressed this phenomenon in his well-known poem, *Bharat Teertho* ('India, the Pilgrimage'), in the following words in Bengali:

Keho nahi jane kaar ahobane kato manusher dhara

Durbar shrote elo kotha hote, samudre holo haara.

Hethaye arjo, hetha anarjo, hethaye dravid cheen—

Shauk, hun dal pathan mogole ek dehe holo leen.

Poschime aaji khuliachhe daar, setha hote sobe aaney upahaar,

Dibe aar nibe, milabe milibe, jabe naa phire—

Ei bharater mahamanaber sagarteerey.

Its free translation in English reads:

> No one knows to whose clarion call
> Peoples responded and came
> In streams
> And joined the ocean!
> Hither lived the Aryans, the non-Aryans,
> The Dravidians, the Chinese, the Scythians,
> The Huns, the Pathans, the Mughals —
> And merged into one body and soul.
> Today, doors to the West are open
> New ideas are flowing,
> Incessant giving and taking,
> Meeting and mingling are on,
> They will not return
> From this supreme Ocean of Humanity
> That is India.

In his seminal essay, 'History of India', written in 1902, Tagore talked of India in civilizational terms where all sorts of invaders had come but most of them never left. On the contrary, over the years, they got 'Indianized' and contributed to the country's art and culture. Tagore wrote: 'This new India belongs to humanity. What right have *we* to say who shall and who shall not find a place therein? Who is this "We"? Bengali, Marathi or Punjabi, Hindu or Mussalman? Only the larger "We" in whom all these — Hindu, Muslim and Englishman, and whosoever else there is — may eventually unite, shall have the right to dictate who is to remain and who is to leave' (Guha 2010: 188-89). According to historian, Sekhar Bandyopadhyay:

> It was in this assimilative power of Indian civilization that India could claim her difference from Europe, because here the idea of 'India' developed more as an inclusive civilizational community, rather than as a political-territorial state. Indeed, the whole of India was never politically unified until the late seventeen century, as the various regional groups continued to maintain their political autonomy and local patriotism, and continuously fought with each other. That dismisses that 'political' history of territorial warfare as a dark history of nightmares. He privileges instead the other more inclusive history of social assimilation which internalized even the invading outsiders. It was in this assimilative spirit, Tagore, argued, that the true essence of Indianness, her national identity was to be found. This essence could not be rationally defined or described, but like life in a body, it needed to be felt or experienced (Bandyopadhyay 2009: xxi-xxii).

Conclusion

There was a basic difference between Tagore and Gandhi. While Tagore used to play with ideas, Gandhi's concern was people. Tagore himself had realized it when he said that Gandhi is 'a lover of men not of ideas' (Engelmeier 2009: 180). Nehru was somewhere in between. Since he became the Prime Minister as well as the Foreign Minister, both ideas and people had to matter to him equally. As it happens with all men of ideas, particularly when they are prolific, in their writings and statements inconsistencies tend to creep in. Tagore was no exception. But therein lies the beauty. He was a thinker par excellence and his footprints are traceable in several aspects of India's policies after independence and foreign policy was one such realm.

Notes

1 Partha S. Ghosh is Senior Fellow at the Nehru Memorial Museum and Library, Teen Murti House, New Delhi – 110011. He is also the Editor of *India Quarterly* (ICWA/Sage). Earlier, he was Professor of South Asian Studies at the School of International Studies, Jawaharlal Nehru University.

2 The national anthem of Sri Lanka was composed by Tagore, in 1938. In Bengali it read *Nama, Nama, Sri Lanka Mata* which was translated into Sinhalese as *Apa Sri Lanka, Nama, Nama, Nama, Nama, Mata Sundar Sri Boroni* by Ananda Samarakun, one of Tagore's favourite students at Santiniketan. It was on Samarakun's request that Tagore wrote and set it to tune.

References

Bandyopadhyay, Sekhar (ed.) *Nationalist Movement in India: A Reader* (New Delhi: OUP, 2009).

Bhattacharya, Sabyasachi (compiled and edited). *The Mahatma and the Poet: Letters and Debates between Gandhi and Tagore 1915-1941* (New Delhi: National Book Trust, 1997).

Engelmeier, Tobias F. *Nation-Building and Foreign Policy in India: An Identity-Strategy Conflict* (New Delhi: Cambridge University Press, 2009).

Ghosh, Partha S. "Domestic Sources of India's Policy of Non-Alignment." *India Quarterly* (New Delhi), 36 (3-4) (1978): 348-362.

Guha, Ramachandra (edited and introduced). *Makers of Modern India* (New Delhi: Penguin Viking, 2010).

Kar, Sisir. *Bengali Books Proscribed under the Raj* (New Delhi: Samskriti, 2009).

Mehra, Ajay K. "India's Experiment with Revolution", *Heidelberg Papers in South Asian and Comparative Politics*, Working Paper No. 40, (September 2008).

Mehta, Rini Bhattacharya. "In the Shadow of the Nations: Dissent and Discourse in Rabindranath Tagore's

Political Writings, 1914-41", *South Asia: Journal of South Asian Studies* (of Australia), 35(1) (2012).

Nandy, Ashis. "Nationalism, Genuine and Spurious: Mourning Two Early Post-Nationalist Strains", *Economic and Political Weekly* (Mumbai), 12 August 2006. Available at http://www.epw.org.in/showArticles.php?root=2006&leaf=08&filename=10421&filetype=pdf

Natarajan, Nalini. *Atlantic Gandhi: The Mahatma Overseas* (New Delhi: Sage, 2012).

Nehru, Jawaharlal. *A Bunch of Old Letters: Written mostly to Jawaharlal Nehru and Some written by Him* (Bombay: Asia Publishing House, 1960).

Rothermund, Dietmar. *Mahatma Gandhi: An Essay in Political Biography* (New Delhi: Manohar, 2009).

Yamaguchi, Hiroichi. *Japan* (New Delhi: National Book Trust, 2006).

Rabindranath Tagore and the British Press (1912-1941)

Kalyan Kundu

The story of the emergence of Rabindranath Tagore in the West is now well known. It was initiated in the summer in 1912 when he sailed for Europe with an intention, in his own words, 'to witness the manifestation of that great awakening of eternal humanity in the heart of Europe'.[1] The other purpose of his visit was to observe the western education system for the benefit of his new experimental educational school in Santiniketan. Thus, on 16 June 1912, at the age of 51, Rabindranath Tagore with his son, daughter-in-law and one of his students arrived in London for the third time, after a gap of 22 years since his last sojourn. His two early visits, one in 1878, as a student at the University College of London and the later visit in 1890 for a month, had neither any significance nor any influence in his life and his future career.

During this third visit, he brought with him two note books filled with his own English translations, or, more appropriately English recreations of some of his Bengali poems. On the day after his arrival, he went to meet one of his acquaintances in London William Rothenstein, an artist and a painter and one of the founders of the India Society[2] to whom he handed over one of his note-books. Rothenstein knew Tagore was a poet; earlier he had read some of his 'works in translation' published in *The Modern Review*. For his own interest and also for the interest of the India Society, Rothenstein avidly received the note-book. Browsing through these poems his initial reaction was exuberant. He made three copies of the note-book and sent one copy each to W. B. Yeats, Andrew Bradley and Stopford Brooke for their assessment. Yeats was struck by the simplicity of expression in Tagore's free verses and found in the poems the work of a 'supreme culture'. A similar reaction was evident from the other two assessors. Yeats read some of those poems in two successive sessions (27 June and 7 July) at Rothenstein's house in Hampstead in the presence of a selected audience of poets, artists and intellectuals. This was followed by a dinner reception arranged by the India Society on 10 July, 1912 in honour of Tagore in a gathering of distinguished public in a restaurant in London where Yeats presided. Welcoming the Indian poet, an ecstatic Yeats said, 'I know of no man in my time who has done anything in the English language to equal these lyrics. Even if I read them in prose translations they are exquisite in style as in thought.' Then he read three poems from the note-book. A brief report of the reception was published in *The Times* - the first report on Rabindranath Tagore to appear in the British press.[3]

From then onward, events moved very fast. The India Society published a selection of 103 verses from the two note books in a limited edition entitled *Gitanjali* (Song Offerings).[4] Yeats wrote the introduction to the book. The India Society edition sold out rapidly within two months. Finally, the high street publisher, Macmillan took the publishing rights of *Gitanjali,* Thus Rabindranath Tagore was fêted by literary Britain. All this happened within the first five months of his arrival in London and Tagore was launched into his new English career.

Gitanjali - Critics' Evaluation and its Success

Reviews of *Gitanjali* appeared in most of the daily newspapers in Britain. A week after its publication the first review of *Gitanjali* came out in *The Times Literary Supplement.* The paper wrote,

> Mr Tagore has translated his poems into English prose, simple and often half-rhythmical, so that their sense is not obscured by an obvious inadequacy of language, and in reading them one feels, not that they are the curiosities of an alien mind, but that they are prophetic of the poetry that might be written in England if our poets could attain the same harmony of emotion and idea That divorce of religion and philosophy which prevails among us is a sign of our failure in both. [...] But this Indian poet, without any obsolete timidity of thought, makes religion and philosophy one. He contemplates the universe as primitive poet might contemplate a pair of lovers, and makes poetry out of it as naturally and simply.[5]

A warm and lengthy tribute of *Gitanjali* appeared from the pen of Evelyn Underhill, one of the distinguished authorities of mysticism, in the weekly journal *The Nation.* The reviewer wrote:

> Only the classics of mystical literature provide a standard by which this handful of 'Song Offerings' can be appraised and understood. These hundred-and-three lyrics, here translated by the author into rhythmical prose of singular beauty pre-suppose as their origin that same personal and first-hand experience of the spiritual order – so changeless and so various, so ineffable and so homely – which is reported to us by the great mystics of every period. [...] Hence, for those interested in the spiritual history of man – the continuance in our own day of that living tradition of intercourse with reality which we owe to the mystical saints - the appearance of these poems is an event of great importance. From the point of view of pure literature, their high quality can hardly be contested; yet it is not mere literary excellence which their author has sought, nor is it here that their deepest interest lies. They are offerings, from finite to infinite – oblations, as their creator holds that all art should be, laid upon the altar of the world.[6]

The critic of *The Daily News and Leader* commended:

> This is indeed, a book of the soul—not the soul of the ascetic, who rejects life, but the soul of the seer and the lover who finds life the visible expression of the eternal. 'Oh East is East and West is West and never the twain shall meet', said Mr. Kipling. In these poems the twain do meet, for they touch that ultimate chord of humanity which is the same by the Thames as it is by the Ganges and they touch it with a mingling of the thought of the West with the rapture and reverie of the East that is unique.[7]

Thus the reviews of *Gitanjali* appearing in British newspapers and journals were extremely favourable. Yeats' introduction to *Gitanjali* helped to establish the poet's identity, necessary in presenting a new author to a new readership. The poems of *Gitanjali* genuinely stirred the soul of the poetry lovers in Britain. The reception of this slim book of verse and its author on its first appearance became phenomenal. It was as if a 'Tagore boom' was established in British book market and Macmillan had to reprint the book eleven times in a year, an unprecedented event for a book of poetry indeed.

Why did *Gitanjali* suddenly strike a chord among British readers? To discuss that we have to look at the literary landscape of Edwardian-Georgian Britain, before the first World War.

The Edwardian era (1901-1910) revealed the predominance of Rudyard Kipling in English literature. His writings were chiefly remembered for his celebration of imperialism. Other representative literary figures of the time were Hilaire Belloc, Henry Newbolt, G. K. Chesterton – all of whom sought to revive a style that distanced itself from Victorian romanticism and aesthetics.

Alongside this group there was a group of poets equally dominating the literary scene; they were old traditionalists, parochial and anti-modern. They included such names as Robert Bridges, Sturge Moore, Walter de la Mare, Earnest Rhys and others. The common features of their work were romanticism, sentimentality and hedonism.

There were some writers such as Bernard Shaw, H. G. Wells, Thomas Hardy and W. B. Yeats, who, through their works, debated and discussed the issues of the time, the amorality, insensitivity repression and philistinism of the English middle class. With the exception of these writers, the overall literary expectation of British readers was at its lowest point. Regrettably both camps failed to contribute any new form, style and ideas to rescue English literature from its stagnant state. Ezra Pound was frustrated observing that the English literary tradition consistently resisted modernism and that he 'should cut out practically all London poetry save Yeats, Eliot and stuff of really unusual interest.'[8] In a similar note Eliot made the same point when he observed that the situation of poetry in 1909 or 1910 was stagnant to a degree difficult for any young poet of today to imagine.[9] This was also noted in the British press. *The TLS* commented in 1912,

> The chief cause of decadence in any art is impoverishment of subject matter; and poetry is always liable to this impoverishment when it has enough intellectual power to pass from its primitive stage of dealing with the particular task to dealing with the general. It must accomplish this transition, if it is to remain living art in a society that is largely concerned with ideas ; for otherwise the poets, for the sake of their art, fall into an obsolete state of mind and fail to interest anyone except the connoisseur of that art. Poetry must conquer the province of ideas if it not to be subdued by them into prose. It must learn to express the emotion stirred by ideas ... and in doing so, it must remain poetry with the old music, imagery, and unhesitating sense of values. That is the problem, which troubles our poetry at present and seems to endanger its very existence...'[10]

Gitanjali fulfilled that expectation of the wider poetry readership in Britain.

The socio-economic condition of Britain during the pre-war period also contributed to the success of *Gitanjali.* 'Great Britain in the period of 1900-1914 provides a fair representation of mass society in an industrial country' on the other hand, 'it also offers us ample evidence not of increasing social homogeneity but a more marked and disturbing social division', wrote the historians.[11] Social homogeneity was lost due to uneven distribution of the wealth. The society was markedly dived into 'masses' and 'classes'. This was evident in the description in Rowntree and in the statistics of Charles Booth. Even upper-class Britain in its uncontrolled affluence seemed unhappy. 'This era of craving for unlimited wealth possessed by affluent English society for their luxury and pleasure with sufficient leisure in hand had developed a feeling of emptiness within. As enjoyment of any kind cannot be satisfied with more enjoyment; increased enjoyment acts as fat in the fire,' explained Nirad Chaudhuri.[12] As a result the spiritual aspiration of the people, both within the *masses* and the *classes*, was at an all-time low. The presence of delicate subtlety and finer sensuality was to be found in Tagore's poems; in addition, a direct and playful approach to the poet's divinity, written in the form of Psalms of David, satisfied the spiritual expectations of the general reading masses. They were the major shareholders of *Gitanjali*'s popularity. One English daily wrote describing the social condition of the time,

> Before the war we saw luxury and freakishness, road-hogging and frolicking, while in slums and other forgotten places there was an ugly and a devastating poverty that rebuked the Empire. In our vulgar satisfaction we had forgotten and rot had set in. It was, then, in the midst of the glistening and vulgar riot that a gentle voice came from the East, and it is a hopeful fact that the poems of Rabindranath Tagore claimed some attention.'.[13]

This was the time when people in Britain, lost and downhearted, were waiting to hear some hope and confidence; ready to welcome even an alien literature. Tagore brought with *Gitanjali* a note of optimism.

Tagore's popularity elevated him to the status of a mystical cult and that mystic seal was something he had to carry with him throughout his British journey. *Gitanjali* generated tremendous enthusiasm within the literary élite in Britain. They were overwhelmed on finding simplicity, serenity and an abundance of Eastern mysticism within the poems—the kind of mysticism that also prevailed in mediaeval Europe in the works of St. Francis of Assisi, Thomas à Kempis or, in modern times in the works of, William Blake, Whitman and Herbart Spencer. About these poems Stopford Brooke wrote to Rothenstein:

> Mysticism of this lofty and profound kind is at root similar all over the world and this accounts for the strange unity of the East and the West in these poems, and makes them ready to find a sympathetic home among the large, quiet and silent group of English people who do not talk against any form of materialism, but think and feel apart in stillness of the eternal matters.[14]

The Gardener and *The Crescent Moon*

Following *Gitanjali,* Macmillan published two other volumes of Tagore's poetry *The Gardener* – a collection of love poems, and *The Crescent Moon* – verses for children, within a space of twelve months. *The Observer* wrote about the poems of the *The Gardener* :

> He does not in his poetry set the themes of life to great music; he speaks them in a soft voice to the heart with all the simplicity and directness in his power, He takes the little intimate things which comprise life and fashions them into pearls which reflect the colour of the sky, the mightiness of love and life.[15]

The critic of *The Scotsman* wrote,

> The rendering (of these poems) has a dignity, a poetic quality, that many an English author might naturally envy. The poems themselves sing the progress of a passionate love affair, and do so with exaltation, a dreaminess and in its feeling.[16]

In spite of the beauty of the poems the critic of an evening paper - *Pall Mall Gazette* wrote:

> The present book of love lyrics is less mature and less profound than the spiritual poems in *Gitanjali* [...] but the simplicity is marred by a too frequent use of the refrain, and the honeyed flavour of the diction grows monotonous.[17]

Within a space of only eleven months, from *Gitanjali* to *The Gardener*, the Tagore readers were already so tired of the monotony of his style and expression that some parodies of his poems appeared in several newspapers.

The Nobel Prize and the Knighthood

Essentially on the merit of *Gitanjali,* Tagore's name was proposed to the Swedish Academy as one of the contenders for the Nobel Prize for literature in 1913 (though during the announcement of the award two other volumes of poetry were published). Although the majority members of the Royal Society of Literature proposed the name of Thomas Hardy as a candidate, Tagore's name was proposed by one of his close friends - Sturge Moore - and his name was finally approved by the Swedish Academy for the award. The news of the award appeared in the British Media on 14 November, 1913.

In spite of the tremendous popularity of *Gitanjali* the decision to award the Nobel Prize for literature to Tagore for the year 1913, surprised the British media. If one took into account his Bengali oeuvre undoubtedly he was an eligible candidate if not the best of all contenders. However, given his only small handful of English works it was felt that the award was to too much and too soon. Even those papers which hailed Tagore and his *Gitanjali* became very critical about the prize. Though some broad sheet dailies were very careful in presenting the news, some tabloids were vociferous. *The Times* wrote,

> The Swedish poets Karfeldt and Heidenstam, and the writer Hallstorm, who are all members of the academy, have expressed their satisfaction with the award and state that the Indian poet's works, although they have only recently became known in the Western world, show an original and poetic vein of great depth and undoubted literary merit.[18]

The general readership was not happy with awarding Tagore the prize and ignoring Hardy. A weekly journal wrote with a touch of sarcasm, 'The latest news of Mr. Tagore is that he is to receive the Nobel Prize for 1913. It is perhaps as well that the committee that administers the funds of the deceased manufacturer should continue to make itself ridiculous'.[19] An extreme speculation came out in a weekly journal *Truth* where their Paris correspondent suggested that Prince William's visit to Calcutta (in 1911) brought the award of the Nobel Prize to Rabindranath Tagore.[20]

The award of the Nobel Prize enhanced further the sales of Tagore's other volumes. One critic of *The New Statesman* spat venom stating that -'Those myriads of Anglo Saxon sheep who, because of the Nobel award, are rushing to buy Mr Tagore's volumes are buying something which, whatever it may be in the original, is in English rather thin and very monotonous. The unjustified boom we have always with us.'[21]

Finally, in 1915, Tagore was awarded the most coveted British accolade, the Knighthood.

A General Review of Tagore's Other Works

After *The Crescent Moon* seven more volumes of Tagore's poems appeared in Britain before his death in 1941. Of these seven volumes, *Fruit Gathering* (1916), *Stray Birds* (1917), *Lovers Gift and Crossings* (1918) and *The Fugitive* (1921) were published by Macmillan and all were translated by the poet himself. In addition to the list a volume of Kabir's poems *One Hundred Poems of Kabir*, translated by Tagore was published by the India Society in 1915 (later on published by Macmillan). In 1925, a small sheaf of verses of Tagore, translated by Thompson in the *Augustan Books of Modern Poetry Series* (Sixpence Series) was published by Ernest Benn, a niche publishing house in London. Finally a long prose poem *The Child* was published by Allen and Unwin in 1931. *The Child* was the only work of Tagore that was first written in English and was originally inspired by watching a passion play in Germany. It was later translated into Bengali, titled *Shishutirtha*. His last book of verse, also produced by Allen and Unwin, in 1932 was *The Golden Boat*, a collection of poems translated by Bhabani Bhattacharya. Besides those individual volumes a collection of his major works *Collected Poems and Plays* was issued by Macmillan in 1936 to celebrate the poet's seventy-fifth birthday. Reviews in the British Press of those books were increasingly unsatisfactory as the translations, unlike *Gitanjali*, were becoming more and more inadequate, there was too much paraphrasing and much of the quality and sharpness of his style was lost. According to Chaudhuri, Tagore failed to develop his English throughout his 'English' career, and that made a substantial contribution to his eclipse: 'Somehow or the other, he developed for his own English a love which was as infatuated as Chesterfield's love for his natural son—both being children of love.'[18] In his latter works, only *Stray Birds*, a collection of 326 aphorisms, received good press, otherwise reviews of his last two books

Lovers Gift and Crossing and *The Fugitive* claimed the works were shallow, wordy and without any substance.

Eighteen months after the publication of *Gitanjali* the *TLS* forecast, that 'although the popularity that caught him up in a flame (a popularity unfailingly registered by the Nobel committee) is likely to fade as rapidly as it was aroused.'[22] Tagore's publisher Macmillan was very much aware that the duration of the Tagore boom would be transitory. So without wasting any time Macmillan published volumes of his work one after another, ignoring the quality of the work. At one stage they published six books within a year. They even published a draft manuscript of his play without the final clearance from the author.[23] 'Strike while the iron is hot'- appeared to be the only aim of the publishers. From Tagore's side there was no resistance. Inevitably Tagore was more concerned to submit the manuscript within the deadline set by his publishers rather than maintain the quality of his works. Moreover, he was then desperately in need of funds for his school. As a result his translations were becoming more and more paraphrased and inadequate. Eight years after the publication of Gitanjali Thompson was bold enough to accuse Tagore in a letter saying, 'As I have told you before, you—with great assistance from your publishers—have been your own worst enemy. You thoroughly deserve to have Rhys' book written about you! (In fact Thompson did not like Rhys' shallow praise of Tagore in his biography) You have been taking the line of least resistance for years, translating from a certain stratum – and that not the least striking one – of your work, and unconsciously mixing the wine of *Gitanjali* with water, till folks have found the draught tasteless. Why, O why? One sin is, you have increasingly paraphrased, instead of translating. You have lost far more of your own force than you need have done.'[24] In a solemn confession Tagore wrote in February 1921 from America, 'In my translations I timidly avoid all difficulties, which has the effect of making them smooth and thin. I know I am misrepresenting myself as a poet to the western readers. But when I began this career of falsifying my own coins I did it in play. Now I am becoming frightened of its enormity and am willing to make a confession of my misdeeds and withdraw into my original vocation as a mere Bengali poet.'[25]

The falling of the standard of Tagore's translation was not the only reason for the decline of his reputation as a poet. Eighteen years after the publication of *Gitanjali,* in 1930, a letter was published in *The Oxford Mail* which could be considered as a representative explanation of the new generation of British readers. The letter read,

> His [Tagore's] poetry has been translated in nearly every European language. Yet how alien it is [. . .] What is a new world to Yeats, is a strange world to most of us. Not that Tagore's verse is difficult or obscure. It is crystal clear; but it is written from another continent. In its abstraction, its constant use of earthly things as spiritual symbols, it is nearer to the poetry of Shelley than to that of any other Western writer. But Shelley is unpopular with our newer poets, and so is Tagore. For Tagore's first concern is with God, a God of whom the East has never for a moment lost sight. So there is a conviction in his approach, a spontaneity in his devotion. But for a long time now our own poets have been on a different road. We simply have not the nerve nowadays to write mystical verse, to start a poem 'Thou art the sky, and thou

art also the nest. O thou beautiful!' And so, although Tagore's poetry has been popular amongst some people for some time, and though the very essence of his English will always give his verse distinction, it would be safe to prophecy that his work will not be at all closely related to the English poetry of his day [...] But if we cannot imitate, we can approve and admire. And then there is much to be admired in a poet, a musician, a teacher and a thinker.[26]

Out of 52 plays and dance-dramas written in Bengali only eight of Tagore's plays were published in translation from Britain, of which only *Chitra* and *Post Office* received the attention of the British play-readers and also received good press. The readers of *Chitra* very much appreciated the lyrical quality and its underlying women's liberation issue rather than its dramatic element. On the other hand the *Post Office*, although it received some criticisms, was the only play of Tagore's to be performed widely in the West and still is sought after by those theatre directors who prefer innovative productions. This was the play first staged by Abbey Theatre, a professional group in Dublin in Ireland, on 17 May 1913, before the publication of the drama in a book form. The drama critic of *The Irish Times* wrote, 'It is not a very imposing play, but it has many attractive and, at the same time, pathetic features.'[27] *The Maharani of Arakan*, a one-act play based on Tagore's short story *Dalia* was the first production of Tagore's play in Britain at the Royal Albert Hall in 1912. A report of that production was published in *The Times* on 31 July, 1912.[28] The main obstacle to appreciating Tagore's play was carefully explained by one of the critics of *The Athenaeum* who wrote, 'Mr Tagore endeavoured to make a developing action, an interplay of characters, the vehicle for presentment of spiritual truths; but in both the burden of the message was disproportionate to the machinery devised for delivering it.'[29] The use of too much symbolism, allegories and metaphors in his plays became progressively incomprehensible to the readers. His last drama *Red Oleander* published in 1925 was severely criticised in the press 'Mr Tagore is too serious a writer to be suspected of publishing absolute nonsense on purpose, so one must suppose that he did it by accident. Presumably he is able to follow the workings of his own brain, and it would be very interesting to hear from him just what it is all about' – a sarcastic comment from the critic of the *Sheffield Daily Telegraph*.[30] It was a real irony that Tagore had to explain the significance and interpretation of this drama to his British readers which was published in the 28 August issue of *The Manchester Guardian*.[31] In writing a play Tagore never followed any standard convention. 'There is no list of dramatic personae, so that the quantities and even sex of the performers were not mentioned' – wrote one of the critics in an evening paper.[32]

Of the 95 short stories by Tagore in Bengali, 34 of these stories in translation were collected in three volumes, all published by Macmillan: *Hungry Stones and Other Stories* (1916, thirteen stories), *Mashi and Other Stories* (1918, fourteen stories) and *Broken Ties and Other Stories* (1924, seven stories). Most of those stories were not translated by Tagore and neither were his three novels in English: *Home and the World* (1919), *The Wreck* (1921) and *Gora* (1923). Most of Tagore's friends loved his short stories. Thompson firmly believed that the original short stories in Bengali were comparable to and occasionally surpassed the merit of most short-story writers from the West. Rhys enthusiastically wrote: 'as we read them we feel at once the touch of the born storyteller.'[33]

Rothenstein wrote, 'I find a book of short stories, which to my delight 'The Postmaster' [...] and the little story will always have a particular value for me.'[34] The critics of the British press, gave a mixed reaction and did not necessarily agree with Tagore's friends. The critic of *The Daily Telegraph* wrote , 'As studies of Bengali life most of these tales are weirdly; as fiction they will not suit – and doubtless were not written to suit all tastes.' [35] A reviewer of *The Manchester Guardian* wrote, 'These stories... may sometimes seem remote and, to a Western sense, lacking in sharpness of edge,'[36] The failure of appreciation by the British readers of those stories was due to their lack of comprehension and basic ignorance of the customs, traditions and cultural situations that surrounded the context of the stories. Hence, most of the reviews were superficial, shallow and negative. Thompson feared that the original stories were chopped and lopped in translation to satisfy Western expectations, and that therefore the finer qualities and sensitivities were lost.[37] Tagore's own explanation was that Indian life was so foreign to British readers that it was likely to tax their imagination too much for perfectly comfortable reading.

The reviews of Tagore's novels in the British press followed almost the same argument as that of his short stories. However, a section of intellectual readers appreciated *Gora*. The tension between the orthodox Hinduism, liberal Brahmoism, and Christianity: the interactions of the educated young men and women; their debates and dialogues amidst a political, psychological and social whirlwind were presented in a lively way in this novel and that was favourably appreciated by the readers. The critic of *The Nation and Athenaeum* wrote, 'Many of his characters are remarkably alive, and the skill with which he makes them grow in subtlety upon the canvas of his book is often admirable.'[38] Another critic wrote about *Gora*, 'We ought not in fairness to apply English standards to a novel written in Bengali for Bengalis any more than to Russians [...] but to the Hindu of today these things (the regeneration of India and the social evils that hamper it) are very living questions.'[39] Some readers also appreciated the underlying theme of women's liberation in *Home and the World*. Rothenstein, profoundly moved by the beauty of *The Home and the World*, wrote to Tagore: 'I don't know when you wrote it but it seems to me to contain your ripest wisdom.'[40]

Among other books of Tagore's published in translation, his memoir *My Reminiscences* was highly praised by readers: 'Your 'Autobiography' especially has been a great pleasure,' commended Yeats.[41] According to *The Nation*: 'If it is safe to prophesy immortality for anything Sir Rabindranath Tagore has written, it is safe to prophesy it for this (My Reminiscences).'[42] From other assorted reports and interviews in the British Press, Tagore's first interview with a special correspondent from *Christian Commonwealth* which appeared on 21 May 1913 was important as he described his own poetical background. Also for the first time, mainstream readers of Britain saw the photograph of Rabindranath Tagore. From the same interview readers were also aware that the poet founded a residential boarding school where he engaged in experimenting with a new system of education as opposed to the stereotypical English educational structure, introduced by the British in India. At the beginning of 1914, Tagore's school was visited by J. Ramsay Macdonald, the future Labour Prime Minister of Britain, who wrote an authentic account of his day in Santiniketan in a long article that appeared in *The Daily Chronicle*.[43]

Barring a handful of intellectual readers, most of the general reading public failed to fathom the highly philosophical treatise of Tagore's essays. The individual essays within *Sadhana –The Realisation of Life* were delivered as lectures in the USA and also in the Quest Society in London and attended by a large audience. This volume of essays received the highest sale of all other essays. But that was not an indication of understanding the deeper philosophy of its writer. The book was published after three volumes of verses so the readers had a natural curiosity about his philosophical works and that perhaps explained the increased sale of *Sadhana*. The reviewer of *The Nation* wrote,

> *Sadhana* is not (a) philosophical treatise, it is a personal statement, which makes free, but not exclusive, use of the philosophic formulae of Indian religion, in the course of expounding its author's vision of life. Its temper is at once individual, national, yet also universal; accepting all the natural links of our closely woven humanity, not as fetters, but as supports to the soul; its temper is at once individual, national, yet also universal; accepting all the natural links of our closely-woven humanity, not as fetters, but as supports to the soul.[44]

Other reviewers found the philosophical aspects in *Sadhana* to be closer to Emerson and Cared, showing that there was not much difference in the thought process between the East and West.[45]

Creative Unity, another collection of Tagore's essays, appealed to many readers due to their clarity and thought provoking elements. The TLS wrote,

> His writings alone show the absurdity of affirming that East and West can never meet; hundreds of western men and women in reading him must feel that here they are in contact with someone with whom they could converse far more fully about the things for which they must care than with the majority of their fellow-countrymen.[46]

The theme of those essays which were written directly in English were more accessible to the readers for their clarity and argument rather than those translated from his Bengali works.

Nationalism, Renunciation of Knighthood and Aftermath

Of the essays that raised serious debate and controversy in the West was *Nationalism*, comprising three individual essays: 'Nationalism in the West', 'Nationalism in Japan' and 'Nationalism in India' and a long poem *Sunset of the Century* (this is in fact five poems written on the last day of the nineteenth century under a common title). It was published during the First World War in August 1917. The war itself provided the whole argument around *Nationalism*. He wrote:

> A nation, in the sense of the political and economic union of a people, is that aspect which a whole population assumes when organised for a mechanical purpose[...] But when with the help of science and the perfecting of organisation this power begins to grow and brings in harvests of wealth, then it crosses its boundaries with amazing rapidity. For then it goads all its neighbouring societies with greed of material prosperity, and consequent mutual jealousy, and by the fear of each other's growth

> into powerfulness. The time comes when it can stop no longer, for the competition grows keener, organisation grows vaster, and selfishness attains supremacy. Trading upon the greed and fear of man, it occupies more and more space in society and at last becomes its ruling force.[47]

According to Tagore the modern war was the consequence of aggressive Western materialism — of science divorced from spirituality — by which these Nations will drag the rest of the world down into the bottom of destruction. Whenever power removes all checks from its path to make its career easy, it triumphantly rides into its ultimate crash of death.

The Times Literary Supplement treated the argument in *Nationalism* sympathetically. In a lengthy review the critic wrote:

> This is not to admit that the poet's view is just. It is not true that the public action of England as a state has been inspired simply by the desire for power and material gain. One hopes that increasingly a great moral ideal is becoming discernible as that which the federation of peoples included in the British Commonwealth are called to realize. On the other hand, it would be vain to deny that the love of power and the love of gain are motives which have had their share in the building up of the British empire. If we do not think the poet's view fair, we may allow that from his standpoint it is explicable.[48]

However, the critic found that Tagore had not given any guidance for practical solutions such as how the poet would like to substitute the present regime of India or how the economic needs of mankind are now to be supplied without a complex industrial organization. Critics of *Nationalism* actually failed to realize that the book was not a series of solutions for the problems of nationalism, but was actually a grievous warning of the evil that would stem from it—evil that one can argue is still present in the world today. A harsh comment on the subject of *Nationalism* came from the critic of *The Spectator* who wrote:

> We need not perhaps take too serious a view of a poet's whimsical and partly humorous incursion into a field with which he is not familiar. Perhaps these lectures are only the Eastern counterpart of a *Sentimental Journey*, the half-smiling, half-pettish records of the likes and dislikes of a man of genius, whose rather irresponsible expression through a pen which can write two languages with equal ease and felicity.[49]

Tagore's condemnation of Western nations as organized political bodies built to satisfy material greed, mechanical organization and imperial aggression was certainly not what his British audiences expected to hear, especially when Europe was at war. As a result, his image of the 'Mystic poet of the East' received a serious knock for the first time. This was aggravated further when he renounced the Knighthood in protest after the Jallianwala Bagh massacre in 1919.[50] His British audience looked upon the renunciation as a deliberate insult to both King and Crown. All these had a cumulative effect of displeasure and dissention amongst the British public. When he visited England again in 1920 his sensitive mind realised that a temporary distancing had been created between him and his close English friends. 'The literary elite which had sat at Tagore's feet in 1912 had come for insight into the universal human heart, not into the affairs of the Punjab or of Bengal,' wrote Ratcliffe.[51]

From this time onward Tagore's role became more like that of a missionary, a proponent of universal humanity, rather than a mystic poet. Frustrated, Tagore left England for Europe with a heavy heart, visiting country after country in Europe devastated by the war, promoting his message of East-West fellowship and seeking support for his newly founded Visva-Bharati University. He was convinced that world peace could only be achieved when East and West would meet together on common ground; where knowledge would flow in two streams—from the East and from the West—and in their unity there would be perceived the oneness of truth that pervaded and sustained the entire Universe. He acted like a one-man peace envoy, received in Europe with 'the kind of welcome that might have been expected for any heavenly visitant in the hell to which man has reduced mankind,' described a columnist of the weekly journal *The Nation and Athenaeum*.[52]

During 1926, his controversial tour in Fascist Italy at the invitation of Mussolini resulted in an international backlash, especially within Europe. The British press followed the event closely. During the months of July and August in 1926 *The Manchester Guardian* reported Tagore's Italian visit and the arguments and counter arguments about Tagore accepting the Fascist dictator's invitation. However, in contrast to his cold reception in 1920–21, in 1926 he was warmly received in England again, reminiscent of the summers of 1912 and 1913. A large number of his photographs appeared in the papers during this visit.

* * *

After the Jallianwala Bagh massacre, Tagore became more and more critical of the British administration in India. Once he had an unreserved faith in the English justice of *Boro Ingrej*, the Great English Heart in England, as opposed to the administrators and clerical English men within the colonies. This faith had now deserted him. While India was passing through political unrest with such incidents like the police atrocities in Solapur for wearing the 'Gandhi cap', Gandhi's Salt March, the arrest of Abdul Ghaffar Khan at the border of Peshawar, the riots in Dhaka, Tagore was in London helplessly watching his country going through turbulent times. During this period he wrote a series of articles in *The Spectator* showing his frustration with the British rule of India. In his article, 'An Appeal to Idealism', he condemned Europe saying:

> Once Asia in her spring time of exuberant life offered the world her spiritual ideals, to-day Europe in the illumination of her intellect has brought her science and also her spirit of service. But unfortunately she has not come to Asia to reveal the generosity of her civilization, but to seek an unlimited field for her pride and power, trying to make these things eternal. She has come with her need and not with her wealth; and therefore she has belied her own mission and used the truth itself for a utilitarian purpose of self-aggrandizement.[53]

In the context of the riots at Dhaka he wrote in a letter to the editor: '[...] that this event at Dacca has alienated, more than anything else in Bengal, the sympathies of those who were still clinging to their faith in British justice. Other happenings had shaken public confidence, but this has struck at its very foundation.'[54] In his article on Gandhi's refusal to attend the Round Table Conference in 1930, he initially did not approve of the decision, but later in the same article he approved the decision.[55] From letters, essays and interviews that appeared in the British Press during 1930, one can gauge anguish and conflict within him. Once he held in the highest esteem the greatness of British civilization from its history and literature. He established a close relationship with the finest English minds of his time. The articles and letters published in *The Spectator* were the testimonies to the aberration of his own convictions. Now he became the severest critic of the administrators of his own country, and he became more isolated.

Crisis in Civilization – The Final Curtain

When another War was looming in Europe, his faith in European civilization had totally crumbled. His last message to Europe came in his eightieth birthday message *Crisis in Civilization*, an abstract of which was published in *The Manchester Guardian.* It said:

> The failure of humanity in the West to preserve the worth of their civilization and the dignity of man which they had taken centuries to build up weighs like a nightmare on my mind. It seems clear to me that this failure is due to men's repudiation of moral values in the guidance of their affairs and to their belief that everything is determined by a mere physical chain of events which could be manipulated by man's cunning or might. The consequences of this belief are proving terrible to man.[56]

It was his last warning to the West. A year before his death, Oxford University conferred on him Doctor of Letters D.Litt. (*honoris causa*). In its four hundred year-old tradition, this was the first time that the conferrer went outside Oxford to give it, rather than the recipient coming in to receive it. This was not so significant for Tagore, but it was significant for Oxford to uphold its own institutional prestige.

From 10 July 1912 until his death on 8 August 1941, Tagore's twenty-nine-year relationship with Britain and the British media followed an uneven path. Although enthusiasm for Tagore from the British public was not always uniform, often praised or rejected — sometimes ebullient, sometimes low — he was in the media spotlight throughout the course of his British journey. The British media never lost sight of Rabindranath Tagore. He was widely reported in newspapers and journals. Reviews of his books, writings, interviews, lectures, news of his exhibition of paintings, travels, meetings with like-minded figures and exchanges with them—whether they be political, philosophical or educational—his ideas, photographs, sketches and cartoons—all appeared in the Press with equal importance.

The media even followed him when he was away from the British Isles. He was reported not only in the mainstream broadsheets and the tabloid press but also in countywide regional papers. Even in papers that didn't fit the remit or ethos of Tagore such as *The Bazaar Exchange and Mart* (an exclusive business paper), *Broad Arrow* (military journal), *Ladies Field* (women's fashion magazine), *Psychic News* (a magazine on parapsychology) and *Municipal Journal of Public Works*—he was intermittently featured. Rarely in history were there instances of a poet from one country being so exhaustively reported about in the media of another. In those twenty-nine years he was praised and revered by a section of intellectual Britain for the clarity of his thought, as a thinker with integrity and honesty and a great visionary of his time.

Notes

1 Rabindranath Tagore. 'Parassey' : *Rabindrarachanaballi*, Vol 22, (Kolkata: Visva-Bharati, 1978) p. 443.

2 The India Society was founded in London in June 1910 by William Rothenstein with the objective, as written in Society's constitution, 'to promote the study and appreciation of Indian art, music and literature, ancient and modern, by means of publications by correspondence with kindred societies and with museums of literary records relating to branches of Indian culture or by such other means not being contrary to the rules, as the Society or the Executive Committee shall time to time determine.' Between 1912 and 1915, the India Society published three works of Tagore, viz. *Gitanjali, Chitra* and *One Hundred Poems of Kabir.*

3 *The Times*: 'Dinner to Mr. Rabindranath Tagore', 13 July 1912, p. 3 c6.

4 Out of the 103 poems included in *Gitanjali,* 53 poems are taken from the Bengali *Gitanjali*; the remaining are taken from his other books of verse—*Naivedya* (16), *Gitimalya* (16), *Kheya* (11), *Sishu* (3), *Chaitali* (1), *Kalpana* (1), *Utsarga* (1), *Smaran* (1) and one song from his play *Achalayatan.* The total comes to 104 as in one translation he had combined two poems into one

5 *The Times Literary Supplement*. 'Mr. Tagore's poems', London, 7 November 1912, p. 492.

6 *The Nation,* 'An Indian Mystic', London, 16 November 1912, pp. 320-22.

7 *The Daily News and Leader*, 'An Indian Poet', London, 21 January 1913, p. 4.

8 D. D. Page (ed.). *The Letters of Ezra Pound 1907-1941*, (London: Faber and Faber, 1971), p. 177. The letter was written to Harriet Monroe on 26 August 1917.

9 C. K. Stead. *The New Poetic: Yeats to Eliot*, (London: Hutchinson, 1964), p. 45,.

10 *The Times Literary Supplement.* 'Mr. Tagore's Poems', London, 7 November 1912, p. 492,.

11 Joseph Rowntree and Charles Booth were social reformers. Charles Booth's book *Life and Labour of People of London* (London, 1908) is relevant here.

12 Nirad C. Chaudhuri. *Atmaghati Rabindranath*, (Kolkata: Mitra and Ghosh, 1992), p. 143.

13 C. F. Lawrence. 'The Light of India', *The Daily Chronicle*, London, 19 May 1915, p. 4.

14 Mary Lago. *The Imperfect Encounter: Letters of William Rothenstein and Rabindranath Tagore 1911–1941*, (Cambridge Mass: p. 41, Harvard University Press, 1972).

15 *The Observer.* 'Life and the Poet', London, 16 October 1913, p. 4 c3.

16 *The Scotsman.* 'The Gardener', Scotland, 16 October 1913, p. 2 c2.

17 *Pall Mall Gazette.* 'Under the Banyan', London,14 October 1913, p. 7 c4.

18 *The Times.* 'The Nobel Literature Prize', 14 November 1913, p. 8 and 'Swedish Tribute to Mr. Tagore', London, 15 November 1913, p. 13.

19 *The New Age.* 'Readers and Writers', London, 20 November 1913, p. 80.

20 *Truth.* 'Notes from Paris- The Prince and the Poet', London, 24 December 1913, pp. 1502-1503.

21 *The New Statesman.* 'Current Literature', London 13 December 1913, p. 309.

22 *TLS.* 'East and the West', 14 May 1914, p. 236.

23 The draft copy of *The King of the Dark Chamber* was printed as a book whilst Tagore was revising the manuscript back in Santiniketan.

24 E. P. Thompson. *Alien Homage*, (Oxford University Press, 1993), p. 47.

25 *ibid,* p. 48.

26 R. E. Warner. 'The Work of Rabindranath Tagore', *Oxford Mail*, Oxford, 27 May 1930, p. 4.

27 *The Irish Times.* 'Abbey Theatre', Ireland, 19 May 1913, p. 9.

28 *The Times.* 'The Indian Dramatic Society', 31 July 1912, p. 9 c6.

29 *The Athenaeum.* 'Drama', 7 November 1914, p. 486.

30 *Sheffield Daily Telegraph.* Sheffield, 23 July 1925, p. 3 c6.

31 *The Manchester Guardian.* 'Red Oleanders - Author's Interpretation', p. 8 c7, (Manchester and London) 28 August 1925.

32 *Dublin Evening Mail.* 'Red Oleanders', Ireland, 22 July 1925, p. 2 c3.

33 Ernest Rhys. *Rabindranath Tagore – Biographical Study*, London, Macmillan, 1915, p. 47.

34 Lago. *The Imperfect Encounter*, p. 41.

35 *The Daily Telegraph.* 'Hungry Stones', London, 29 November 1916, p. 4.

36 *The Manchester Guardian.* Special Supplement, 'Sir Rabindranath Tagore's stories'. 26 April 1916, p. 3.

37 Thompson. *Alien Homage*, p. 22.

38 *The Nation and the Athenaeum.* 'The World of Books', London, 9 February 1924, p. 669.

39 *TLS.* 'New Novels – *Gora*, 28 February 1924, p. 126.

40 Lago. *The Imperfect Encounter*, p. 255.

41 Bikash Chakravarty (ed.). *Poets to a Poet 1912-1940*, (Kolkata: Visva-Bharati, 1998), p. 154.

42 *The Nation.* 'Memories of a Poet', 25 August 1917, p. 536-7.

43 J. Ramsey Macdonald. 'Mr Rabindranath Tagore's School', *The Daily Chronicle*, 12 January 1914, p. 6.

44 *The Nation*. 'The Circle and the Centre', 13 December 1913, p. 499.

45 *The Manchester Guardian*. 'Poet and Philosopher', 6 January 1914, p. 4.

46 *TLS*. 'Man the Creator', 31 May 1917, p. 259.

47 Rabindranath Tagore, *Nationalism*, (Macmillan, London, 1917). The present citation is from *The English Writings of Rabindranath Tagore*, Sisir Kumar Das (ed.), (New Delhi: Sahitya Akademi, 1996), p. 421.

48 *TLS*. 'The protest of a Seer', 13 September 1917, p. 435.

49 *The Spectator*. 'The Neo-Hindu in America and Japan', London, 13 October 1917, pp. 386-387.

50 The Jallianwala Bagh massacre, (also known as the Amritsar massacre) took place on 13 April 1919 when Brigadier-General Reginald Dyer ordered his army to fire upon a crowd of non-violent protesters, along with Baishakhi pilgrims, gathered in the Jallianwala Bagh garden in Amritsar, Punjab to protest the arrest of two leaders despite a curfew which had been recently declared. The dead numbered between 370 and 1,000, or possibly more. This brutality stunned the entire nation.

51 Thompson. *Alien Homage*, p. 31

52 *The Nation and The Athenaeum*. 'A League of Spirit', London, 9 April 1921, p. 49-50.

53 Rabindranath Tagore. 'India: an Appeal to Idealism', *The Spectator*, London, 7 June 1930, p. 932,.

54 Rabindranath Tagore. 'Great Britain and India', *The Spectator*, London, 30 August 1930, p. 280.

55 Rabindranath Tagore. 'The Round table Conference', *The Spectator*, London 15 November 1930, p. 724.

56 *The Manchester Guardian*. 'Dr. Tagore and the War', Manchester and London, 1 April 1941, p. 6.

Rabindranath Tagore: Encounters with the French

Supriya Roy

In the Rabindra-Bhavana Archives, apart from files containing the correspondence between Rabindranath and intellectuals of France, there is one file named simply, France. This file contains letters from various individuals in France. Mostly written in French, these are from his admirers – students, and people who spontaneously offered him their homage. I have chosen one letter from this file written by 20 year old Andre Gillaumeau from Yerres. He writes (translated):

> Prestigious Poet, Your personality is that pronounced, your genius that universal, that I, an obscure young Frenchman, hesitate to pronounce your name. I am only 20 years old, the age of dreams and all hopes. But how to dream in our occidental civilization where action is the queen. Always acting, never dreaming. Here poetry is banned from life. Here a dreamer is considered a fool by the people who surround him. The minds are active and never contemplative or speculative. In Europe one is afraid to show one's feelings. A civilisation should be disinterested. It should not aim to improve the destiny of man but go in search of the idea, the truth, without necessarily ending up in a material benefit for man. My soul is in need of other things than logic and reason. Will I find it in your philosophy? Great Tagore, these words, I have written them with an occidental soul, I hope for an answer. I resign from the world of thought, I kill myself morally, I refuse to think, till I have your answer…Tagore—answer me![1]

This letter, in a way, sums up the problem faced by a section of the people of France at that period after the war who felt Rabindranath could give them an answer.

On the other hand, in a letter to Rabindranath shortly after the Indian poet spent a few days in Villeneuve with Romain Rolland, the French writer wrote:

> "Never forget the magic power of your name and all it represents for an elite of Europe. Undoubtedly, it knows you only partially, by imperfect translations. But Duhamel has just told me these last days that there is something more essential than the direct effect of books—it is their 'fragrance'—these mysterious emanations of a God-pervaded soul. There exist a multitude of poets whom one reads and whose art is admired. But very few radiate this enchantment which awakens and exalts the forces of life. You are one of these good genii."[2]

What was this magic power and how did it affect the French?

The Nobel Prize awarded to Rabindranath Tagore, an Asiatic, literally forced the European reading public to acknowledge the existence of a culture based on traditions not their own and made them realize that outside the Western sphere of influence new forces were stirring of which they, now, had to take notice. Jean Guehenon, writing in *La Revue de Paris* in 1919, said,

> That the very name of a poet who in his country enjoyed such a reputation should have been almost ignored by the whole of Europe until these last few years, goes to prove the limits of human glory. It also proves the narrowness of our civilization and points out—whatever one may say—its provincialism.... The knowledge that these ideals are different from ours, at least makes us aware of the relativity of our European concepts. We do not sufficiently realize that millions of human beings are fed on different ideas from ours, yet live.[3]

It began as a re-appraisal of many concepts that were deeply rooted in Western tradition. India suddenly ceased to be a merely political abstraction. It was a country that brought forth a poet of genius comparable to their own Dante, Shakespeare and Goethe.

In France it was the intellectual elite, or at least the most progressive part of it, that went over to him wholeheartedly and unhesitatingly. Romain Rolland, André Gide, the Comtesse de Noailles, Paul Valery and many other outstanding writers and scholars in France celebrated him as one of the greatest poets of the age and spontaneously responded to his message.

There arose at that time a very real need for a re-valuation of standards all over Europe; and this need was brought about by events of a political as well as of a moral nature. Rabindranath, in his own distinctive language, expressed the need of the hour better than any other writer at that time. Whether the majority intuitively grasped the meaning of his message, one cannot say. But the response of intellectuals was undoubtedly due to the awareness that Rabindranath succeeded in saying the things that are in our minds, as Gilbert Murray said, but which we cannot quite bring out.

In the years that followed the war, Europeans wished to replace the merely material considerations which had guided them for so long by a spiritual and moral revaluation of the universe. Rabindranath's arrival on the continent took place at the right moment. He came to Europe in 1920-21, in the years following the war when people were ready to receive his message, driven as they were towards religion by their despair and their sense of frustration. They felt that a prophet was among them bringing a message of hope and fulfilment.

Dwarakanath Tagore in Paris

Incidentally, Rabindranath's grandfather Dwarakanath went to Paris in 1945-46 just a few decades before him. His appearance in Paris created a great sensation. Dwarakanath was a handsome man, and he lived in a truly magnificent oriental style while at Paris. The King, Louis Philippe, honoured him at his court at a grand evening party. The room was hung with Indian shawls, then the height of ambition of every French lady. Imagine their delight when the Indian Prince placed a shawl on the shoulders of each lady as she left the room!'[4]

Early Visits

Rabindranath's first three visits to France were in the form of stopovers on his way to England. However brief these visits were, he has kept lively accounts of them in letters and in diary. On the 9th of October 1878, a young lad of seventeen who was making his first trip abroad, reached Paris.

> What a spectacular city! To be in that wilderness of towering palaces is to be overwhelmed. It seems that in Paris the poor do not exist. Where is the need for such massive ostentatious buildings for puny humans, I wondered. We went to a hotel so colossal in every respect that, as in overly loose garments, I did not feel comfortable staying in it. I was reduced to insignificance in this room. Monuments, fountains, gardens buildings, cobbled roads, carriages, horses and the commotion of people among other things rendered one speechless.
>
> On reaching Paris, we went to a Turkish bath...I realized that having a Turkish bath is the same as sending your body to the laundry![5]

The two brothers then went to the Paris Exposition which housed almost 53000 exhibits and attracted more than 16 million visitors. 'I do not remember the Paris Exhibition in any orderly manner, but rather as a heap of confusion,' he wrote back home.

His second visit was in 1890; he was accompanying his brother Satyendranath and a friend, Loken Palit. Before leaving for Europe, he wrote a long essay on why he was making this voyage. This essay was later used as an Introduction to his *Diary of a Traveller in Europe*[6]. He begins the essay by saying:

> For a very long time I had nursed a desire to plunge myself right into the middle of European civilization and experience in the throb of my blood and nerves the blow, the whirlpool, the madness of it all—its tidal wave of dance and music, its roaring laughter and applause and its froth-like enthusiasm, its electrifying speed and its mighty flow.
>
> For ages I have been sitting by the seashore building and unbuilding sand houses and wondering while many of my friends, one by one, have churned the oceans, each according to his strength, and come back; such is their enthusiasm that even after returning to dry shore they cannot stop brandishing their limbs.

He is able to explain the problem with the people of his own country and why he needed this stimulus from the West. He continues,

> We are age-old Indians, quite ancient and very tired. I often experience in myself that vast

> racial antiquity. When I concentrate and probe deeply within me all I can see is thought, rest and otherworldliness. As though there was a long vacation both within and without. It is as though we completed our work in the office at the dawn of the world, which is why in the heat of the afternoon when everyone else is busy working, we have drawn the shutters and rest in certitude; we have drawn our salaries and have now retired managing our household on a pension. This suits us fine.

His impression of Paris had not changed much within these 12 years. This time, too, he arrived by train from Brindisi. While still on the train, they entered France. His diary notes: 'Now we are in France. On our right, a stream flows foaming along. Like the French, it is swift, restless, impassioned, fond of laughter, chattering. But more transparent and child-like.'[7]

They spent the night at hotel Terminus. They spent just a day in Paris. In the morning, they visited the Eiffel Tower from where he sent postcards to his wife and other family members. He bought some books at a bookstore. In the evening they went to a hippodrome to see a pantomime, the equestrian spectacle, *Jeanne d'Arc.*

When Rabindranath left for Europe in 1912, it was twenty years since his last trip abroad. He undertook the trip because he wished to be in touch with the wider world and felt a need for the stimulation of contact with western artists and intellectuals. Before leaving for the West, he promised his students and the teachers of his school that he would write to them regularly about his travels. He intended to have these letters published in a book;[8] as an introduction, he wrote a long essay on his thoughts before leaving. He wrote that there was a platitude prevalent in Bengal that European civilization was materialistic—that there was no spirituality in it. According to him, 'Whenever we find something propitious in human society, at the root of it there is a spiritual power...If we find any human progress in Europe, we have to know that at the root of this progress is the soul—certainly not something created by insensate matter.' All these years he had been searching in the Vedas, the Upanishads, Buddhist scriptures, Sanskrit classics, the message of the medieval saints and in the history of India for the soul of India; now, in this pilgrimage, he was going to search for the soul of Europe. On 14 April that year, the luxury liner, *Titanic,* had hit an iceberg and had capsized. Reports had come that in order to save the lives of women and children, a couple of thousand men, including millionaires, had chosen the watery grave. Referring to this incident, he asked, 'Is there no relation between self-sacrifice and spirituality? Is this not a sign of the power of the spirit?...Does not spiritual strength give valour and prowess to man?'[9] It was this spirit that he was searching for in his trip to Europe.

Alexandra David-Neel

On his third trip in 1912, he was accompanied by his son and daughter-in-law. They hired a motor-cab and made a whirlwind tour of the sights of Paris. He had planned to spend a longer time in Paris and had come armed with letters of introduction to various important people from Alexandra David-Neel. But that was not to be. Alexandra David-Neel (1868-) a student of Sylvain Lévi studied

Philosophy and Oriental languages. She then left Paris to acquire further knowledge on occultism spending more than 30 years in India and other parts of Asia. She was the first European woman to be granted permission to enter Tibet. She was probably one of the earliest encounters Rabindranath had with the French. She came to Calcutta in 1912. She visited the Tagore family home and was aware of Rabindranath's thoughts and works. She visited Santiniketan and was struck by the educational methods adopted in the poet's school. She writes,

> The aim of the founder was not simply to give to the boarders of Santiniketan a modern scientific instruction…it was to form a generation of Indians who, while possessing extensive scientific knowledge, would preserve the traditional spiritual attitude of ancient India. I was able to taste at Santiniketan the atmosphere quite particular of composure in which one bathes. Classes were given under the shade of big trees and they were seen afterwards dispersing in the fields, a book in hand. The frame was ideal to facilitate both study and reflection.[10]

1920: The First Important Trip to France

In 1920, Rabindranath, accompanied by his son, Rathindranath and daughter-in-law, Pratima left for Europe and America. As was his wont, he tried to justify this trip to the West: '…I often wonder why I am going on this trip. It is not for the pleasure of wandering—that I know, but I do not know for certain the reasons for my going. There is one thing that comes to mind and it is this—during the process of churning, butter is separated from the buttermilk; after the churning in Europe, the intellectuals and thinkers would not remain invisibly merged in society. In these times, they will surface and perhaps, it will be easier to locate them. It is not only to see them—today they are engrossed in their thinking and I wish to come into contact with those thoughts.'[11]

On 6 August, Rabindranath arrived in Paris. They were received by Sudhir Rudra who had arranged rooms at the Hotel Metropolitan on Rue Cambon. The first few days in Paris were lonesome for the poet; he had no opportunity to meet anyone. The hotel was gloomy yet the weather in Paris was much better than in England. During their first week, they went to the grand opera where *Faust* was being staged. Rabindranath liked it; in fact he felt it was better than any of the operas he had previously seen in America or England.

Autour du Monde, Boulogne sur Seine

Soon enough they shifted to Autour du Monde as guests of Albert Kahn, the Jewish philanthropist. Rabindranath felt comfortable right away. In one letter he informs C. F. Andrews: "The sun is shining bright and the spirit of exhilaration is in the atmosphere." In his next letter, he writes: "We are in a delightful country, in a delightful place in France, meeting with people who are so human."[12]

Albert Kahn invited important people to his house to meet Rabindranath. Most of them were academicians, writers or those who appreciate the arts and Rabindranath enjoyed the stimulating conversation he had with them. Rathindranath, in his diary, gives us an idea of the kind of people he met and also what they discussed. Lifelong friendships and associations were founded here, with people like the Karpèles sisters, Sylvain Lévi, Albert Kahn and Comtesse de Noailles.

Andrée and Suzanne Karpelès

The Karpelès sisters came to visit them everyday. Andrée, the elder sister, was a post-Impressionist artist and student of Abanindranath Tagore. The two sisters had been to Calcutta and had known Rabindranath. Suzanne, a student of Sylvain Lévi, soon became the poet's secretary and helped with his mail. We must remember that in 1913, Rabindranath was awarded the Nobel Prize and on this visit, he was a celebrity already. Suzanne has put on record her experience of those days; we also get to share a few interesting anecdotes. The first day she brought the poet over to Autour du Monde, which was at quite a distance, the poet had a lot of luggage and she was quite apprehensive of the amount she would have to pay the cab driver. Imagine, then, her surprise when the cab driver took out from under his seat a copy of the day's *L'Humanité* which had a photograph of the poet on the front page; he asked how could he take his fare from this person![13]

The Karpelès sisters grew very close to Rabindranath and his family. Andrée (1885-1956) started painting a portrait of the poet in oil. It was Andrée who introduced them to the works of Impressionists and post-Impressionists in Paris. She took them to an art dealer's gallery in Place de Madeleine where an exhibition of these works was on. Rabindranath saw the works of Cezanne, Manet, Renoir, Van Gogh, Rodin and others for the first time. Rathindranath, his son, comments in his diary: 'The effect on us was overwhelming. Van Gogh especially impressed me." This sister became so close to the family that she visited Santiniketan in 1922 and taught at the new art school for some months. She became a life-long friend. When she left Santiniketan, a farewell was arranged for her where Rabindranath composed a song for her. In this song, Andree Karpeles lives on in Santiniketan. This song, *Bhara thak smritisudhay vidayer patrakhani...* is still sung at all farewell events in Bengal:

> May your cup of parting stay filled with the nectar of memories
>
> At the celebration of our re-union, bring it back.
>
> Let tears of sadness in the core of silence
>
> Nurture in hiding a new song of the heart.
>
> In the path you will take, you shall be alone—
>
> Your only guide in the darkness, a shining thread within.
>
> The Muse of separation in the heart's lotus field
>
> Shall secretly pour her nectar all day. [translated][14]

Suzanne (1890-1969) played an important role in the re-vitalization of Buddhism in Cambodia.

With a triple degree in Pali, Sanskrit and Tibetan from the University of Paris, she went to Southeast Asia to further her studies. She also taught at Calcutta University. She would sit with the poet and translate the poems of *Lipika* into French.

St John Perse and Andre Gide

The first Frenchman to hear the poems of *Gitanjali*, a few years before the war was the poet, Saint John Perse (1887-1975) who was awarded the Nobel in 1960. So impressed was he with the poems that he sent a copy of the *Gitanjali* to his friend André Gide. Within the year 1912, Gide translated the poems into French. Even before he was awarded the Nobel, Rabindranath became familiar to a handful of enlightened persons in France.

In the introduction to *L'Offrand Lyrique*, Gide has discussed not only *Gitanjali*, but *The Gardener*, *The Crescent Moon* and *Sadhana*, which includes his poetry and philosophical writings. Gide writes:

> What then is the secret of this quivering joy which flows in streams, sparkles like water and warms like daylight? What is this truth which simultaneously nourishes and intoxicates the soul? Is it the fruit of Brahmanic philosophy? Is it the cult of Visnu? No, it is the *love* in this philosophy, it is the *love* in this religion. Because, as he says in the preface of the book where he has assembled his thoughts:
>
> For western scholars the great religious scriptures of India seem to possess nearly a retrospective and archaeological interest; but to us they are of living importance. [*Sadhana*]
>
> What I admire here, what fills me with tears and laughter, is the passionate life in this poetry– which one would have thought to be so intellectual and abstract – something that shivers, throbs like a sentence in Pascal's *Mystery of Jesus*– but here with shivering joy.
>
> Let all the strains of joy mingle in my last song–
> of the joy that makes the earth flow over
> in the riotous excess of the grass,
> the joy that sets the twin brothers, life and death,
> dancing over the wide world, the joy that
> sweeps in with the tempest, shaking and waking
> all life with laughter, the joy that sits still
> with its tears on the open red lotus of pain,
> and the joy that throws everything it has upon the dust,
> and knows not a word. [*Gitanjali*, Poem 58][15]

It fell on Musée Guimet's 'Friends of the East' group to arrange for the felicitation of the Nobel prize winning poet. Suzanne was at that time the Secretary of this group and she recounts how people would line up to ask for invitations to the event. People from all walks of life wanted a glimpse of this prophet-like poet.

Comtesse de Brémont

Another frequent visitor to Rabindranath was the Comtesse Anna de Brémont (1864-1922). American Anna, at the age of 17, married the Comte de Brémont, a French medical doctor. Widowed in 1882, she began a career as a journalist and writer. Her books include a volume of sonnets and poems. She was translating *The Fugitive* into French and would discuss this with the poet.

Professor Sylvain Lévi

Professor Sylvain Lévi (1853-1935), the renowned Orientalist would visit Rabindranath twice a week. Rabindranath was greatly attracted to this simple, kind-hearted professor who was so dear to his students. In a letter to C.F. Andrews he wrote, 'He is a great scholar as you know, but his heart is larger even than his intellect and his learning. His philosophy has not been able to wither his soul. His mind has the translucent simplicity of greatness and his heart is overflowing with trustful generosity which will never acknowledge disillusionment. His students come to love the subject he teaches them because they love him.'[16] The two would have long conversations. Rathindranath writes in his diary: 'Father has been trying to impress upon Monsieur Lévi, Monsieur Kahn and others of the necessity of interchange of thought between France and India. In India we get western culture only through English literature—we have no continental culture. Europe is suffering at the present moment because its culture has rejected the east altogether. She must get Eastern idealism and faith back; otherwise it would be quite fatal for her. Father has suggested interchange of scholars.' [17]Sylvain Lévi, it may be recalled, joined Visva-Bharati as its first Visiting Professor.

Albert Kahn

Albert Kahn, who was Rabindranath's host in France, not only this year but on all his subsequent visits to France was a banker and philanthropist. He was convinced that a knowledge of foreign societies encourages respect and peaceful relations among peoples. From 1898 onwards, armed with the necessary financial resources, he set up the series of bursaries he called *Autour du Monde*—around the World. Realizing that his era was to witness great changes, he began to build up an iconographic memory of societies, environments and lifestyles around the world. From 1909 to 1931 he commissioned photographers and film cameramen to record life in over 50 countries. The images were held in the Archive of the Planet. According to the poet's son, '...it was fully worthwhile to come to France to meet this one man—Monsieur Kahn.'

On 20 August 1920, Rabindranath wrote to C.F. Andrews, describing his host,

> He is thinking great thoughts, carrying out great ideas, living the abstemious life of a *tapaswi*. He is devoting his immense resources for the good of humanity—not through any comfortable and respectable path of sentimental charity, but by discovering and systematically attacking the root of the

> evil from which human beings are suffering. It requires a stupendous amount of organised information in all branches of knowledge and human activity—this is the kind of work about which he has thought deeply and for long, made plans, and has set a huge host of workers to carry them out. It is an amazingly comprehensive scheme for which an extraordinary power of thinking and organising is needed, above all is required that spiritual love of ideal which does not depend for its fire upon repeated supplies of fuel of immediate success. It is an inspiration for me to know this man and to talk to him. I feel clearly when I am in touch with men of his type that the ultimate reality for man's life is his life in the world of ideas, where he is emancipated from the gravitational pull of the dust and where he realises the spirit.

One evening, Albert Kahn took Rabindranath to Restorante Prevoste for a chocolate. A young man, recognizing the Indian poet, stood up and undaunted by his presence, left his own chocolate and started to recite a poem from Gide's translation of *Gitanjali,* 'Where the mind is without fear...' Rabindranath thoroughly enjoyed the incident. Such incidents were common during his visit when people on the street recognized him instantly and paid him homage.

One day, Monsieur Kahn offered to take them on a motorcar drive through the battlefields of Rheims. He could not sleep that night. Rabindranath wrote: '...I was taken to Rheims and other devastated regions of France in a motor-car. It was a most saddening sight. It will take a tremendous effort and also an immense lapse of time to make this a thing of the past. When the spiritual ideal is lost, when human relationship is completely broken up, then individuals freed from the creative bond of wholeness find a fearful joy in destruction.' [Paris, 7th September 1920]

Comtesse Anna de Noailles

Another colourful personality Rabindranath encountered was the Comtesse de Noailles (1876-1933), a poet in her own right. She was friendly with the intellectual, literary and artistic elite of the day including Marcel Proust, André Gide, Paul Valery, Jean Cocteau and Pierre Loti. In 1906, her image was sculpted by Rodin. She was known for her beauty—men of all ages succumbed to her charm and beauty. She came to meet the Indian poet, sure of her victory over him. Suzanne Karpelès has shared with us her eye-witness account of that day. The Comtesse started working her charm on Rabindranath who responded by raising his consciousness to a higher level; the more the charm applied, the higher he rose, frustrating her again and again! Finally, Rabindranath reminded her that they were both poets and the significance of their meeting should not be forgotten. Immediately, Anna spoke in a voice shorn of all coquetry—she said she would like to hear his poetry. After that, she too, read out some of her poems.

She told Rabindranath how on the day the war was declared with Germany—she happened to be with Clemenceau, prime minister of France. He was so depressed, he told her that he did not care to talk but would like to read some poems with her. She opened the newly published volume of *Gitanjali* and the war was ushered in by their reading together from the poems of Tagore.

Henri Bergson

Another of Albert Kahn's friends who visited regularly was the French philosopher, Henri Bergson (1859-1941). He was awarded the 1927 Nobel Prize in Literature 'in recognition of his rich and vitalizing ideas and the brilliant skill with which they have been presented.' Rabindranath was familiar with the philosophical writings of Bergson. In a letter to C.F. Andrews, Rabindranath writes, 'The great philosopher Bergson came to see me and we had a most delightful talk. He has read my *Personality* and what he said about my work was beyond my expectation.'[18]

At times, Bergson's philosophical beliefs influenced Rabindranath's poetry—though there was a clear distinction in their thoughts. Writing about the poem, '*Chanchala*' from the *Balaka* volume of poems, his biographer, Prabhatkumar Mukherjee comments,

> Taking Bergson's poetic elements out of his theory of mobility, he has coloured it with his own spiritual aesthetics to create a poem of exquisite beauty. But when the poet goes beyond his poetry to discuss his theory—there he does not agree with Bergson. Endless speed, endless progress is not possible in science, nor proven by religious doctrine—hence those who imagine Rabindranath's and Bergson's philosophy to be in the same category do injustice to both.[19]

The poem Prabhat Mukherjee writes about is '*Chanchala*' or "The Restless One":

> O great river
>
> Unseen, silent, your stream
>
> Continuous, without cease
>
> Flows on forever.
>
> The void throbs to the beat of your Fierce bodiless pace,
>
> At the shock of your immaterial current's race.
>
> Masses of foamy matter arise;
>
> The sky weeps in fire-laden clouds.
>
> Keen shafts of light radiate in streams of colour
>
> From the spreading darkness.
>
> Suns, moons and stars
>
> Like bubbles dissolve
>
> In ranks upon ranks,

Revolving and dying….

Look, how the stream surges into sound.

The boat quivers.

Leave on shore what you have saved on shore:

Do not look back any more.

Let the call of what lies ahead

Draw you forward

On this vast stream

From the clamour behind

To unplumbed darkness, boundless light.[20]

Sudhindranath Dutta, a renowned poet and critic had once written to Rabindranath on this matter of Bergson's influence in his poetry. He wrote:

> …A Bengali philosopher had once tried to seek common ground between 'Balaka' and Bergson's theory in a lecture. As far as I could recall, I did not agree with his conclusions then and such exercises appear ridiculous to me even now. But as your new books reach me one after another in succession, I realise that your poetry even while being a sworn enemy of the invigorating prattle of Bergson's élan vital remains nevertheless the best evidence of creative impulse reaching out to the unknown beyond. Your theory of becoming is more eternal and ceaseless, more self-originated and self-contained than Bergson's.[21]

Romain Rolland

In 1916, on his first trip to Japan, Rabindranath saw the beautiful civilization of Japan being destroyed by fierce nationalism and militarism, that had been learnt from the West. He delivered at the Imperial University of Tokyo, a lecture denouncing nationalism at the very height of war-fever in Japan. The lectures delivered in Japan soon reached Europe. The humane personality of Rabindranath appealed with poignant force to Romain Rolland, who, through his sister, translated them with an introduction in the midst of the struggle of nations. He declared that a new voice had arisen in the East proclaiming peace and goodwill to mankind and called upon Europe to listen to it with humility.

Romain Rolland (1866-1944) was a French dramatist, novelist, art historian who was awarded the Nobel Prize in Literature in 1915. In his memoirs, *On the Edges of Time,* Rathindranath recalls: 'Strangely enough, whenever Father mentioned his desire to meet Romain Rolland our French friends, somehow or other, avoided the subject. By his attitude towards the defeated Germans,

Rolland had become a *persona non grata* in France and nobody cared to help us to find him, although we discovered that he was then living in Paris…'[22]

Rabindranath had been corresponding with Rolland for many years. In April 1919, Rolland invited him, along with many other celebrated thinkers, to sign an anti-national Declaration of the Independence of the Spirit, which Rabindranath did. They were kindred spirits and recognized each other's qualities. In 1925, Rabindranath wrote,

> Human civilizations have their genesis in individuals, and they also have their protectors in them. One of the few proofs that the present day is not utterly barren of them is the life and work of Romain Rolland. And that the present day needs him most is proved by the scourging he has received from it, which is a true recognition of his greatness by his fellow beings.[23]

Paris, 1921

In 1920, after spending about six weeks in France, he left for Holland and Belgium and then spent some months in USA. On 16 April 1921, Rabindranath took his first air flight from Croyden to Paris on board "Goliath", a Farman plane, carrying 12 passengers to Paris. Albert Kahn received them at the airport and filmed the descent of the plane.[24]

Rabindranath spent a hectic fortnight in Paris before leaving for Strasbourg. He gave a lecture on 'Indian Folk Religion' at the Musée Guimet. He read a paper, 'East and West' at the Conference at the Palais de Justice before Mr. Kahn's Committee. At the University in Strasbourg he delivered his lecture, 'Message of the Forest' before an immense crowd. Professor Lévi introduced Rabindranath:

> In entertaining, tonight, Rabindranath Tagore, the University of Strasbourg do render homage not only to a poet of genius and a genius marking the millenium of a great nation, the French University of Strasbourg entertains a sister university of India. For the last twenty years Tagore has been dreaming to dedicate a university to his country, where India, long confined in her isolation would enter into contact with the thought and civilization of the whole world…[25]

An interest in India seemed to have surfaced among the intellectual elite of France. Henri Barbusse (1873-1935), the writer and socialist leader had founded *Clarte*, a periodical for new writers in literature. He was looking for essays on India by Indian writers and wrote to Rabindranath for help. Rabindranath wrote to Pramatha Chaudhury who knew French, to arrange to send these writings. Rabindranath seemed also to have brought together quite a few couples in matrimony because of their mutual interest in the poetry of the Indian poet! Professor Le Brun met his future wife through Rabindranath's poetry. The case of Alfred Foucher and his wife Eugenie had a similar story. Eugenie was writing a dissertation on Rabindranath and often went to Alfred for guidance. Alfred Foucher was professor at Sorbonne and President of the Institute of Indian Civilization. Their frequent meetings led to the altar!

Rabindranath enjoyed the company of these French scholars.

French Studies in Visva-Bharati

On his return from Europe, Rabindranath introduced the study of French language and literature in Visva-Bharati. The teacher appointed was a Parsee gentleman named H.P. Morriswala. His teaching was augmented by French visitors who were only too happy to take a few classes. Paul Richard, husband of Mirra Richard (later to become Mother of Sri Aurobindo Ashram) stayed for a few weeks and taught conversational French. Later, when Sylvain Lévi joined as Visiting Professor, his wife, Desirée, taught French language.

The tercentenary birth anniversary of Moliere was celebrated in Santiniketan. Morriswala spoke on the life and works of Moliere. Sylvain Lévi recited a sonnet by the French poet and read an extract of his play. Rabindranath, who presided over the occasion, spoke on his experience of reading Moliere. His brother, Jyotirindranath had translated Moliere's plays into Bangla from the French.

On another day, Rabindranath translated into Bangla an English translation of Racine's play, *Berenice* which he read out to his students.

We may recall his speech, 'The Centre of Indian Culture' where he had said:

> Let me state clearly that I have no distrust of any culture because of its foreign character. On the contrary, I believe that the shock of outside forces is necessary for maintaining the vitality of our intellect…European culture has come to us not only with its knowledge but with its speed. Even when our assimilation is imperfect and aberrations follow, it is rousing our intellectual life from the inertia of formal habits. The contradictions it offers to our traditions makes our consciousness glow.[26]

In 1924, on his way to South America, he stayed for a week as Kahn's guest in Paris. In 1926, after his controversial visit to fascist Italy, Rabindranath toured Europe. He visited Rolland at Villeneuve and stayed in Paris, again as Kahn's guest for a few days. He wished to keep his visit private and met only a few friends. He met Sylvain Lévi and his wife Desirée. He also met Professor Jules Bloch, a linguist who studied Indian languages.

Paris, 1930

This visit to France in 1930 was a milestone in the life of Rabindranath Tagore. In Paris Rabindranath's career as a visual artist was launched at the ultra-modern *La Theatre Pigalle.* He arrived in Marseilles with a box of 400 paintings which he hoped to show in Europe. He was a guest of Albert Kahn but this time at Villa Dunure, a beautiful seaside villa at Cap Martin. Victoria Ocampo, his Argentine hostess who was in Paris, came over to meet him and on seeing the paintings, promised to arrange for an exhibition in Paris.

Paris Diary

On 20 April, Rabindranath Tagore arrived in Paris; he was received at the station by Victoria Ocampo and Dal Hogman, husband of Andrée Karpelès. Rabindranath stayed at Mozart Gardens. After lunch, Ocampo brought Georges Henri Rivière of the Ethnographic Museum, who was most impressed with the paintings and promised all help with the exhibition. The next day, Victoria Ocampo brought in Paul Valery (1871-1945), French poet and polymath, to see his pictures and who, after seeing them said, 'They had to be exposed, that they would be a lesson for French artists.'[27] On 23 April, Rabindranath painted all morning. In the afternoon, André Gide came to meet the Poet. André Lhote (1885-1962), French sculptor and painter, also influential as a teacher and writer of art, came to select pictures for the exhibition; he chose 120 paintings.

It was decided that the exhibition would take place in the exhibiting rooms of Le Thiatre Pigalle, Rothschild's new ultramodern theatre. The catalogue was prepared and Comtesse de Noailles wrote the Preface.

On 28 April, Sylvain and Madame Lévi visited Rabindranath and on 30 April, Ezra Pound, the poet, visited Rabindranath. At 4:30. that evening, Rabindranath visited Sorbonne and met leading professors of the university and Indian students at Sorbonne.

On 2 May, after lunch, he met Drieu La Rochelle, (1893-1945), a French writer. At 3:30 pm. the exhibition of Rabindranath's paintings was opened under the patronage of the *Association française des amis de l'Orient.*.

An interesting letter written by Andrée Karpelès to Rathindranath who could not be present gives an eye-witness account of the exhibition. Excerpts from that letter will give the reader an idea of how Paris reacted to the paintings of Rabindranath.

> We arrived at the exhibition at the opening -- at 3 pm and there were already people carefully looking at the pictures. The rooms -- well lighted up by invisible lights and with grey velvet on the walls, are in the cellar. "*Tout Paris*" knows this very modern theater, but this was my first and surprised glimpse of it: a queer staircase leading to a cellar transformed in [to] a steamer deck (a bar) and on the other side leading to the exhibition rooms. In these, I, at once forgot the bar, the modernism [sic], etc. and just was struck by Gurudeva's pictures, hung up, they strike still more from far away than when I admired them first. The purity of their outline, their strength, the sureness of their values, and intensity of their meaning carry their message very far, they can support the test of different lights, of distance, of new surroundings.
>
> There was one very big room and two smaller ones, not crambed [sic] full; the pictures were at their best there, well selected, well hung, well isolated.

The crowd intensified more and more; of course when Gurudeva entered, clad in white (like a huge statue overlooking, from the stairs, the crowd of dwarfs), people turned from the pictures to gaze at him, and the word of 'Prophet' was on many lips.

He was soon surrounded. La Comtesse de Noailles with a large black skirt, wide like a Marwaree one, made a suitable figure to be photographed with him; she was vivid, nervous, passionate, and the catalogue containing her poetical introduction was a success for her also.

People of all corners of the world were in the room, from China came a young poet, from Ceylon there were de Fonseka and his wife; my Peruvian friends bubbling with admiration; two 'saris' were there but not the 'sari' I wished for, draping Pratima; all parts of America and England were represented; I don't know by what miracle Gretchen Green suddenly arrived, straight from New-York! I heard Italian ladys [sic] going in raptures over the colours and trying to guess the secrets of this 'new art'. Our friends and publisher Monsieur Roches said: 'mais ce n'est pas l'ouvre d'un seul homme, c'est l'ouvre de tout un peuple!' The landscapes seemed to attract much attention; they were compared and preferred to Victor Hugo's landscapes. The name of Gauguin was often pronounced also. People need a point of comparaison [sic], a lifebuoy for their mind -- one is always afraid of something entirely new; that might take you away from the safe shore of every-day impressions to the dangerous shores of the Unknown.

Some artists said: 'textures, and leather work, and frescoes ought to be made after these decorations". Some others were sorry there were not more pictures in the style of the simple serene drawings with one pure perfect outline. H. Jviman(?) was there, very interested; I caught a glimpse of the different great Sorbonne Professors whose judgment I do not care for; there was Goloubeff also; all reacted in different ways, but none were indifferent, visitors did not just come out of snobism [sic] attracted by the artist's name, nor out of mere curiosity; they remained a long time; saw the pictures over and over again, discussed them with friends, etc.-- one felt in the air the electricity of human minds excited, the whole atmosphere was stirred with interest, with life -- This is quite a new school, said a lady, "Nothing like the Calcutta School or Abanindranath's work.'

- No it's quite new, I answered, it's the first result of the 'Santiniketan School'.

- 'So all his life the poet has been a painter, said another; it's painted by someone who has such a sure hand; who mastered the workmanship and enslaved the lines.'

Dal and I stayed till the end, half past six; it was so crowded and so hot that your father retired in a little side-room; people begged to be introduced etc. ...

As it was, there was a little too much photographing going on with the brutal noise of magnesium, but on the whole it is *wonderful* how quickly the whole thing has been organized, the pictures framed and the necessary rumours spread in time, and in so short a time ...[28]

The exhibition was a huge success and was followed by exhibitions all over Europe and England. The reviews were encouraging. Henri Bidou, writing for the Paris Press, said:

> ...A latent genius was asleep; that is made plain by the sureness of the design, the beauty of the tones, the liveliness of every detail, and the sense of ornament. For almost a lifetime, this genius has been kept in the shadows, for the highly developed faculties of the conscious mind left no room for the expression of this hidden force. One fine day it revealed itself and the poet felt that another person was being manifested in him, but the new minister has not changed the laws of the State.[29]

Rabindranath often said, that his poetry and songs, he would leave for his countrymen, but his paintings belonged to the West. In a letter to Nirmalkumari Mahalanobis, close associate and confidante, he had written in 1930:

> When I write poetry, the Bengali language emerges spontaneously along with the thoughts evoked in them. But when I paint pictures, neither the lines nor the colours belong to any region. This belongs to whoever appreciates it - it is not necessarily of Bengal only because I happen to be a Bengali. Therefore of my own accord I hand over these pictures to the West. My countrymen have perhaps come to know that I belong to no particular race—this is the reason they are averse to me—they do not hesitate to malign me. That I am not hundred per cent Bengali, that I am equally of Europe, too - may this be proved through my paintings.[30]

Rabindranath Remembered

Rabindranath did not visit France after 1930. He did correspond with some of his French friends like Sylvain Lévi and Romain Rolland. In the early thirties, Gide wrote for the Golden Book of Tagore a note on the *Gitanjali:* 'The incomparable poetic purity of the small book was radiant in my eyes with such a brilliance that I consider it a matter of great honour to bring a reflection of it to France. Throughout the war this fixed star has continued to shine above all our political and religious discords and has poured on the world a tranquil light of love, confidence and peace.'[31]

During Rabindranath's centenary year, France celebrated the Indian poet rather grandly. The Bibliotheque Nationale de France had an exhibition of translations of his works as well as books and articles written on him. Many periodicals brought out special Tagore numbers. Prithvindranath Mukherjee brought out a book, *Pharasider chokhe Rabindranath* (Rabindranath in the Eyes of the French). This book contained Bangla translations of 18 essays by various French scholars on Rabindranath. In an essay by St John Perse, he said:

> Rabindranath was worried about the dreadful consequences of materialism and the future of humanity dependent on a civilisation that relied wholly on industrial labour. He spoke of a twin crisis of collective humanity in need of international fraternity and the individual in need of total perfection.
>
> Where lies the path to salvation? To which refuge shall we turn when, according to Rabindranath our only goal should be the preservation of only the spiritual amongst the deepest human impulses?

> The Anglo-Saxon theory of application did not reassure him. He did not trust the French love for reason either, where he failed to discern an extreme form of human urge. But he knew that the French soul had been enriched by a long legacy of moral philosophers and lovers of humanity, that the French society was built upon a spontaneous liberalism and a part of the French soul empathises easily with the joys and sufferings of humanity at large.
>
> He had a deep regard for the humanism and liberal consciousness of the French indologists.[32]

Even though not on such a scale, but Rabindranath's 150th birth anniversary did not pass unnoticed by the French. There were cultural events, and special TV and radio programmes. The Government of India sponsored an exhibition of his paintings, *The Last Harvest*, in Paris in 2012.

Rabindranath's works are still being translated afresh in French.

Tagore as Envoy

In 1921, Rabindranath founded his university, Visva-Bharati. One of the objectives of this university was: 'To seek to realize in a common fellowship of study the meeting of the East and West, and thus ultimately to strengthen the fundamental conditions of world peace through the establishment of free communication of ideas between the two hemispheres.'

He realized the need for men to understand values other than those which have come to be identified with one's own way of life. One needed to know about other cultures and other ways of life and it was with this in mind that he founded his university. It was the same ideals that made him visit different countries to spread this message.

The idea of Visva-Bharati encapsulates a viable blueprint for the survival of humankind in harmony, which we have as a legacy from Rabindranath.

Rabindranath was a wayfaring poet and was one of the most widely travelled persons of his age. Travel was not very comfortable in those days and his yearning to see the world and meet the great minds of his times proved strong enough to warrant such frequent trips. It was growing on him, more and more that he had a mission in the world. On one level, he was a Bengali and longed for his nest in Santiniketan but on another level, he was an Internationalist, far ahead of his times and he soared where few could reach. He believed it was possible to create a world of peace and harmony and brotherhood and he did and said all that he could to work towards that goal. He was unafraid to speak out clearly what he believed in. Nearly a hundred years ago, Rabindranath was a lone visionary who went from country to country with the message of universal brotherhood.

The Poet Writes to Andre

I had started with Andre Gillaumeau's letter to Rabindranath. We had found no reply in the file. Did Rabindranath respond to this vulnerable, sensitive young man? We hoped our poet would not disappoint Andre. We, at Rabindra-Bhavana, were so involved with Andre's letter, we decided to write to him at the address he had given. To our surprise, a letter arrived from France. This is what Andre said:

> "Your letter has finally reached me, after some detours. ...
>
> I remember very clearly that letter and it is with surprise and happiness that I traced it, as well as with nostalgia. How far is all this; I was 20 years, now I am 74.
>
> I remember very well to have received an answer, written by a student a disciple, but I do not have it anymore. The essence of the answer was: you have to work on mutual understanding of the various civilizations.
>
> As far as the problem of action and contemplation is concerned...I would not be able, yet, at my age to give it a solution. I am not sure if there is one, perhaps it is not necessary also. I had a career as an officer; this has served me as far as action is concerned, mainly in the period under consideration. But I am convinced that there is more true prayer in a second of contemplation than in a day of action.
>
> Thank you, Madame, to have sent me that letter. It made me half a century younger."[33]

We, on our side, felt vindicated in our belief in Rabindranath. Andre did survive, and with his sensitivity intact. Another life had been touched by our poet.

Notes

1 Andre Gillaumeau to Rabindranath, Yerres, 27 February 1937 (Rabindra Bhavana Archives)

2 Rolland to Rabindranath: Villeneuve, 8 July 1926 (Rabindra-Bhavana Archives, Visva-Bharati)

3 Jean Guehenon: "Le Message de l'Orient—Rabindranath Tagore" in *La Revue de Paris,* September 1, 1919.

4 Krishna Kripalani. *Dwarkanath Tagore: A Forgotten Pioneer.* (New Delhi: National Book Trust, 1980).

5 Rabindranath Tagore. *Letters from a Sojourner in Europe*, edited by Supriya Roy. (Kolkata: Visva-Bharati, 2008) [Tagore Travelogues- translation of *Yurop Prabashir Patra*] Translated by Manjari Chakravarti; pp. 33-34

6 Translated from *Yurop Jatrir Diary. (*Kolkata: Visva-Bharati, 1986). Translated by Supriya Roy, not published.

7 Translated by Supriya Roy from *Yurop Jatrir Diary.*

8 *Pather Sanchay,* Kolkata: 1346 B.S.

9 Translated by Supriya Roy from "Jatrar Purvapatra" in *Pather Sanchay. (*Kolkata: Visva-Bharati, 1406 B.S).

10 *L'Inde Où J'ai Vecu*, Paris:1986.

11 Translated from "Vilatjatrir Patra", *Santiniketan Patrika*, Jyaishtha, 1327 B.S. Translated by Supriya Roy. (Unpublished).

12 Paris, 13 August 1920 and 20 August 1920.

13 Translated from the Bengali translation, "Pharasider Chokhe Rabindranath" by Suzanne Karpelès published in *Visver Chokhe Rabindranath* (Rabindranath through the Eyes of the World), compiled by Prithvindranath Mukherjee. (Kolkata: Rupa, 1994), pp. 182-87

14 Translated by Santosankar Dasgupta for this article (unpublished).

15 Translated by Chinmoy Guha in *Gitanjali: Song Offerings. (*New Delhi: UBSPD, 2003), p. 277.

16 Letter to C.F. Andrews , Paris, 28 August 1920.

17 Rathindranath Tagore *The Diaries of Rathindranath Tagore,* edited by Supriya Roy. (Kolkata: Karigar, 2013), pp. 99-100.

18 Rabindranath to CF Andrews, Paris: 20 August 1920.

19 Prabhatkumar Mukherjee, *Rabindra-Jivani,* vol. 2. (Kolkata: Visva-Bharati, 1395 B.S.),. p. 486.

20 Translated by Supriya Chaudhuri in *Rabindranath Tagore: Selected Poems*, edited by Sukanta Chaudhuri. (New Delhi: Oxford, 2004), pp. 221-24.

21 Translated by the editor for this essay (unpublished). Sudhindranath Dutta to Rabindranath: 26 September 1935.

22 Rathindranath Tagore. *On the Edges of Time, (*Kolkata: Visva-Bharati, 1981).

23 "To Romain Rolland: an appreciation" (5 October 1925) in *Modern Review*, January 1926; p. 49.

24 This film can be seen on You-tube over the Internet.

25 "Rabindranath Tagore in Strasbourg", *Modern Review,* July 1921, pp. 95-96.

26 *The Centre of Indian Culture. (*Kolkata: Visva-Bharati, 2001), pp. 33-34.

27 Andree Karpeles to Rathindranath, Paris, 23 April 1930.

28 Letter from Andrée Karpelès to Rathindranath, undated but written around the 2nd or 3rd of May 1930, Rabindra-bhavana Archives, VB.

29 Newspaper clippings, Rabindra-Bhavana Library, VB.

30 Rabindranath to Nirmalkumari Mahalanobis, Berlin: 28 August 1930.

31 *The Golden Book of Tagore*, ed. Ramananda Chatterjee. (Kolkata: The Golden Book Committee,1990) reprint, p.88.

32 In the book by Prithvindranath Mukherjee, *Visver Chokhe Rabindranath, (*Kolkata: Rupa, 1994), pp.104-05.

33 Paris, 23 April 1991.

Albert Kahn Collection

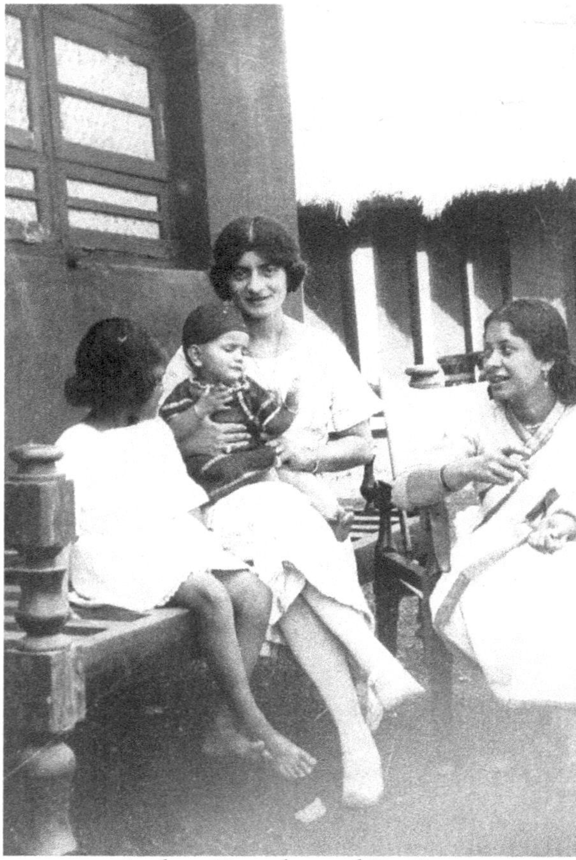

Andree Karpeles and Pratima

Comtesse Noaille

In Albert Kahn's garden

Albert Kahn

RT 1878

Paris 1930

RT by Andree Karpeles

RT with Sylvain Lévi

Sylvain and Desiree Lévi in Santiniketan

Villeneuve, 1926 With Rolland

Tagore and Italy

Stefano Beggiora

Let me dream that from the fire-bath
the immortal soul of Italy will come out
clothed in quenchless light[1]

Cultural Relations between Bengali Intellectuals and Italy between late 19th and Early 20th Century.

The historical period in which the intellectual genius of Rabindranath Tagore flourished is a crucial one for modern India. The period between the end of the 19th and the first decades of the 20th century was a season of transition and a time of social and political changes, most of a dramatic nature for both Bengal and the whole Indian subcontinent. Social, religious and new political and ideological trends on a regional and national level were about to open the gates to Independence. In this scenario, Tagore was a landmark. A holistic intellectual and an activist and social reformer, he transcended reality by virtue of his poetics and thought, and eventually became immortal.

Tagore contributed to Bengali internationalism, a culturally active school that worked to establish a fruitful relationship with the West on political as well as cultural bases. Its principles were cosmopolitanism, humanitarianism and pacifism. The India of the early twentieth century appeared to Bengali intellectuals as a country profoundly aware of its cultural richness and ready to assume its responsibilities in the progress of humanity.

Swami Vivekananda (1863-1902), another prominent representative of Bengali internationalism, extolled ideals of tolerance, mutual respect and charitable compassion. The millenarian Indian tradition had much to teach to the modern West by means of inter-faith dialogue. This was thought as functional to the redemption of the human being (Dasgupta, 7).

The geopolitical layout of the world at that time, the impact of major technological developments and the political and economic hegemony of the West contributed to the colonial subjugation of India. Claiming self-governance and expanding international relations were thus pivotal assets on the path towards Independence. In this particular period, India, following the Bengali Renaissance, sought a direct, unmediated relationship with the rest of the world thus overcoming the barriers erected by colonialism. Particular attention was paid to those contexts (global or regional) that promoted self-determination, unity, independence and affirmation of identity. Such politics was basically an *ante litteram* 'state-making' or 'nation building.' Back then, however, such initiatives were developing in a moment of acute crisis although they were moving away from imperialist or colonialist ideologies and ethnic bias.

Although the 1917 Great October Socialist Revolution in Russia was the first to cause a global crisis in applying Marxism in practice and thereby paving the way for communism and world socialism, the United States, Germany, Ireland and Italy also became arenas for exploration and confrontation (Dasgupta, 59). Yet a close engagement with India as a free country was unlikely in the West, especially in the political scenario of that time. Any opening was deemed to be looked upon as anti-British. Conversely, cultural exchanges were encouraged. In fact, India found rather receptive milieus. Personalities like Rabindranath Tagore, Mohandas K. Gandhi, Subhas Chandra Bose and many others had the occasion to visit Europe and Italy not only to evaluate possible alliances and to explore political strategies functional to Indian nationalism, but also to test, compare and share their perspectives and concerns on human values across and beyond national borders. In the period between the two world wars, there were numerous contacts and friendly relations between members of the Bengali cultural elites and Italian academics, intellectuals and politicians (Prayer 1994, 1-15).

The history of Italy served as a model for many Bengali intellectuals. Conscious of this, Rabindranath Tagore, more than others, was willing to promote cultural exchanges. Since 1870, Italy had become a constant interlocutor for Indian nationalism. Its role in the construction process of a European culture through the Renaissance had been a crucial one. Such notions, however, reached India only after being filtered by the British. The liberalism of the educated classes in India had been influenced by Victorian culture. Hence the importance of Italian history, whose artists and intellectuals were a source of the greatest inspiration. The Italy of Dante, Michelangelo, Machiavelli (author of *The Prince*, a work that is considered the Italian *Arthashastra* of Kautilya) was one that had fought for independence, national unity and dignity. Furthermore Italy, just like India, was characterised by great linguistic and cultural diversity, was politically divided into principalities and kingdoms, and had suffered foreign occupation. The great changes of the Italian Risorgimento were a source of inspiration for the Bengali Renaissance. Bengali intellectuals and reformers read and quoted the writings and speeches of Italian 'freedom fighters', learning from their self-abnegation and total commitment to the national cause. In particular, the works of Giuseppe Mazzini were translated into several languages. In Bengal, Mazzini was considered a useful example to the national cause (Ray, 11). His admirers, each with a different political agenda, exalted him and tried to emulate his Republican ethics, while the religious element was expected to serve the purpose of making nationalism universal and immortal. The latter aspect was particularly cherished in India and received the heartfelt approval of Mahatma Gandhi. Moreover, Mazzini's hope for a 'new religion' showed a certain degree of fascination towards eastern traditions.

Few decades later, his call for unity and his appeal to put an end to autocratic governments attracted the attention of Rabindranath Tagore. On the basis of such fascination, it is safe to argue that the *mahakabi* approached Italy in an effort to explore the local political situation as well as to promote cultural exchange. Like his contemporaries and predecessors in Bengal, Tagore looked at Italy as a leading force in the development of new ideas in the heart of Europe. Further to Risorgimento, Tagore, in his earliest studies, explored Roman classicism, Roman Catholicism, the arts and liberal thought that made Italy great in history. In particular, Tagore was keen to meet Senator Benedetto Croce, one of the greatest Italian intellectuals, philosophers and liberal thinkers of the twentieth century. Croce's works enjoyed great fame both in Europe and in Bengal, and Tagore had read his essays on aesthetics and literary criticism.

As for Tagore's standing, his name became a famous one only after his winning the Nobel Prize in 1913. From then to the mid-1920s, thirty works of the poet were to be translated in Italian. Carabba of Lanciano took the responsibility for such an important publishing venture. Some of the works of Tagore - e.g. *Sadhana. Reale concezione della vita* (Sādhanā, The Realization of Life), *Gitanjali* (Gītānjali, Song Offerings), *Il giardiniere* (The Gardener), *Uccelli migranti* (Stray Birds), *Citra: dramma in un atto* (Citrāngadā/Chitra: A Play in One Act), *Il re della camera buia* (Rājā/ The King of the Dark Chamber), *La luna crescente* (The Crescent Moon) - were already familiar to Italian intellectuals before the 1920s (Beggiora B, 133-39). They had been published along with the most important writings of Niccolò Machiavelli, Leon Battista Alberti, Aristotle, Seneca, and the poets Giacomo Leopardi and Ugo Foscolo. It was the launch of his most recent works in Italian that facilitated the visit of the *mahakabi* to Italy.

Unfortunately, during the gap between the two world wars, Italy was fast moving towards one of the darkest pages of its modern history. The rise of Fascism, seemingly a fervent culture motivated by nationalism and the myth of the glory of Rome, deceived many. Tagore had the chance to meet Benito Mussolini and leading figures of the fascist orientalist school. Unwillingly, he was dragged into the political arena of a Europe in turmoil. In response to such exploitation, Tagore, once again, proved himself capable of representing Bengali internationalism at its best and showed responsibility, depth of spirit and sound ethics.

The Journeys of Rabindranath Tagore in Italy

The fortune of Tagore in the West depended on a number of factors. A fine gentleman, educated, with the charisma of a sage, Rabindranath Tagore somehow satisfied the hunger for exoticism in certain British cultural circles. But it was his poetry, closer to British taste than that of any other Indian intellectual, which contributed to his affirmation. India, however, in the lyrics of Tagore, was still a distant geo-cultural reality whereas his introspective vision was overall perceived as far too personal. In other words, when the West could not understand Tagore, a loss of fascination took place. Ezra Pound and W.B. Yeats were among the first who praised the literary genius of Tagore. Yet soon after celebrating his poems and granting him access to the peaks of success and fame in old Europe, they drastically changed their mind, especially Pound (Hurwitz, 53-63).

Though trying to show appropriate modesty, Tagore rejoiced in his fame. This explains his surprise towards a consensus that was unjustifiably whimsical. This element is extremely useful to introduce the present work. The Western perception of India and India-ness were superficial and instrumental to historical contingencies. This explains why the figure of Tagore – though discussed and translated in several publications – is still little known to the lay public. To some, Tagore was truly a master. Yet despite the admiration and consensus he was able to gain, the Fascist regime manipulated him. The glory of the poet and his celebration ended abruptly as soon as Tagore distanced himself from the abuses of the Fascist regime. What was happening in Italy was not news. India and her culture had been manipulated and misrepresented since antiquity. Orientalism contributed to the creation of suitable icons, often not corresponding to the reality, in order to praise what was convenient.

The very first visit of Rabindranath Tagore in Italy dates back to 1878, when he passed a few hours in Brindisi, during a stopover on his way to Great Britain, which gave the poet the opportunity to enjoy the Italian countryside and its people. When he arrived in Italy for his first official visit, in 1925, his fame was already great (Formichi, 14).

The visit took long to be arranged since various diplomatic parties needed to come to an agreement. Back then, Italian academics had important relations with Kalidas Nag, a young Bengali scholar who had moved to Paris around 1920 and became the pupil of Sylvain Lévi, the French Orientalist and Indologist who was famous for his studies on Buddhism (Radhakrishnan, 433). Nag, in 1921, visited Italy and was in touch with Italian Indologists, most notably Carlo Formichi and Giuseppe Tucci. Between 1921 and 1924, Nag was instrumental in initiating cultural contacts between Italian scholars and Kolkata intellectuals. Eventually, he built a bridge between Tagore and the elite of Viśva Bhāratī and Italy (Formichi, 49). On the one hand, Nag made it possible for the circulation of the unpublished works of the poet. On the other, Formichi worked in order to obtain an official invitation for Tagore to visit Rome, where he was expected to deliver a series of lectures.

The invitation came in 1925 not directly from the Italian Government but from an academic institution (Formichi, 12). In the original programme, during a twenty-five day sojourn on his way back from South America, Tagore would have visited Genoa, Florence, Rome, Turin, Milan and Venice. The poet, though honoured, was trying not to get involved in political issues, especially considering that, from the 1920s, Italy was going through a time of profound crisis and political turmoil. This is revealed by a certain embarrassment in the diplomatic circles with Fascism. To prevent accidents and to avoid jeopardising his reputation at an international level, Tagore tried to avoid public expressions of esteem by Mussolini, who himself was very keen to meet the poet (Beggiora A 50-51).

The tour was abruptly suspended. Tagore fell ill and part of the programme was cancelled. This is the reason why he quickly reached Venice, via Milan, somehow evading a pompous welcome ceremony in the capital. The soul of the poet was overwhelmed by the warm welcome of the people of Milan and Venice, and the circumstances that forced him to quickly close the tour were looked at as a missed opportunity (Formichi, 29). This, however, paved the way for a second official visit to

Italy, which was due to take place the following year.

Shortly before the arrival in Italy of Tagore, there had been some facts of considerable political importance. In October 1922 there was the famous march on Rome following which King Victor Emmanuel III handed power to Benito Mussolini and his National Fascist Party. Previously there had been a great deal of political unrest, details of which cannot be summarised here. It should be noted, however, that in those years the Italian economy was very unstable and there was an alarming increase in levels of public disorder. The parliamentary system was unable to cope with the crisis. Mussolini was able to quickly take control using his private militia, the Blackshirts, who consistently exercised pressure on all political opponents of il Duce. Basic civil rights were withdrawn, e.g. freedom of speech and press. In short, all information had to pass through the filter of Fascist censure. The next elections saw an overwhelming victory of Fascism, this being the result of coercion, propaganda, intimidation and violence.

This season was characterised by political murders (e.g. Giacomo Matteotti in 1924 and Giovanni Amendola in 1925) and the exile of several intellectuals, liberal thinkers and Socialists. Some of them, from France and Switzerland, started to openly criticise Fascism and set the bases for resistance. Il Duce was aware that the potential damage that might result from disastrous international relationships. Fascist propaganda started to work and foreign journalists and cultural delegations were often invited to witness the important cultural and technological achievements of Fascist Italy. The regime was determined to promote its work in an international arena, especially after the brutal events that characterised its debut, in order to divert attention away from the 'Matteotti accident' and the mass persecutions carried out by Fascist action squads.

Tagore was unaware of the drama that was taking place in Italy. Yet, once *in loco*, he was able to gain a clearer picture. Besides, the totalitarian turn of Fascism was yet to come and, until 1926, Tagore probably did not have the means to understand the arguments of Mussolini. In fact, both Italy and India were struggling with the problem of national rebirth. Such processes involved moral imperatives. However, for the poet this was supposed not to be detrimental of individual freedom. Il Duce was aware of Tagore's mind. This was even clearer when, during a conference, Tagore warned European powers about the use of science for the manufacture of instruments of destruction and about false pacifism, probably an allusion to the Italian situation in its domestic policy and the plans of colonialism (Prayer 1996, 19). Although the Italian Government did not appreciate the speech, Mussolini, interested in a dialogue with Tagore, sought the approval of the poet and observed that the suspension of democracy was only in response to a specific historical necessity and that for this reason his dictatorship would only be of a temporary nature.

In 1925, Mussolini expressed his deep appreciation for the work and thought of Tagore. He sent books to the Viśva Bhāratī University in India through those who then were the more distinguished Italian orientalists: Carlo Formichi and Giuseppe Tucci. The following year, Tagore decided to accept the invitation from Mussolini himself and when he arrived in Rome, he was received by il Duce and King Victor Emmanuel. Here, he witnessed a theatrical representation of one of his dramas, *Chitra*, and a *lectio magistralis,* accompanied with all the honours the regime could display. He had the opportunity of discussing philosophical problems with Senator Benedetto Croce in a personal conversation, but under the strict surveillance of the regime. But it was only when he was out of the country that he freely met other anti-fascist exiles. In Switzerland, Tagore realised that Fascist propaganda had manipulated his visit (Radhakrishnan, 485). He was made aware by some refugees about the painful reality of the regime in Italy, and was informed about the brutal murder of Matteotti. The poet was so moved by what he learnt that shortly thereafter he published an open letter in the *Manchester Guardian*, a British newspaper, where he formally condemned the regime and complained about its manipulative propaganda. This resulted in a drastic turnaround of the Fascist press. The Italian media pathetically tried to destroy the reputation of the poet, who just a moment earlier was celebrated so unreservedly.

Some Considerations on the Relationship between Tagore, Mussolini and the Italian Indologists

The cultural exchange between the Viśva Bhāratī entourage and Italy was of enormous importance. Carlo Formichi, an Indologist, was interested in capturing the attention of Indian government authorities. As soon as he knew about the interest of Mussolini in Tagore, he approached the poet and became the key interlocutor of Italy at Viśva Bhāratī. In order to please Tagore, some five hundred books of Italian history, literature and art (including on Fascist ideology) were sent to Viśva Bhāratī. Tagore replied as under (Kundu):

'I assure you that such expression of sympathy, coming as it does from the representative of the Italian people, will open a channel of communication of cultural relations between your country and ours, which has every possibility to produce an event of great historical relevance.'

Tagore's visit in 1926 was the opportunity to set the foundation for closer cooperation between the two countries. This was mainly achieved by favouring an increase in cultural exchanges by means of seminars, conferences and exchanges of scholars and students. Programmes were developed to encourage the study of the language and culture of the partner country (Formichi 241 and ff). Tagore looked at Italy as a privileged interlocutor in that he recognized some similarities in the history and political circumstances of the two countries. Yet, most of all, Tagore was interested in the exchange of ideas and meetings between individuals from different cultural backgrounds. This, however, proved to be extremely difficult, especially in the aftermath of Tagore's letter on Fascism in the *Manchester Guardian*.

For the Italian Government, the episode was a serious turnaround on the part of the poet. Formichi himself felt betrayed and his relations with India suffered a setback. Tagore was accused by Formichi of not fulfilling his promise to avoid political discourses. Tucci, more diplomatically, always considered Tagore as a *guru* and spent years trying to repair the frayed relations between Italy and India. As an orientalist and explorer, he became the spokesperson of il Duce in the Indian subcontinent.

On the basis of recent analysis of historical documents (e.g. epistolary and statements released by witnesses), it is quite likely that Tagore had been deceived by the regime. This explains why he was initially welcomed, celebrated and loved, and, then,all of a sudden, the trend reversed. Further, on getting to know the real face of Fascism, Tagore noticed that his statements had being conveniently tailored by Formichi before being passed on to the Fascist media. This added to a series of incidents that marred the poet's journey to Italy and eventually led to a worsening of relations between Italian Indologists and Tagore as well as Viśva Bhāratī intellectuals. What appears clear today is that Formichi and Tucci tried in all possible ways to isolate the poet from his usual circle of friends of Bengali intellectuals in order to bring him as close as possible to il Duce.

The meeting between the poet and the dictator was one characterised by mutual appreciation. It was this unexpected courtesy that compelled pacifists and anti-fascist circles in Europe to ask the poet to release a categorical condemnation of the regime. Tagore, however, was under the strict surveillance of Formichi and his entourage and could not, by any means, meet with anti-fascists. During the two meetings with Mussolini, the poet understood the dictator and his charismatic leadership as a model of 'creative force'. This dynamism was reflected in his title, i.e. il Duce (lit. 'the guide') of the Italian people. Led by Mussolini, Italians were supposed to take pride in their heritage and in the rediscovery of the glory of a majestic past. This sacred fire was meant to create a sentiment of national regeneration.

Once in Switzerland, Tagore was contacted by Romain Rolland (Dutta and Robinson, 228-31). There they were joined by another French writer, Georges Duhamel, who tried to convince the poet of the actual situation in Italy. He thus came to know about the persecutions and of the many expatriates living in Switzerland and France who had to flee in order to survive the darkest age of modern Italy. It would only be after the testimony of Giacinta Salvadori, wife of historian and anti-fascist in exile, Prof. Salvadori, that Tagore wrote the open and uncompromising letter of condemnation of the regime (Prayer, 1994, 54).

From the document we read (Dutta and Robinson, 332)[2]:

'My mind is passing through a conflict. I have my love and gratitude for the people of Italy. I deeply appreciate their feeling of admiration for me, which is so genuine and generous. On the other hand, what Italy revealed in Fascism alienates itself from the ideal picture of that great country which I should love to cherish my heart.'

The Bath of Fire and the Legacy

'Italy clothed in quenchless light' – an article that appeared in the Italian newspaper *La Tribuna* on June 2, 1926 along with Tagore's autograph (signed May 31st) – Italy would emerge thus only after supreme sacrifice, an act of purification through the process of the renewal of the people and the nation: *the bath of fire* of history.

It seems incredible to think how prophetic the words of Tagore were, especially in view of the events to come. The Second World War, the implementation of racial laws and the civil war between Nazi-Fascists and the Resistance are part of the national drama. This apotheosis of violence and abuse was, for Tagore, the opposite of the concept of culture and civilization. Not only did he firmly denounce the British colonial policy (and colonialism in general), he also witnessed how the positive aspects of Western culture were superseded by the horror of world conflicts. In fact, Tagore had the opportunity to closely observe the years of World War 1 and the prelude to the second world conflict. Italy did not hesitate to become involved in both.

The spiritual testament of the poet can be read in his last work, *Crisis of Civilization* (*Sabhyatār Sankat*, 1941). This is a strong act of accusation against the Raj, culpable of having failed to respect human dignity and to have inaugurated in India a police-state, a pathetic parody of civilization that did not deserve respect. This discourse applied to other regimes, in Europe or elsewhere.

In *Crisis of Civilization*, Tagore gives voice to his criticism of colonialism and the administration of the British Raj in India, which he always judged deleterious. However, he does not reject Western culture (e.g. literature, arts, etc.), of which he was an admirer. The main problem of the time was nationalism and patriotism. This led people to believe that there could not be any respect between opposite nations. Such was the case for the several freedom fighters in India, including those in the Congress Party. Tagore was not of this opinion. The brutal plunder, the abuse and finally the crimes committed by the British in his country did not represent a culture clash between two worlds.

The sympathy of some Bengali intellectuals for fascism was limited to idealism, although none of them would appreciate the ideology of dominion promoted by Mussolini. Subhas Chandra Bose himself personally had the opportunity to firmly reject such an insinuation. However, it is plausible that the positive inclination of Bengali internationalism towards the political stances of Fascist Italy had deep roots in the way they culturally viewed nationalism. Tagore probably combined the authoritarianism of Benito Mussolini with the necessity to have a strong personality guiding the political life of a nation (Prayer 1996, 24-25). The ideal of power and virility in India was traditionally linked to the inner qualities of man, to his ability to dominate the tensions of the mind and to purify the body through *sādhanā*, the spiritual discipline. In the theories of Mazzini as well as in fascism, the idea emerges that duty is above right. This is moral philosophy, rather than politics. However, this very doctrine brought unity and self-determination only after civil war and foreign occupation. Fascism was successful in mobilising the masses and to inculcate a spirit of discipline and responsibility towards the nation. The future of the country was supposed to be conducive to major social reforms, great infrastructure, land distribution, industrialisation and scientific progress. Similar ideas were supported by Tagore, although Mahatma Gandhi did not completely agree with them (Wolpert 119-20). This general atmosphere -- and the fact that Italy was a dedicated partner – impressed the Bengali elite and India in general. In short, Italy was a model in the process of affirming cultural tradition as well as indigenous politics. Those were the years when the Swaraj movement was successfully spreading across India and gave voice to Indian nationalism and patriotism.

An example of this trend is the sacralisation of Mother India, which is celebrated in Bankim Chandra Chattopadhyay's *Vande Mātaram* hymn (from the novel *Anandamath*). Tagore made of it the first national anthem which was sung on the occasion of the first session of the Indian National Congress in 1886. The poet was able to capture the feelings of the people of India, i.e. deep affection for their homeland and proud acknowledgment of her roots and diversity (See also Das, 824).

Tagore, however, was profoundly aware of the crisis caused by the colonial exploitation of the homeland. He did not want idolatry of the flag as the dynamic principle of nationalism. Moreover, the freedom of India and her people was not intended in a purely political sense, but as spiritual emancipation, i.e. the freedom of conscience that leads to knowledge and the highest values (Radice). This is the reason why he favoured the awakening of human conscience above all other virtue. Independence would then result naturally.

By virtue of this reasoning, Tagore detached himself from the political arena and focussed on his artistic and intellectual activities. He unambiguously distanced himself from nationalism as a principle. Inevitably, many patriots, that once looked at Tagore as a source of inspiration, felt betrayed and accused him of cowardice and connivance with the English. But Tagore was beyond all this. During his travels in old Europe, and perhaps thanks to his visits to Italy, he developed the idea that nationalism as a principle was a dangerous and potentially harmful concept. He wrote of how the flags, insignia, eagles and other wild rampant beasts, symbol of the European superpowers, were mere simulacra, empty idols around which the blood and the consent of peoples coagulated (Sen, 113-16). This was confirmed by his visits to Japan first and the United States later, when he saw that nationalism was an opportunity of advancement otherwise lost by these cultures. In retrospect, Tagore had sincere admiration for Italy, hence his sadness in criticising the thencurrent political situation. With World War II coming, the times were not mature for a rapprochement. The farsightedness of the poet and his speeches were judged dangerous and looked at with suspicion both abroad and among Indian freedom fighters. For the very same reason, during those difficult years, his fame in the West suffered a setback. At first celebrated, European cultural circles became all of a sudden cooler to his presence and campaign to raise funds and support for his university in Śāntiniketan.

In Italy too Tagore saw the risks of war and nationalism. In fact, he did his best to promote the philosophy of internationalism, universalism and spiritual unity (Rigon, XII, XIV). For Italy in particular and, for the West in general, World War II was to turn into the most dramatic event of the 20th century. War, however, is part of a larger drama, i.e. a human tragedy. In the post-war years, with the birth of the Republic and the years of economic development, Italy is described and discussed in a way that does not take into account the shock caused by a profound process of social change. What followed, *la ricostruzione* (lit. recovery) was a dramatic period. The post-war economic boom led to the inevitable transformation and disappearance of a predominantly peasant culture. As poet and intellectual Pierpaolo Pasolini noted, it was an 'anthropological mutation.' This phenomenon refers to the profound mutations that were taking place with the advent of consumerism. This change, in the original formulation of Pasolini, is one of the most dramatic phenomena in contemporary Italy. It consists of the eradication of traditional Italian culture. In *Gli Italiani non sono più quelli* ('Italians are no longer themselves')[3], Pasolini argued that the middle class completely changed its values and adopted a hedonistic and consumerist approach to life. The power of consumerism contributed substantially to the eradication of traditional indigenous values, paving the way to a hybrid 'mass culture.' This new system is discussed as amoral in that there is no space for consciousness, spirituality, religion, or even patriotism.

India too, now a global power, is going through a rapid process of socio-cultural transformation while at the same time it is proudly clinging to a tradition of millennia. The example of Pasolini seems thus appropriate. The Italian poet, along with other writers, visited India in 1960 on the occasion of a conference commemorating the centenary of the birth of Tagore. His first trip to India is essentially visual, emotional[4]. His nonconformist gaze, little inclined to celebrations, is initially superficial towards both the event and the figure of Tagore. He did not realise that the answers to the contradictions of contemporary society, which he so poignantly described, had already been addressed by Tagore who proved himself a key thinker in social theory. The path indicated by Tagore is the story of Viśva Bhāratī, as a pedagogical centre for the investigation and promotion of culture at all levels. It is along these lines that Italy and India must continue to work together, thus fulfilling Tagore's ideals, especially in terms of youth education. When Rabindranath Tagore came to Italy, some noted that his most inspired speeches were not those delivered on public celebrations but improvised ones, when he was addressing students or pupils informally in schools. This is the memory of him we love to keep (Assagioli, 60-62).

'*Io mi sento giovane come voi, più giovane di voi; lo studente è come un uccello che vola nell'aria e si posa ovunque – così è il poeta, ed per questo che studente e poeta sono fatti per intendersi* (I feel as young as you, younger than you. The student is like a bird flying through the air and landing everywhere - so is the poet, and that's why students and poets naturally understand each other).'

Notes

1 Official greeting of Tagore to Italy on the date of his arrival in Naples, May 31st, 1926 (Das, 989). The Italian newspaper *La Tribuna*, June 2nd, 1926, titled: 'L'Italia ammantata di luce inestinguibile' (Prayer 1996, 16; Dasgupta, 80).

2 Letter to Charles Freer Andrews, Hotel Bristol, Vienna (Austria), 10 July 1926.

3 In the newspaper *Corriere della Sera*, June 10th 1974.

4 On this occasion he wrote *The Scent of India*, translated by David Price, (London: Olive Press, 1984), original title: *L'odore dell'India*; I ed. (Milano: Longanesi, 1962).

References

Aslan, Odette. *Rabindranath Tagore*, (Paris: Seghers (Ligugé, Aubin), 1961).

Assagioli, Roberto. 'Impressioni sulla visita di Tagore in Italia' [*Ultra*, 4, 1926: 43-44, 46-47], *Esoterismo e fascismo: storia, interpretazioni, documenti*, ed. Gianfranco de Turris, Roma: Edizioni Mediterranee, 2006), pp. 60-62.

Beggiora, Stefano (A). 'La vita di Tagore', *Tagore sommo poeta dell'India moderna*, ed. Monia Marchetto, (San Marino, Rimini: Il Cerchio, 2012), pp. 31-54.

--- (B), 'Opere di Tagore tradotte e pubblicate in Italia', *Tagore sommo poeta dell'India moderna*, (ed.) Monia Marchetto, (San Marino, Rimini: II Cerchio, 2012), pp. 133-39.

Das, K. Sisir (ed.) *The English Writings of Rabindranath Tagore: a miscellany. Vol III*, (New Delhi: Sahitya Akademi, 1994).

Dasgupta, Tapati. *Social thought of Rabindranath Tagore: a historical analysis*, (New Delhi: Shakti Malik/ Abhinav Publications, 1993). Dutta, Krishna and Andrew Robinson (eds.), *Selected Letters of Rabindranath Tagore*, (Cambridge: Cambridge University Press, 1997).

Formichi, Carlo, *India e Indiani*, (Milano: Alps, 1929).

Hurwitz, M. Harold, 'Ezra Pound and Rabindranath Tagore', *American Literature*, Vol. 36, n. 1, 1964: 53-63.

Kundu, Kalyan, 'Mussolini and Tagore', *Parabaas*, Kolkata, May 2009. http://www.parabaas.com/rabindranath/articles/pKalyan.html, (accessed April 4, 2012)

Pasolini, Pier Paolo, 'Gli italiani non sono più quelli', *Corriere della sera*, June 10th 1074. (II ed. 'Studio della rivoluzione antropologica in Italia').

Prayer, Mario. *In Search of an Entente. India and Italy from the X to XX Century. A Survey*, (New Delhi: Munshiram Manoharlal Publishers, 1994).

--- 'Internazionalismo e nazionalismo culturale. Gli intellettuali bengalesi e l'Italia negli anni Venti e Trenta', spec. issue *Rivista degli Studi Orientali*, LXIX, 1996: 1-113.

Radhakrishnan, Servapalli ed., *Rabindranath Tagore, A Centenary Volume 1861-1961*, (New Delhi: Sahitya Academi, 1992) (First pub. 1961).

Radice, William, 'Tagore's Poetic Greatness', *Parabaas*, Kolkata, May 2003. <http://www.parabaas.com/rabindranath/articles/pRadice.html, (accessed April 4, 2012).

Ray, K. Mohit ed. *Studies on Rabindhranath Tagore*, (New Delhi: Atlantic Publishers & Distributors, 2004).

Rigon, Marino ed. Tagore/Balaka, Milano: Guanda,1975.

Sen, Amartya, *The Argumentative Indian: Writings on Indian History Culture and Identity*, (London: Allen Lane/Penguin, 2005) (Italian ed. *L'altra India*, Milano: Mondadori, 2005).

Tagore, Mussolini and the Italian Press*

Kalyan Kundu

On 20 July 1926, Rabindranath Tagore wrote to Leonard Elmhirst from Vienna: '[…] it was a sinister design on the part of my evil fate to have brought me into any relation with this man (Mussolini) who may be altogether a fraud and no real personality… and I regret that I have allowed myself to pay this last visit to Italy.'[1] The letter was written in the context of his Italian tour in 1926.

Surprisingly, four years after that visit, Tagore wanted to reconcile with Mussolini and wanted to resume a good relationship with the Italian Government. He expressed such an intention in a letter written in November 1930, from New York that read: 'Your Excellency, I am personally indebted to you for the lavish generosity you showed to me in your hospitality when I was your guest in Italy and I earnestly hope that the misunderstanding which unfortunately caused a barrier between me and the great people you represent, will not remain permanent, and that this expression of my gratitude to you and your nation will be accepted.'[2]

Such an act of ambivalence and contradiction on the part of the Poet often baffles the scholars. While seeking a probable explanation, one has to examine the entire perspective of Tagore's Italian tour in the turmoil of the socio-political scenario in the Italy of the 1920s. It is a fascinating account of a high drama, of an imperfect encounter, of desperation and dissension. The Italian media also played a substantial role in creating an international backlash.

Political Backdrop in Italy in the 1920s

Before October 1922, the year when Mussolini came into power, the economy of Italy was unstable and there was an alarming growth of public disorder, especially the rise of crime and murder. Italian politics was going through a crisis. The Prime Minister, Luigi Facta, stepped down. Mussolini saw that the parliamentary system was crumbling and failing to tackle the crisis. Although the Socialists were one of the most dominant political forces in the country, the intelligentsia failed to accept them as a ruling party. That was the opportune moment for Mussolini to take control of the situation. On 27 October 1922, he organised his infamous march from Milan to Rome with his followers and his semi-military corps, the 'Blackshirts' (*Squadristi*). The progress of the Blackshirts was unhindered, shops looted on the way, telephone exchanges destroyed and government offices occupied; any resistance was harshly treated. The police was a silent spectator. The King Vittorio Emmanuelle III

---

* An abridged version of the above article was first published in an American Webzine, *Parabaas*

declared a state of emergency across the country and ordered the army to impose martial law. Facta had already resigned, another ex-premier Salandra declined to take charge of the country. The King was not keen to offer the post to Giolitti, another ex-premier, in whom he had no confidence. Seeing the popularity of Mussolini, the King took an undemocratic decision – asked Mussolini to form the government, despite cabinet's opposition. Mussolini, whose belligerent followers were causing a menace countrywide, became the 27th prime minister of Italy at the age of 39. On 29 October 1922, the frenzied, jubilant followers of the Prime Minister ransacked the socialist party offices everywhere, bookshops were wrecked or burnt down, and opposition members were humiliated and tortured. The only consolation was that the Italian stock market became stable on Mussolini's appointment.

Mussolini quickly took charge of the situation. From the very beginning he was very cautious about his foreign policy. Mussolini was convinced that his domestic image would not be enhanced unless countries outside Italy had full confidence in him. He repeatedly mentioned to foreign agencies, journalists and dignitaries that fascism was a special ideology only appropriate for Italy and 'not for export.' (This was also mentioned to Tagore).

In 1924, Mussolini and his militia used violence, false propaganda, and vote-rigging to win his first election with a good majority. Some of the socialist leaders such as Amendola and Matteotti openly challenged the fairness of the election. Matteotti also prepared a dossier of evidence describing details of the election's irregularities and sent copies to England and Belgium. As a result, they were eventually assassinated by the Blackshirts. A mass persecution took place and many intellectuals, namely, Professors Salvadori and Salvemini, conductor Toscanini, Angelica Balabanoff, an expert of European socialism, and many others left the country and took refuge elsewhere in Europe.

The appalling treatment of these socialists, especially the assassination of Matteotti and Amendola, had a detrimental effect on Mussolini's image in the rest of the Europe. Adverse criticism found its way to some socialist papers published in France. As a result, Mussolini was looking for a means to divert international attention away from the Matteotti incident.

It was against this political backdrop that Tagore came to Italy in 1925 to begin his first Italian tour on his voyage back from Argentina. Incidentally, although it was his first sojourn in Italy, the poet did touch the Italian shore in 1878 and 1890 on his way to England.

Tagore's First Italian Tour, 1925

In 1920-21, Tagore travelled extensively through Europe, especially the countries devastated by the war. During that time he repeatedly stressed to his European listeners about the need for the meeting of the East and the West through a common fellowship in the hope of promoting world peace. His role, at that time, was that of a cultural ambassador and a one-man peace envoy.

He himself took the initiative through founding his own University—Visva-Bharati at Santiniketan—which he envisaged would be a meeting place for intellectual cooperation between the East and the West. *Yatra Viswam Bhabati Ek Needam* meaning where the world meets in one nest—was the motto of the university. He was looking for visiting professors from Europe for his academy and that was another reason for his tour in Europe in 1920/21.

Tagore came to know about the Italian Professor, Carlo Formichi, from one of his students, Kalidas Nag. Formichi was a great scholar of Indology. At that time, Formichi was the Professor of Sanskrit at the University of Rome. Nag was convinced that the erudite presence of Formichi would be of great benefit to Visva-Bharati. In May 1921, Tagore's son contacted Prof Formichi informing him of his father's intention to visit Italy and to discuss the possibility of the Italian professor joining Visva-Bharati.[3]

Formichi was also aware of Tagore's literary excellence and was looking forward to meeting him and his visit. However, while the Professor was eagerly awaiting Tagore in Florence in May 1921, the tour was suddenly abandoned for some unknown reason. Formichi was disappointed about not meeting Tagore and kept trying other methods to bring Tagore to Italy. He was keen that this visit take place for three main reasons: First, at that point, Tagore had been to most European countries except, of course, Italy; second, he was a genuine admirer of the Poet and was keen to meet him; third, with the help of Tagore and through his contacts, he wanted to fulfil his passion for visiting India.

It took another three years for Prof. Formichi to finally persuade a non-governmental academic organisation Circolo *Filologico Milanese* (Philological Society of Milan) to raise enough funds to bring Tagore to Italy. Two other societies, one in Florence and the other in Turin, also agreed to share the costs of the trip, on condition that Tagore would deliver lectures in both those places. However, the Philological Society was the main sponsor of Tagore's Italian tour in 1925, and they sent an official cable to Tagore at Santiniketan confirming his invitation to Italy. Tagore accepted the invitation. It was all arranged that he would visit Italy on his way back from Peru, where he was originally destined in order to attend the centenary celebrations of Peru's independence.

However, on his way to Peru, while in Argentina, Tagore became seriously ill and he had to cancel his Peruvian trip and returned to Europe after nearly two months of recuperation in Argentina. When he disembarked at Genoa on 19 January, 1925 in the middle of the deep winter, accompanied by his Secretary Leonard Elmhirst, he had not yet fully recovered and was very fatigued. The only people in the quayside to receive Tagore were his family, Prof. Formichi and a representative of the Philological Society. That was the first meeting of the Poet and Prof. Formichi.

From Genoa, Tagore travelled to Milan (21 January, 1925). On the very afternoon of his arrival, he visited *La Scala* to watch Verdi's opera *La Traviata.* The next day, he attended a local meeting with poets, intellectuals and artists arranged at the hotel. The main agenda of the meeting was to discuss the possibility of an exchange programme of scholars and intellectuals between India and Italy. In the evening, he addressed a public lecture organised by the Philological Society. There was no particular title of his lecture.[4] He mainly reminisced about his first Italian experience forty-seven years ago and was very distressed to find the 'the dark misery lowering over Europe'. He openly condemned his listeners: 'Do you realise how the mark of ugliness is everywhere apparent, in your cities, in your commerce, the same monotonous dreadful mask, so that nowhere is there a living expression of the spirit of love?' Reports of that lecture came out in several Italian papers. Unfortunately his lecture upset a section of militant fascists. The spiritual note in Tagore's speech on humanism failed to impress those Italians who were struggling hard to overcome their economic and social decadence. A professor of Iowa University even criticised: '[...] the present fascist government which is operating without check of an intelligent Italian public opinion, international altruism as preached by Tagore cannot live [...]'.[5] Tagore was unaware of this reaction in the media during his stay in Italy.

On the following day, Tagore was received at Peoples' Theatre (Teatro del Popolo) in Milan; gave an hour-long sitting for his portrait to the famous painter Rietti at the Duke's Palace. However, in the evening, he fell ill again and had to cancel all appointments. The Italian medical team did not take any chances, advising Tagore to return to India. His proposed tour in Florence and Turin were called off. Eventually, he went to Brindisi, after a short detour in Venice, to embark on the liner heading for India. The original tour was scheduled for twenty-five days, but he returned to India after only fifteen.

Before leaving Milan, Tagore invited Prof. Formichi to join Visva-Bharati as a visiting professor at Visva-Bharati's expense and requested him to find funds to set up an exchange programme for an Italian scholar and also funds for Italian books and resources for his University. Prof. Formichi accepted the invitation.

This abrupt end of Tagore's tour caused some embarrassment to the organisers. Tagore was equally upset. So he promised his Italian friends and admirers that as soon as his health improved, he would make a second trip to Italy. This he promised in a beautiful poem, which was immediately translated into Italian by Formichi and circulated in the major Italian papers.[6] Tagore sailed back to India on 2nd February 1925.

In 1925, Tagore had no plans to visit Rome or to meet Mussolini although Mussolini was aware of Tagore's presence in Milan.

The Second Italian Tour, 1926

Prof. Formichi received the official appointment letter from Visva-Bharati in July 1925. He was delighted as his visit to India became a reality; at the same time he was a little apprehensive as he was unable to source funding for an Italian scholar for the proposed exchange programme. Furthermore, he was unable to purchase the Italian books he promised Tagore. The situation was a little embarrassing for Formichi as he was aware that other European visiting professors who preceded him, donated substantial amounts of resources to Tagore's university. Being desperate for funds and finding no other alternatives, he ultimately wrote a letter directly to Mussolini explaining everything and requesting support.[7] It was more than likely that it was Prof. Formichi's first contact with Mussolini.

Until that point, Mussolini had no issues with or policies towards India. His only reference to India was found in an article published in *Il Popolo d'Italia* where he predicted that the Moplah rebellion would be the end of British rule in India. For a few years, the above article by Mussolini was his only commentary on India available in Italy.[8]

However, when Formichi's letter arrived at his desk for his personal attention, immediately a sinister plan was triggered in his mind. He became seriously interested in Tagore. Perhaps he assumed that any positive remarks and praise by Tagore, a symbol of peace and human liberty with towering international fame and respect (at least during that period) for his fascist regime would be highly valued by the outside world. Suddenly, Tagore became a big catch to Mussolini. There were other advantages too. Tagore did not know the language and was not very much aware of Italian politics. So when Formichi's application came to his desk for his personal attention, Mussolini generously approved the application. Mussolini was confident his approval would act as bait and Tagore was bound to get reeled in.

Prof. Formichi's worries were over. He joined Visva-Bharati in November 1925 for a four-month contract and brought with him the entire library of Italian classics, and the service of Prof. Guiseppie Tucci, another distinguished scholar of Indology for a year, all paid for by the Italian government.

Tagore was overwhelmed at Mussolini's gesture and immediately cabled Mussolini expressing his gratitude. He wrote: '…I assure you that such expression of sympathy, coming as it does from the representative of the Italian people, will open a channel of communication of cultural relations between your country and ours, which has every possibility to produce an event of great historical relevance.'[9]

Tagore wanted to show his gratitude personally by going to Rome and asked Prof. Formichi to arrange his trip. A second visit was due in any case as Tagore wanted to keep the promise he had made to his Italian friends on his first visit. Prof. Formichi informed the Ministry of Foreign Affairs of the poet's intention of visiting Italy again. Now, Mussolini's plan was off the ground. The Government extended its hospitality to Tagore and his entourage.

Tagore arrived in Rome on 30 May 1926 with his son, daughter-in-law and his adopted granddaughter. The following day, accompanied by Prof. Formichi, he went to meet Mussolini at his office in Palazzo Chigi. For the very first time, the Poet and the Dictator came face to face.

In his opening conversation Mussolini said, 'I am one of those many Italians that have read all your books; of course those that have been translated in our language.'[10]

Mussolini's biographer believed that his literary and cultural awareness was not as much as he claimed. There is little confirmation that he knew more of Shakespeare and Moliére than he could have learnt from the encyclopaedia or indeed from the mere titles of their plays.[11] However, Tagore was visibly impressed by this statement. He thanked Mussolini for the generous gift and for the services of Prof. Tucci at Visva-Bharati. Mussolini then enquired how long the poet intended to stay in Rome.

Learning that it was the Poet's intention to visit Florence the following week, Mussolini insisted on his staying at least a fortnight in Rome, to take complete rest and to enjoy some sight-seeing. Tagore expressed his gratitude further and during their extended dialogue mentioned that he had not yet decided on the subject of his public lecture. Mussolini immediately insisted: 'speak on art, speak on art.'

In this context, Mussolini's biographer wrote: 'He liked to stress the great importance of art and he himself, inevitably, claimed to be a seminal influence upon contemporary artists; but in private he was ready to confess that he did not understand pictures and inwardly he resented that Italy had been held back from political greatness by the illusory and corrupting pursuit of aesthetic values.'[12]

However, the conversation was extremely cordial and lasted for half an hour. Prof. Formichi was translating Mussolini's words into English but not Tagore's words into Italian as Mussolini could understand English when spoken slowly.

At the end, the Prime Minister said: 'Just let me know what you would like to do best and I shall be more than happy to arrange it for you.'

On his way back to the hotel, when Formichi asked Tagore about his first impressions of Mussolini, 'Without any doubt, Tagore replied, his is a great personality. There is such a massive vigour in that head, that it reminds one of Michelangelo's chisel. Moreover, there is a simplicity in the man which makes it hard to believe that he is really the cruel tyrant many indulge in depicting.'[13] Parts of this statement was reported in the press.

Again Mussolini's biographer wrote : '...His personal magnetism worked best with those who saw him rarely; nevertheless, he could always impress a visitor when he tried and all the fascist leaders remembered how they had at times fallen under a real spell...'[14] and Tagore was no exception.

The above is only a small summary of the first meeting between the two, which was also highlighted in some of the Italian Press. When a reporter of a well-circulated fascist paper, *La Tribuna,* asked the poet to write a few words about the new Italy, he wrote, 'Let me dream that from the fire-bath the immortal soul of Italy will come out clothed in quenchless light.' This statement was also quoted in several of the daily papers.

Though most of the reports in the Italian Press were friendly and sympathetic to Tagore's ideals, there were critics too. *La Voce Ripublicana* wrote on 4 June: 'The idea of a 'mixture of diverse elements' is constantly occurring in Tagore's speeches. He wants to unify in one civilisation the characteristics of East and the West. But it is not sufficient to speak in English, to stay at a grand hotel, to study western texts, to give interviews, and to have a secretary engaged in writing letters on a machine, to know the spirit of the West. To appreciate our nature it is necessary to come into contact with our world and our life. The civilization of Europe is essentially dynamic, while the civilisation of India is essentially static and dualistic, and Tagore's idea of a meeting of the two is absolutely utopian.'[15] An ex-professor of history and the senator Alessandro Chiappelli also contradicted Tagore's idea of the East-West fellowship. He maintained that the *dhyanyoga* of the East had no relevance in the West.[16]

During the next two weeks (1-13 June 1926), Tagore visited most of Rome's landmarks, met the King, had lunch with the British Ambassador, was received by the Governor of Rome at the Coliseum, and gave several interviews to newspaper reporters, though some of his quotes were wrongly interpreted in the press. As Tagore did not know the language, he found no discrepancies in what was written. He was heartily received at the University of Rome, gave several lectures of which the most important one was that presented at the Union of Intellectuals: *The Meaning of Art*, attended by Mussolini himself.[17]

In the afternoon, after his meeting with the British Ambassador, Tagore told Formichi, 'As long as Mussolini lives, Italy can be said to be safe. Now I know what I shall answer when, after crossing the Italian border, I hear people speak ill of your country.' [18]

It was an irony that Tagore himself had to speak ill of Italy after crossing the border!

Tagore met Mussolini for the second (and last) time before his departure from Rome on 13 June. This time, the meeting was longer and heartier. As usual, Tagore discussed his ideas of an East-West fellowship. He said, 'Italy possesses a great personality, and therefore she, as a nation, is most suited to promote a rapprochement between both the Asian and the European civilisations.' Mussolini agreed. Tagore also added, 'Your Excellency, you are the most slandered man in the world.' Mussolini, with an innocent smile replied, 'I know, but what can I do?' [19]

Then the Poet expressed his intention to meet the great Italian philosopher, Benedetto Croce. Before Mussolini could respond, Prof. Formichi objected in a louder tone: Impossible! Impossible! Mussolini, cunning as ever, immediately stopped him and asked to arrange the meeting. Next, Mussolini showed Tagore the blueprint of his plan to build an 'International City' in Rome.[20] Tagore was again highly impressed and finally he requested of the Duce a signed photograph.

The meeting was over, though, at present, there is still no record available of this second meeting with the Italian Press.

The next day, Tagore left Rome for Florence followed by his visit to Turin. On 22 June before leaving the border of Italy, the poet sent a cable to Mussolini, 'I take leave with hearty thanks to Your Excellency, and to the Italian people represented by Your Excellency, for your generous hospitality and kind sentiments towards me.'[21]

With that note, Tagore and his party left Italy by train for Switzerland never to return Italy again!

Post-Italian Tour

In Switzerland, Tagore spent a few days in the company of Romain Rolland at Villeneuve. There he had a long conversation with Rolland, who was very critical of the fascist regime in Italy.[22] Rolland tried to give Tagore a more truthful picture of the growing repression, violence and atrocities that took place in the fascist state and eventually condemned his tour.

There was no doubt that Tagore himself felt a kind of uneasiness, especially during those latter stages of his Italian tour. But he was so thrilled by his reception and intrigued by Mussolini, that he was unwilling to believe he had made any mistake in coming to Italy. Rolland also made Tagore aware of some of the oblique comments in the European press on his Italian visit. He tried to bring Tagore out of his ambivalence, stressing the fact that freedom of expression had ceased in Italy.

Listening to Rolland, Tagore found himself in a great dilemma and eventually he agreed (though under pressure) to publish a write-up in the form of an interview, deploring some of the facts he did not like in the tour. Rolland brought his friend Duhamel from Paris, with a view to publicising the interview in the French press.

But that arrangement also went wrong. In his article, Tagore could not dismiss from his mind the magnificent reception he had received from his Italian host. A frustrated Rolland and Duhamel finally requested Tagore not to say or write anything before he heard the truth from those now living in exile.

A meeting was arranged with Tagore and Prof. Salvadori in Zurich and Modigliani in Vienna. In Zurich, Prof. Salvadori could not meet the Poet because of his illness but his wife took his place.

In her opening conversation, Signora Salvadori asked, 'You, who are so good, why did you come to Italy, now the land of violence and persecution?'

Tagore explained the background of his tour. At one stage he said, 'I had no opportunities to study the genesis or the activities of the fascist movement and I did not express any opinion about it. In fact, in most of my interviews, I was careful to explain that I was not competent to say anything either for or against fascism, not having studied it… About Mussolini himself I must, however, say that he did interest me as an artist… Mussolini struck me as a masterful personality… The people with whom I came into contact in Italy were almost unanimous in assuring me that Mussolini had saved Italy from anarchy and utter ruin.'

Signora Salvadori emphasised … 'It is not true. This is the opinion of people who are in favour of Fascism … People who hold contrary views were not allowed to see you … It is not true that Mussolini had saved Italy from financial ruin. The financial position of Italy was better before Mussolini came into power. Look at the Lira, it was 70 to the pound, and it is now 130 … Foreigners do not know, and merely repeat conventional tales. But what makes us unhappy is that you have unintentionally helped to support Fascism. We know it is unintentional, for you are too good to do so.' [23]

Signora Salvadori related to the poet only those facts which she had personally witnessed. The whole conversation lasted for an hour and was later published in *The Manchester Guardian* in October 1926. Indeed, he heard similar stories when he met with Mr Modigliani, the attorney of the

Matteoti Trial, and Angelia Balabanoff in Vienna.

Tagore was clearly perturbed by what he had heard and developed the same mental agitation and distress as he had during the Amritsar massacre, which took place in India seven years beforehand.

From Vienna he wrote a letter to his friend Charles Andrews in India, explaining in detail the background of his Italian tour, how he was trapped by the interviewers, that he had never praised fascism at any stage of his tour but admitted that he had praised Mussolini as an artist. The letter was elaborate but courteous, and was eventually published in *The Manchester Guardian* on 5 August 1926.[24] At one point he wrote: 'The methods and the principles of fascism concern all humanity, and it is absurd to imagine that I could ever support a movement which ruthlessly suppresses freedom of expression, enforces observances that are against individual conscience, and walks through a bloodstained path of violence and stealthy crime. I have said over and over again that the aggressive spirit of nationalism and imperialism, religiously cultivated by most of the nations of the west, is a menace to the whole world .... It would be most foolish, if it were not almost criminal, for me to express my admiration for a political ideal which openly declares its loyalty to brute force as a motive power of civilization.'

Prof. Formichi was taken aback to see the Poet's sudden change of attitude towards Italy: he also sent a counter article in the same paper three weeks later stating his side of the story.

Rolland did not like the softer tone of Tagore's letter. However, he selected the salient points from that letter, re-edited and translated it and distributed it amongst the European press. And, with this, Mussolini's plan to acquire international legitimacy of his fascist regime using Tagore in his political propaganda came to an unfortunate end.

The Italian people had a glimmer of Tagore's letter to Andrews and his interview two weeks after. The Bologna paper *Assalto* reported (28 August): 'That Tagore, who came to Italy twice and inflicted us on his very heavy poetic lucubration, is an old actor who is worthy of our highest contempt...This guru is kept by various governments. He is paid so much at each lecture... this viscid, insinuating individual, who is as honeyed as his words and poems, came to Italy as he was invited, paid and helped by the government. He exalted Italy, glorified fascism, and sang the praises of Mussolini... As soon as he crossed the border, this old man with an unsound soul, who impressed the public with his long black tunic and his white beard, talked behind the back of Italy, Fascism and its great leader, who is endlessly greater than him... He approximately behaved like prostitutes who always swear they are in love with their latest customer. Today we claim we do not like Tagore as a poet anymore because he is emasculated and without backbone. He disgusts us as a man because he is false, dishonest and shameless.'[25]

An abusive and distasteful accusation indeed!

Postscript

Now the question is as to how far Prof. Formichi and Prof. Tucci were involved in Mussolini's ploy. Researchers believe that Mussolini engaged those two Italian professors as confederates to help him in his heinous intention of using Tagore as a mouthpiece for the propaganda of his regime. That is not substantiated by the evidence in hand. Tagore personally invited Formichi to join Visva-Bharati as a visiting professor at Visva-Bharati's expense. Prof. Tucci joined Visva-Bharati at Formichi's recommendation, although Tucci's salary for one year was paid by the Italian Government.

It must be mentioned that, in Mussolini's Italy, people had three options before them for their survival: the first option for those people who refused to compromise with the government and its policy, was to silently leave the country, taking refuge in another country. It was not an easy option to leave one's homeland, foregoing one's establishment and professional income. Yet Prof Salvemini, Prof. Salvadori, Angelica Balabanoff, Modigliani and others chose this option as they were not ready to compromise their principles with the policy of the government. The second option was living within the country but detaching oneself from the government and remaining neutral with eyes and ears closed, as did Benedetto Croce, Duke Scotti and other intellectuals, who were the silent minority. It was not quite easy either to maintain this neutral stand. The last option was striking a compromise with the government for the sake of maintaining a safe existence, with or without having faith on Fascism. Prof. Formichi, Tucci and the professors and directors of several universities, civil servants the middle-class intellectuals, people who felicitated the poet were amongst this majority group.

Prof. Formici was a supporter of fascist policy, but he had a genuine admiration for the poet. Hence, sometimes, he had to do a balancing act pleasing his master on one hand and at the same time looking after the comfort and reputation of his visitor. This had been often mistaken by Tagore's associates that Formichi was trying to please his master and helping in his plan.

Although it was apparent that Tagore, overwhelmed with the gift and also the service of the professors, accepted Mussolini's invitation to visit Italy as his guest. But that was not absolutely right. Tagore had already made up his mind to visit Italy a second time to keep his promise. One will be surprised to learn that Tagore made another abortive attempt to visit Europe in the summer of 1925 (six months after his first visit) —before Formichi joined Visva-Bharati— one which again had to be cancelled on health grounds. However Mussolini's gift had only concretized his decision.

Another common allegation about Prof. Formichi was that, being Tagore's Italian interpreter throughout his stay in Italy, he closely guarded Tagore from the people whose beliefs opposed fascist ideals.

Perhaps this allegation was, to some extent, true. Then again, the poet did not show much interest to meet contemporary writers and artistes or find out about their literature and arts during the crucial period of the country. Only sporadically did he mention the names of some of the contemporary writers in the course of his interviews. But he never insisted on meeting any one of them. The only person he was keen to meet was Benedetto Croce. However, the meeting took place under rather extraordinary circumstances. On the other hand, the political situation in Italy was not conducive for the common people to speak out against the atrocities of the government. It should be noted that during his long and secret meeting with Croce, Italian politics was not discussed at any time.[26]

Finally, the researchers believe that Professor Formichi edited Tagore's interviews, comments and speeches for the Press. Tagore's praise for Mussolini and his government were highlighted; whereas Tagore's message of humanism or any comment about the regime were suppressed and not published in papers. As Tagore did not know the language he did not realise the misrepresentation of his statements during his stay in Italy.

This can be explained in several ways. The first and most possible explanation was that, being a fascist himself, Formichi knew the psyche of his people. He did not want the Milan incident in 1925 to be repeated. He did not want Tagore to be involved in any kind of controversy and strictly edited his statements before translating and handing it over to the press. Often, the first pages of the national newspapers were scrutinized and approved by the Prime Minister's Press Office before being given clearance for printing. All the newspapers were directed to send one copy of the paper to the Prime Minister's residence every morning. Even if Prof. Formichi refrained from editing the words of Tagore, the report would have been censored by the press office of the Prime Minister anyway. Equally possible is that Formichi had translated Tagore's words quite faithfully. The press, having the Prime Minister's office in mind, took it upon themselves to edit out portions which were not in tune with the Fascist ideology. As there are no complete transcripts of Tagore's interviews available to compare with the published versions in the newspapers, it is difficult to assume what exactly had happened.

* * *

In 1930 when Tagore came to New York after visiting Russia, he met Prof. Formichi again. Formichi was then the visiting professor of the Italian Chair at Berkeley. During the course of the conversation, Tagore expressed his intention to resolve any misunderstandings he had with Mussolini. Formichi suggested that the Poet write a letter to Mussolini directly explaining his account. The Poet made a draft and instead of sending it to Mussolini directly, sent it to his son in London for his approval. Part of the letter has been mentioned at the beginning of this article. In that letter (21 November, 1930) he wrote at the end: 'The politics of a country is its own; its culture belongs to all humanity. My mission is to acknowledge all that has eternal value in the self-expression of any country.' Those lines, at the end of that letter, summed up Tagore's lifelong conviction and that was perhaps the reason why he wanted to reconcile with Mussolini once again. Mussolini's office received that letter but there was no further response from *Il Duce.*

It was true that Tagore was affected by the strange and captivating illusion of Mussolini, but that illusion lasted until Mussolini invaded Abyssinia (now, Ethiopia) and other Mediterranean states. Tagore never once praised fascism at any point during his Italian tour. As Mussolini and fascism are more or less used synonymously, he was frequently misquoted and misrepresented in the Italian media.

Notes

1. Krishna Dutta and Andrew Robinson (Eds.): 'The Tagore-Elmhirst Correspondence', *Purabi: A Miscellany in Memory of Rabindranath Tagore 1941-1991* (London), p 98, 1991.
2. Krishna Dutta and Andrew Robinson (Eds.): *Selected letters of Rabindranath Tagore*, Letter no. 247, (Cambridge University Press, UK, 1997), p. 394, Letters.
3. The letter was written on 8 May, 1921 by Prof. Formichi to Rathindranath :

   'Dear sir- I dare say your father's works are known and highly admired here by every cultivated person. We are all very happy to hear that he intends spending a week in Rome. I am sure it would be a great success. Let me know, please, as soon as you can, whether I may tell our Rector to announce and prepare everything for the lecture. Your father shall find here in Rome more sincere admirers than he can think of. Make me acquainted, please, with the day of your arrival and the hotel you will live in. Your very sincerely, Carlo Formichi. '

   Source: Formichi File, Rabindra-Bhavan, Santiniketan, India.
4. A typed copy of this lecture dated '22 January, Milan' is still available in the Italy folder at Dartington archive. The lecture was published later entitled 'Voice of Humanity' at *Visva-Bharati Quarterly*, Vol.III, part 1, pp 1-10, 1925.
5. Sudhindranath Bose, *Forward*, 22 July, 1925. Cited from *Rabindrajibani o Rabindra Samity Prabeshak* by Prabhat Kumar Mukhopadhyay, Vol III, Visva-Bharati (India), p. 211, 1952,
6. The original Bengali poem is the last entry of Tagore's book of verse *Purabi* with the title 'Italia'. The English rendition is merely a paraphrase done by the poet himself.

   To Italy

   I said to thee: 'O Queen,
   Like many other lovers, who have brought their gifts to thy feet,
   I have come, as a lark at the gate of thy Dawn, only to sing to thee
   and then to go.'
   From thy window thou speakest to me through thy veil:
   'Now it is winter, my sky is dark, my garden without flowers.'

   I said to thee: 'O Queen,
   I have brought my reed from the eastern shore
   Waiting to play to the light of the dark eyes,
   Open to me thy veil.'
   Thou saidst in answer, 'Go back my impatient poet,

For I have not yet decked myself in colours,
In the sweet month of May, When I sit on my flower throne,
I shall ask thee to my side.'

I said to thee: 'O Queen,
My journey has borne rich fruit in thy words of hope:
Floated on the breeze of spring the magic of the call to me
will burst in flowers in our faraway forest.
I shall seek back my path to thy window
On some sunny day, drunken with the fragrance of roses
and humming with bees' wings.
Today, while I take my leave and go back, I sing:
'Victory to thee!'

7. For the entire text of the letter see Mario Prayer: 'On Tagore', *Rabindra-Vikshya*, Vol. 39, (Kolkata: Visva-Bharti, India, 2001), pp. 31-50. Translated from Carlo Formichi's book *India e Indiani*, published by Alps, Milan, 1929.

8. Mario Prayer. 'Italian Fascists Regime and Nationalist India, 1921-45,' *International Studies*, Vol 28, (New Delhi, India), pp. 249-271, 1991.

9. Mario Prayer. 'Carlo Formichi, India e Indiani', *Rabindra-Vikshya*, Vol 40, (Kolkata: Visva-Bharti, 2001), p. 21.

10. Mario Prayer. 'Carlo Formichi, India e Indiani', *Rabindra-Vikshya*, Vol 40, (Kolkata: Visva-Bharti, 2001), p. 25.

11. Denis Mack Smith. *Mussolini*, Granada, (London), p. 152, 1983.

12. Ibid, p. 155.

13. Mario Prayer. 'Carlo Formichi, India e Indiani', *Rabindra-Vikshya*, Vol 40, (Kolkata: Visva-Bharti, 2001), p. 25.

14. Denis Mack Smith : *Mussolini*, (London: Granada, 1983), pp. 127-8.

15. P.C.Mahalanobis (Notes): 'Our Founder-President in Italy, 1926', Visva-Bharati Bulletin, *Visva-Bharati Quarterly*, (Kolkata, India). p. 283, October 1926.

16. Alessandra Chiappelli:'Tagore, l'Oriente e l'Italia nuova', *Il Giornale d'Italia*, p2c4, 8 June, 1926.

17. For the entire text of the lecture 'The Meaning of Art' see Sisir Kumar Das (Ed), *The English Writings of Rabindranath Tagore*, Vol. 3, (Delhi: Sahitya Akademi), pp. 580-588.

18. Mario Prayer: 'Carlo Formichi, India e Indiani', *Rabindra-Vikshya*, Vol 40, (Kolkata: Visva-Bharti, 2001), p 26.

19. Ibid, p 30.

20. 'The International City' was the original idea of the American architect Hedrick Christian Anderson. He had planned an international centre in Rome where the entire legacy of knowledge and culture from all

the nations of all times was to be preserved. He called this a 'Laboratory of Civilisation'. A team of about forty architects designed an elaborate plan to build a series of townships such as the Temple of Arts, Conservatory of Music, International Institutes of Letters and Science, Palace of Scientific Research, etc. This Township was planned to be built in Austria beside the Tiber, a site where Roman civilisation was established a thousand years ago. This township would be more spectacular and unique than St. Peters or even the Coliseum. There would be the world's tallest marble obelisk, thirty-six metres high, weighing 770 tonnes, on the top of which there would be a bust in the semblance of Mussolini. Tagore was told about this plan because Mussolini had approved this project. Tagore was so impressed by this plan that the same afternoon he went to Anderson's studio to see the blueprint of the project, accompanied by Rathindranath and Signor and Signora Rapicavoli. However, this project never materialised. After using a hundred tonnes of metal there was no money left and the project was abandoned.

21. Mario Prayer: 'Carlo Formichi, India e Indiani', *Rabindra-Vikshya*, Vol 40, (Kolkata: Visva-Bharti, 2001), p 34.

22. For details of Rolland-Tagore dialogue see Sisir Kumar Das (Ed.), *The English Writings of Rabindranath Tagore*, Vol. 3, Sahitya Akademi, (New Delhi, India), pp. 890-99, 1996.

23. For details of Tagore--Salvadori dialogue see Sisir Kumar Das(ed.), *The English Writings of Rabindranath Tagore*, Vol. 3, pp. 899-903, Sahitya Akademi, (New Delhi, India) 1996.

24. For the entire letter see, Kalyan Kundu et.al (Ed,), 'Imagining Tagore- Rabindranath and the British Press (1912-1941)', pp 418-21, Sahitya Samsad, (Kolkata, India), 2000.

25. *Asalto*, 28 august, 1926, [cited from Gaetano Salvemini, *Tagore e Mussolini*; 'Esperienze e Studi Socialisti in onore di U.G.Mondolfo', Florence: *La nuova* Italia, 1957, pp. 191-206.

26. For details of Tagore-Croce conversation see Sisir Kumar Das (ed.), *The English Writings of Rabindranath Tagore*, Vol. 3, Sahitya Akademi, (New Delhi, India), 1996, pp 888-89.

The Pedagogy of Rabindranath Tagore

Sandra Chistolini

In this article, pedagogy is intended to mean the study of the concept of education as it attests to its fundamental role in the forming of a human being. In this definition, we place the humanistic-literary production, considered a classic of Indian teaching, written by Rabindranath Tagore, a poet, dramatist, musician and thinker, who was able to articulate a worldwide idea of education and which has become the heritage of mankind.

The number of documents by Tagore preserved in the Italian catalogue is remarkable compared to that found in national bibliographies in English, German, French, Spanish and Portuguese. Specifically, in the Italian national bibliographic catalogue, as of 26 October 2012, 700 records with Author Rabindranath Tagore appear for the period 1914-2012. *Gitanjali* is one of the books, which has been translated endlessly from 1914 to date. This article cites reference to works, in Italian if available, for the subject being developed herein. Over the last several years, Italy has also seen the appearance of versions in Hindi, Bengali and Urdu, in addition to the noted translations into European languages. The renewed interest in making some writings available for reading in non-European languages can probably be attributed to Indian immigration to Italy. The movement of the population from one continent to another is a fact of great social import that affects the education of the young people. Schools that address new attention to the study of one of the founding fathers of awareness of modern India support the formation of the Indian identity outside India and allow children born in Italy, of families with a history of migration, to approach the knowledge of their culture with pride. The awareness of the depth of one's traditions lends national value to the roots, with which a family can identify from one generation to the next. In the words of Tagore, the national spirit opens up to a universal dimension. Poetry, theatre and music speak to the heart of each person, always and everywhere. Time and space seem to cancel out in the infinite worlds in which children have their first direct human experience. The writings of this great master lead one to discover what unites us in a shared message of peace and unity. He educates towards the meeting of meanings in which everyone can recognise his or her own spirit, easily forget about belonging to a limited geography and savour, instead, the joy of a planetary symbiosis.

The topics mentioned in this introduction are developed in three sections, each of which is composed of two subsections. The first section specifies the educational perspective chosen as the scientific-disciplinary orientation of this article. In the second section, the idea of school, which matured until the foundation of Santiniketan, is presented from a critical viewpoint. The third section illustrates the conception of Tagore's educational system.

Perspective of the Study

In the first place, we would like to address Tagore's educational production and reflect on some teachings that spring from his writings. His thoughts on education and projects for the formation of the human being, free from social and political restraints, were the foundation for teaching experiences that became significant points of reference. This becomes relevant to the study and criticism of possible alternatives to schools set up with the idea of guaranteeing a bureaucratic, administrative system, more than the idea of being occupied with preparing the person to live together with others, in peace and justice.

We are faced with a life dedicated to teaching with the learned use of the arts, which, in literature, poetry, theatre and music, find reliable paths for manifesting a common mutual feeling. Through conferences, debates, letters, schools and contact with different people, Tagore informed the whole world about the deep-seated truth of his teachings. The origin of feeling sinks its roots into Eastern philosophy. He unveils a Weltanschauung in which the spiritual dimension of man prevails: only by entering into one's own spirituality can one know the profound thought manifested in the actual act of thinking.

For Tagore, leading man to self-knowledge means acknowledging the symbiosis that exists between a human being, the outside world and the Divine. Not so much the search for knowledge as knowledge itself is an act that leads to self-perfection. Man may have a yearning for the heights. His desire becomes a concrete action when he manages to consciously gain possession of his own greatness; this means crossing the limit of limits.

In the second place, the intention of this work is to awaken, in the Indian thinker, the perspective that extends to the eternity of man and his spirit. By continuing along his path, in which the mystic element remains constant, he arrives at defining the meaning of life, beyond the space-time factor. In every writing, the intensity of self-meditation and the world shines through. An intensity that has no need of sophisticated, complex speech. Simple words communicate this to the reader who is sensitive to universal problems of birth, life and death, which are always revived, even if in a new form. A story by Tagore can be reflected on alone or in groups; the teachings for an entire school year can be developed. Although aimed at man without boundaries or limits, that is, not restricted to a single country and a particular culture, this thinker offers starting points for producing educational material that can be traced back to individual scholastic situations, thanks to the creative impulse of the teacher. On one hand, Tagore might appear excessively far from man's sufferings and the common teacher, but on the other hand, through his own educational experience, he himself has shown that 'new school' projects can be realised, despite social conditions, despite cultural prejudices, despite the scarcity of resources and, probably, despite the strong bureaucratization of educational systems.

Historical context. Born in the second half of the 19th century and having lived long enough to see the war in the first person, he did not hesitate to raise his voice in favour of peace among peoples.

The roots of peace originate in man himself, who reaches a clear awareness of his role in society and his universal mission. As for the social role, one can think and act to change structures that oppress persons, starting from the sense of freedom that has always existed in each of us and is understood as a universal mission, a duty towards the permanent evolution of man. Progress concerns everyone, at all times and in all places. What has been introduced in the past exists in the present and shall return in the future, according to a sometimes incomprehensible process, but structurally common to mankind. Man belongs to that process and, vice versa, the process belongs to man: actually, the one is the other, the other is the one.

Rabindranath Tagore was born in Calcutta on May 6, 1861 and died in Bolpur (Bengali) on 7 August 1941. He was a man anchored to his people and his culture; from both, he drew the symbols and meanings that entwine his tales and lyrics. In his eyes, India sacrificed the authentic tradition that risks being suffocated by the West. The diversity of the populations is to be used to advantage, not to create divisions, so that each country maintains the awareness of its cultural history and, hence, religious history. Recognition of the difference is not an assertion of superiority of one State over another, it is just the self-awareness that only by admitting that we are different, can unity be attained. Thinking about education is almost presented as a logical consequence of what is expressed through an artistic production. When one understands the meaning of life and the function of man in the vortex of history, one may be able to begin thinking about how to have others participate in one's own knowledge and how to create a better world, inhabited by new men. This is probably one of the reasons why the greatest thinkers of all times have enriched their literary production to also include educational works.

Those who believe in great ideals, in the regeneration of consciousness and in changing society cannot avoid dealing with education. Unfortunately, not everyone ends up proving, at his or her own expense, the practical validity of the principles so loftily supported on a theoretical plane. This is only part of the case of Tagore, who accrued a certain amount of field experience in the educational situations born out of his inspiration. Although he was not a teacher by profession, employed in a school full-time, his nearness to students and teachers was continuous throughout his lifetime and, as we can grasp from the worldwide diffusion of his works, he will remain perpetually remembered. What writings by Tagore can we find in literature today? Little, if we expect indications of the paths that lead to knowledge and perfection, without a shadow of a doubt. Much, if we approach them with serene, critical thought. It is not by chance that works talking about overcoming prejudice based on race, religion and sex are being reread today. It is rather a confirmation that the realisation of the idea of a united world is considered an important stage. One must keep from creating a fence that excludes others, boasting claims of excellence. On the basis of this approach, a well thought-out choice was made that attributes a specific significance to Tagore: he was a poet, an internationally recognised composer, whose writings are understood through translations and, even more so, in the new communication with the Indian cultures that are alive in Europe. These features define the teaching concept of the thinker who educates through his own arts. Without having planned a teaching system, according to a western structure, he remained faithful to the educational style of his culture: shape man for who he is and aim at infinity.

Disciples and faithful followers. Tagore does not fall within a study of charismatic phenomena and is not like a guru to whom certain abilities or talents deemed magnetic can be attributed, and that attract different types of people, fascinated by the mystery. We asked ourselves about a possible tie with Tagore, who also spoke of love, peace and universality, with the various types of masters that still populate India today. Tagore's educational message is of another kind. It represents a unique cultural testimony of great value and is a point of pride for India. His teachings bring people together and allow them to understand how the search for unity can help resolve conflicts. He does not ask that one belong to a religion, a belief or a faith, but, rather, demonstrates the importance of knowing how to distinguish what is right for each human being.

The School of Tagore

We have reached the second section of this article, in which we are trying to gain knowledge of Tagore's idea about childhood, and understand the basis on which he started up the teaching experiment for which he is remembered. In pedagogy, the debate on teaching about peace continues to be very heated, not only as an opposition to war, but, above all, as educating toward the value of life, respect for the environment and love for the incarnate truth in man. Santiniketan is the ideal place, where many of us would like to have lived, at least for a short time in our lives, in order to be able to say that the transformation of the relationship of man with nature is possible, starting with small daily things. In our opinion, Tagore's teaching example is to be sought through trusting in the possibility of liberating the best part of the adult, starting from the child. Below we are presenting three basic concepts in modern pedagogy that were well-analysed by the thinker: the concept of family education; the concept of the child educating the adult; and the concept of cultural independence in forming a free and critical consciousness.

Indian cultural tradition. According to some scholars, Tagore is not to be considered a true educator (Delfino 18) and any and every comparison with professional teachers would be inappropriate. However, in a paper written jointly with Antiseri and Reale, the Italian educationalists, Laeng, notes how Tagore represents a compulsory step towards understanding the Indian educational culture (3: 746-747). The poet has not written a pedagogic set of rules to be set up against the educational methods used in schools of his time (*Per aprire nuove vie alla scuola* 7), nor did he study to be a teacher; he did not even accrue much experience in teaching. During a conference held at the J.J. Rousseau Institute in Geneva in 1921, he stated that he was driven by sensitivity for other persons and trusted in the instincts of children (Pieczynska 79-80). His thoughts on educating children and young people, his stories and his experience of life, as well as the literature produced about his works render Tagore a relevant personality in the history of pedagogic thought. In their historical context, the most divergent positions, with respect to traditional literary precepts of reality, are also legitimate and worthy of attention. Apart from the official classification, which neither removes nor adds much to the greatness of Tagore's literary work, new value should be attributed to the fact that this man communicated a message of solidarity and remains alive in the spirit of his school.

The idea of Tagore's school is derived from the observation of the ill effects produced as much from the foreign domination in India, tending to impose abstract models, as from the conclusion that the school, even in the West, is unilateral and does not manage to properly educate children. The pupils do not internalise the highest moral, civil, religious values and do not have the necessary techniques available to face life with feeling and action. Moreover, Tagore, as a pupil, had personally experienced what he defines as 'the oppression of the school system' to the point of convincing his family to remove him from school. After this, one of his older siblings took over the instruction of young Rabindranath (Delfino 8). The official school environment was described by the poet as being in complete disagreement with an education full of the freedom and joy that, within his family, had allowed him to develop his intellectual and artistic talents (*Antologia di scritti pedagogici* 109). Hence, his dream of all children having schooling like the one he had had in his family.

The school model conceived by Tagore arose out of both the criticism of formal English education, a prototype for the European system, and the recovery of Indian tradition. The criticism concerns attempts to plan a school and organise teaching that might be the most advantageous in one point on the globe, but that cannot, just for this reason, be considered valid and effective for everyone everywhere. In other words, the English school system may work magnificently for that country, being an integral part of it, and is a manifestation of the cultural-historical progression made by man in that society. However, things become complicated if one expects to transport, in its entirety, a model from one country to another, for example, from Europe to India, in the belief that it will function as it is, without taking into account the cultural-historic beliefs that have influenced it. Today, the concept seems quite simple and practically of public domain. In actual fact, if we look around, we find innumerable cases of such expectations: a model is often deemed a model par excellence and every action blindly follows it, without criticism of any kind. In good faith, one may forget that what goes well today many not go well tomorrow, and that what is best for one person, may be the worst for another. Tagore found that the way out for India was to be found in India itself, that is, in the ancient Indian principles that inspired the education of the youth (*La civiltà occidentale e l'India* 35). From here, the return to the tradition of the Indian population, totally different from all European cultures.

To have an elementary school girl wear the dress of a grown woman is simply ridiculous, apart from being silly, unless it is for a masquerade, which, as such, falls outside the norms of ordinary life. The exception is not a rule, even if the rule may allow for exceptions. One understands that the negative criticism of the school led by others originated in Tagore as a valorisation of what is criticised. Referring again to the example of the dress, this may have been acquired from a grown woman and judged as very beautiful, but can become unexpectedly ugly, if worn by a little girl, or comical if used for a masquerade. The Bengali thinker had visited English schools in Great Britain and had seen the application of the programmes and teaching-methodological organisation in Indian society. The inconsistencies must have been so self-evident to him, as to spare no efforts to found a school for Indian children and youth.

When Indian education is modelled on English education, the result is robots, and schools are the factories of these robots. The lessons are empty and have no connection to the life of the country; the teacher guarantees the functioning of the factory, by transmitting knowledge without soul; families experience the situation with mesmerised astonishment, they do not understand and are happy not to understand, because this is the means of approaching the culture of the moment. In speaking about the education imposed from abroad, Tagore writes, 'What is called school in this country today is actually a factory and the teacher is part of it. At ten thirty in the morning, the factory opens at the sound of a bell and the teacher begins to speak; the machines begin to work. The teacher stops talking at four in the afternoon, when the factory closes and the pupils go home, taking some typed pages of knowledge with them. Later, this knowledge is controlled through an exam and then classified' (*La civiltà occidentale e l'India* 41).

***Santiniketan**: House of Peace.* In imitation of the school imported from Europe, Tagore opposed a school that satisfied certain conditions that took into account the social changes that had occurred in India. It was not a matter of going backward, imitating, exclusively, the instruction given by gurus, who morally and spiritually shaped young people. Tagore proposed, in short: a) starting up varied, living courses able to nourish the heart as well as the intellect; b) teaching that avoids clashes or divisions that negatively affect the mental development the youth; c) worthy consideration of the mind that should be cared for so as not to degrade it to something unreal, heavy and abstract, only a focal point during school activities; d) the recovery of ideals from the Indian culture that motivate youth to participate in building up their own country; e) attract the interest of persons towards Indian education; f) replace moral education, generally understood as severe and sacrificing, with the ancient Brahmacharya that protects youth in their development, guides them amicably and educates towards discipline, not only and solely through severity; g) help nature without which neither development nor learning is conceivable (*La civiltà occidentale e l'India* 42-47).

In 1901, in Bolpur, a village in the foothills of the Himalayas, about one hundred kilometres from Calcutta, Tagore founded his new school: Santiniketan, which means 'House of Peace'[2] (*Santiniketan, the Bolpur school of Rabindranath Tagore*). This was a time when India was frantically trying to liberate itself; various movements sprang up for nationalising education. In 1906, the 'National council of education' was organised; it asked Tagore to prepare a constitution for the new Indian school.

Over the following years, Santiniketan became an international university bearing the name Visva-Bharati, i.e. "universal voice". The principal aim of this cultural centre was to offer a place for searching for the truth, in which everyone was involved in human heredity and the common ideals that artists, philosophers and saints left for everyone, not only for the population where they originated. All humanity may know the beauty, greatness and depth of the fabrications of those who left traces of themselves in history. The universality conceived by Tagore is a spiritual temple, where youth from all countries and cultures are educated to mutual understanding (Pieczynska 141-147). In this international setting, Tagore sees, looming ahead, the meeting of oriental heritage with that of the West and the birth of a worldwide culture (Pieczynska 145)[2]

Other educational experiences are alive at the Institute of Rural Reconstruction and at the Siksha-Satra centre, both natural continuations of Santiniketan. Above all, at Siksha-Satra, a 'place of free education', as described by Elmhirst, a collaborator of Tagore (*Rabindranath Tagore, pioneer in education* 66-84), an interesting example of school-community has been created, where young people are employed in both intellectual and practical activities. The style of living is based on cooperation and everyone is educated towards community service. The young people become promoters of the education of illiterate adults inhabiting the surrounding villages. This way models for exchanging knowledge and abilities are implemented between youth and adults and between students and teachers (*Antologia di scritti pedagogici* 179). The first tasks of a child entering Siksha-Satra are handicrafts and caring for the home. Through the former, the child develops manual agility; through the latter a sense of community participation is acquired.

The figure of the teacher changes appearance: from the teaching desk, this figure moves to a position of the daily sharing of the lives, work and learning of the youths. More than being given instruction, in this 'Home-School' each becomes aware of his/her self-training. In this regard, Elmhirst writes, 'If education is to mean something, it is necessary to safely include the supply of the means for experiencing each phase of life of an adult in its embryonic form. The school must be a laboratory, not just for absorbing knowledge or developing well-sheltered nurseries, but for allowing one to emerge, venture in the realm of economy, self-preservation, self- discipline and self-governing, and self-expression in the world of spiritual abstraction and human well-being' (*Antologia di scritti pedagogici* 183).[15] In Tagore we find educational principles devised into forms of educational experience for promoting the intellectual, artistic and moral growth of the youths. As for Santiniketan, to talk about 'school' may be inappropriate, if one looks at it as one would look at one of our urban schools. For Tagore, Santiniketan is a living environment where everyone contributes towards creating the most suitable atmosphere for on-going learning and permanent self-instruction. Students and educators live, work, study and meditate in Santiniketan according to the rules of community life. Since the school is open-air and far from the city, it fosters contact to nature and there is also ample freedom left to the young people as regards attending lessons. The peacefulness of the surroundings makes learning easier. If the students find no interest in, for example, attentively listening to music, Tagore suggests the indirect encouragement of the students, thus giving them another way to improve their sensitivity. Observing the adults who are playing can also be a means of training towards musical sensitivity (*Antologia di scritti pedagogici* 109-116). At the Ashram described by Tagore, one lives in such an ideal state that it is difficult to imagine recreating a similar ambience in our schools, even the open ones. Tagore himself confesses that he was sometimes unable to realise his entire unpretentious school project, due to the requirements of society.

Setting up a school as a happy island, far from social conditionings, above all those deemed more negative, does not appear to be an easy or useful undertaking for the social insertion of youths. Nonetheless, the approach to some educational experiences, which place education to values and ideals at the centre of interest, assuredly constitutes a positive element of comparison.

The Human Relationship with Nature and Society

In this last section of the article, we show how Tagore's idea of nature is partially similar to that of Rousseau[1] and the educational reform that, in the early nineteen hundreds, founded the active school and progressive education. In Italy, scholars, such as Delfino, Ottonello, Poli and Stilo, have studied the teaching profile and sketched the humanistic aspects of this philosophy. These short comparisons confirm the unique, original character of education as described up to this point. In Tagore's pedagogy the constant reference to the spiritual dimension of existence and the harsh criticism of consumerism contribute to giving a notable weight to the issue of education. Finally, the eschatological prospect of life confers its own depth to Tagore's thoughts on education, which is understood through the fervent adhesion to a vision of the world that is religious, first, and then becomes philosophical, literary and artistic.

Students teachers parents. In Santiniketan students practice the Brahmacharya; the eventual cultivation of the earth for their own nourishment; discussions with teachers, even during walks; the study of music and the stars during evening hours; and listening to legends and historical tales. Youths are educated according to nature, free from the conventions and habits that make adults slaves to rules. In the new school, unstructured knowledge is accrued about trees, flowers, fruit and everything with which they come into direct contact. This attention to nature, as an authentic state for true education, brings us to *Emilio* by Rousseau and teachers in the new school of the early nineteen hundreds.

When the child approaches the world for the first time, it does not need books; Tagore found that the lack of suitable teaching materials does not constitute a problem at the beginning of one's education. The more austere and poor the environment, the more the child learns to build what it needs. Its intelligence draws benefit from this and the teacher himself should create and improvise what is required (*Antologia di scritti pedagogici* 65-69). Tagore cites the case of a young teacher who managed to attract the full attention of the class, not because he possessed any especially refined knowledge, but just because:

> '[...] his teaching was personal, he was the source himself, so he was living material and easily assimilated by living human nature. The real reason for his success lay in his intense interest in life, ideas, everything that surrounded him as well as the young people with whom he came into contact. Inspiration did not come from books, but from the direct communication of his ultra sensitive spirit with the world. The seasons had the same effect on him as they had on plants; he seemed to hear the invisible messages of nature in his blood, messages that constantly sail across space, swim in the air, flicker in the sky and tremble in the roots of grass below the earth. The literature he studied had nothing to do with libraries: he had the power to see ideas in front of him, just like he saw his friends, with a precise form of breath of life. So the young people in our school had the good fortune to receive lessons from a live teacher and not from textbooks' (*Per aprire nuove vie alla scuola* 23-24).

As a matter of fact, how many teachers have experienced that their most successful lessons were the improvised ones, because they were sincerely felt by themselves and the students? The improvisation we are talking about does not exclude the preparation of the teachers. It is rather the capacity of the teacher to follow and allow the free development of the child to emerge (*Antologia di scritti pedagogici* 53). Tagore himself abstains from giving specific methods and pre-established projects. He offers guidelines, which, in all of his works, are based on philosophical, pedagogic and psychological consistency. He asks teachers to exercise the duties that were once entrusted to the guru. Although unable to have a perfect guru available, it would be enough to have available

> '[...] teachers capable of placing economic considerations aside and see the interior value, and one of these teachers would be elevated to the condition of guru. Once he has become a guru, he will realise that he can only attain credit if he manages to infuse life as he has lived it into his students, ignite their minds with his knowledge, and make them happy through his affection. By giving his students things they cannot buy or sell and are, actually, priceless, he will gain a devotion that does not depend at all on fear of punishment, but a rather deep devotion to be called religious and genuine enough to be defined as natural. If he is also compelled to accept a salary to be able to live, he will give special value to his task, by offering more than that for which he is being paid' (*La civiltà occidentale e l'India* 52-53).

Parents are asked to not settle with sending their child to school; their task is to build, within the child, the humanity that unites him with all other men. Social class should not influence education to human values. As a criticism, Tagore observes that the process has been inverted today: first one learns that one belongs to a well-off family, and then, when it is not already too late, one builds up the human essence. Therefore, parents should take more care of raising their children with positive feelings towards others and think less about raising them according to fashion. Of course, a child needs to be clothed, but that does not mean it has extreme need of the mostly expensive, elegant clothes. To dress in fashion is a social convention, sometimes respected more by parents than by children; at other time imposed by children on their parents.

> The child who is decked with prince's robes
>
> and who has jewelled chains round his neck
>
> loses all pleasure in his play;
>
> his dress hampers him at every step.
>
> In fear that it may be frayed, or stained with dust,
>
> he keeps himself from the world,
>
> and is afraid even to move.
>
> Mother, it is no gain, thy bondage of finery,
>
> if it keeps one shut off from the healthful dust of the earth,
>
> if it rob one of the right of entrance to the great fair of common human life -(*Gitanjali* 26).

Once again, Tagore refers to the education of inner values, rather than the search for empty outward appearances.

As for the writer's position concerning punishment, he suggests replacing it with atonement. In fact, punishment leads to feeling of revenge of the offended person against his offender, who bears feelings of resentment and humiliation. Vice versa, with atonement, he who has caused the offence may voluntarily find a way to remedy the injury inflicted (*La civiltà occidentale e l'India* 49). Structurally, Tagore's school has no benches or teacher's desk, since, in India, they are deemed non-essential for teaching and since sitting on the ground is much less formal. The search for the essential should guide youth in understanding that civilisation is not measured on the possession of blackboards and pencils, but on factors such as simplicity, naturalness and spiritual peace (*La civiltà occidentale e l'India* 50-51).

Within this well-pictured scenario, in which pupils, teachers, parents, directors and inspectors, school structures and educational principles are adapted to the Indian society of the early nineteen hundreds, one cannot take Tagore's educational project literally. The examples and methods are clearly to be related to each specific country and each social-cultural development. The advice to focus on essentiality, the search for human values, reference to the spiritual dimension, the approach to nature and paying attention to inspiring messages of universal feelings that may come from it represent educational aims that can be pursued in every school.

The aim of education. The guiding ideas that allow Tagore to think about an innovative type of education come from the mantram, that is, the meditation text that he usually recited from the time of his youth: the text says, 'Satyam Jnanam Anantam Brahma, Anandarupamamritam yadvibhati, Shantam Shivam Advaitam', which means, 'Brahma is Truth, He is Wisdom, He is Infinite; He is revealed in endless forms of Delight or Joy; He is Peace, He is Goodness, He is One' (*Rabindranath Tagore pioneer in education* 85). Recited, at first, mechanically, the text moved forward, unveiling its intrinsic meaning into the mind of the poet, who finally reached the spiritual conquest of Mukti, or Freedom. The path he followed is the same one he indicates to others, primarily educators. According to Tagore, the most habitual educational processes lose sight of the formation of the mind, which should be guided towards the realisation of our deepest spiritual relations with the Supreme Truth. Science, philosophy, art and literature are fully developed when one reaches a sense of separation of what is spiritual from what is material. The inner life in man can thus expand itself in vast atmospheres, in which new spaces are captured or created by the imagination. As an individual, man possesses the unity of the living organism, distinct in itself. As a social being, man belongs to humanity, and as a spiritual being, he belongs to all-embracing reality.

Education is meant to preserve the physical man, helping him to find the paths of peace. The lack of mutual social and worldwide participation and the lack of cooperation among men are to be overcome through education to a sense of individual and collective responsibility. The religion of spiritual harmony is what Tagore wishes to make others feel and understand as an element that drives educational processes. The product of the latter will be the man of the future.

In *Personality*, published in 1917, Tagore explains what correct education should pursue: 'The object of education is to give man the unity of truth. When life was simple, all the different elements of man were in complete harmony; but when there came the separation of the intellect from the spiritual and the physical, the school education puts the entire emphasis on intellect and the physical side of man,' completely neglecting the spirit: aiming solely at supplying facts, the education at school accentuates the imbalance of the faculties.

'I believe in a spiritual world, not as anything separate from this world - but as its innermost truth. With the breath we draw we must always feel this truth, that we are living in God'; and further, 'I believe that the the object of education is the freedom of the mind which can only be achieved through the path of freedom, though freedom has its risks and responsibility as life itself has. I am certain, though it would seem that men as a whole have forgotten it, that young people are living beings, more alive than the adults who have built the shell of their habits. So it is indispensable for their spiritual health and their development that they have not only schools where they receive lessons, but also a world, whose dominant spirit is personal love' (*Per aprire nuove vie alla scuola* 15 and 27) (N.B. text freely translated).

For this belief, Tagore asks everyone not to stop at the appearance of external facts. Only the search for universal harmony leads to human freedom. The author's Indian culture resurfaces in his philosophy and teaching: harmony with the universe allows one to feel a sense of completeness and reawakening, that sense of freedom, which persists in a divine dimension in that same universe (*Antologia di scritti pedagogici* 99). Education is achieved in freedom from ignorance and prejudices that create barriers against populations understanding each other. For this dimension to reawaken in minds and hearts, people must be given new faith, and education must be given great ideals (*Antologia di scritti pedagogici* 32).

Although Tagore's school model requires special conditions in order to be realised, the suggestions derived from reading the literature on his works on education and, even more, on the conception of life, are truly countless (*Sàdhanà* 37-38). It does not seem out of place to weigh and compare an education completely aimed at fulfilling an ideal with an education that is totally lacking in such. There are risks in the one as in the other case. An ongoing comparison with the historic dimension could probably and partially furnish tools useful for drawing up a most accurate outline for an educational project, limited to the time and study in which it is expected to be completed. Thus the choice is no longer between the existence, or not, of an ideal in education, but rather in the continuous search for the sense to be given to the entire education process. The ideal itself sinks its roots into that cosmic Weltanschauung, so well expressed by Tagore in his teachings.

Conclusion

What remains of Tagore lies in the proof of his works that make up the analytical element and criticism, from which teaching can be derived. In order to be scientifically correct, each verification of what the poet taught must be conducted on actual facts, which, for the moment, cannot be directly detected through comparative surveys between the Indian and Italian school systems. In Italy, Tagore appears in some books of the elementary school and compulsory school, in some teaching manuals for teachers' training and in university curricula (Chistolini). For children and adolescents he is the poet of love, for teachers he is a compulsory step towards understanding the spirit of India, recognised as the spirit of every human being. There are many psychological and pedagogic approaches, in the field of adolescence, and sociological approaches, within the caste system, and they may all be the object of a careful study between the literary plane, free of order and classification, and the scientific plane, with an overabundance of explanations. Original pedagogic and teaching suggestions flow from the tales, poetry, songs and music, from theatre and film productions, which may fall within the subject-matter of intercultural education and education to peace for teachers' training.

Notes

1. In *Emilio,* Rousseau describes adolescence as a second birth, in which relevance is placed on the changes to the youth, who is putting childhood behind him to enter into a new human condition. Similar attention is found in *The holidays have come*, a work in which Tagore talks about the age of fourteen being a time of growth, in which the adolescent needs not only larger clothes, but also love and understanding, especially from a female figure.

2. Under the title *Santinicheton (or Shantinichetan)*, a text composed of 17 parts was originally published and contained talks and meditations by Tagore that range from 1908 to 1916. Thereafter, in 1935-1936, the 17 pamphlets, jointly with some other talks, were published in two volumes. The Italian text includes translations from the English, such as: *Shantinichetan*, *Some letters from India*, *Asrama education* and Pearson's book.

References

Chistolini, Sandra. *Tagore Aurobindo Krishnamurti. Unità dell'uomo e universalità dell'educazione.* (Roma: Euroma-La Goliardica, 1990).

Delfino, Giuseppe. *L'ideale educativo di Rabindranath Tagore.* (Genova: Edizioni Pedagogiche, 1972).

Ottonello, Giacomo. *Rabindranath Tagore. Principi educativi e scuole nuove in Oriente.* (Roma: Armando, 1977).

Pearson, William W. (Ed.) *Santiniketan, the Bolpur school of Rabindranath Tagore.* (New York: The Macmillan Company, 1916).

Pieczynska, Emma (Reichenbach). *Tagore éducateur*. Neuchàtel, (Paris: Delachaux & Niestlé, 1922).

Poli, Flavio. *Tagore e Santiniketan*. (Bologna: EMI Editrice Missionaria Italiana, 1982).

Reale, Giovanni, Dario Antiseri, Mauro Laeng. 3 Vols. *Filosofia e pedagogia dalle origini ad oggi*. (Brescia: La Scuola, 1986).

Rousseau, Jean-Jacques. *Emilio o dell'educazione*. (Scandicci (Firenze): La Nuova Italia, 1995).

Stilo, Salvatore. *L'umanesimo integrale di Tagore*. (Torino: Reggio Calabria, 1982).

Tagore, Rabindranath. *Antologia di scritti pedagogici*. (Brescia: La Scuola, 1975).

Tagore, Rabindranath. *Asrama education*. Santiniketan, (Kolkata: Santiniketan Press, 1958).

Tagore, Rabindranath. *Gitanjali (Offerta di canti)*. (Lanciano: Carabba, 1914).

Tagore, Rabindranath. *La civiltà occidentale e l'India*. (Torino: Boringhieri,1961).

Tagore, Rabindranath. *Le vacanze sono arrivate: Le opere. Poesia, teatro, prosa*. (Milano: Club degli Editori, 1966).

Tagore, Rabindranath. *Per aprire nuove vie alla scuola*. Fatti e documenti raccolti da A.C. (Milano: Libreria Ed. Milanese, n.d.).

Tagore, Rabindranath. *Personality. Lectures delivered in America. (*London: Macmillan and Co., 1917).

Tagore, Rabindranath. *Rabindranath Tagore pioneer in education. Essays and exchanges between Rabindranath Tagore and Leonard K. Elmhirst*. (London: J. Murray, 1961).

Tagore, Rabindranath. *Sàdhanà. Reale concezione della vita*. (Lanciano: Carabba, 1915).

Tagore, Rabindranath. *Santinicheton. (*Vicenza: Esca, 1978).

Tagore, Rabindranath. *Some letters from India*. Santiniketan: India, 1923.

Tagore's Connections with Belgium

Philippe Falisse

Rabindranath Tagore's anguish for the Belgians was acutely expressed when, in 1914, at the outset of the First World War, he wrote 'The Boatman'. The poem strongly depicts the tragedy of the war, the helplessness of its victims and the belief in a saviour who will finally come and bring peace and happiness. He had told his friend C.F. Andrews: 'In 'The Boatman', the woman in the silent courtyard 'who sits in the dust and waits' represents Belgium.'

Rabindranath visited Belgium only once. He had sailed to Europe in May 1920, and had already toured England, France and Holland when he arrived in Belgium, on 2 October 1920, to 'deliver a lecture at the Royal Artistic Circle' in Antwerp. The Flemish newspaper 'Ons Vaderland' reported that the poet had taken 'The Message of the Forest' as his subject for that occasion. 'The lecture had been organised by Dr. de Groot and was open to the general public. There was no entry fee.' [1]

On 3 October, Rabindranath wrote to C.F. Andrews: 'I am writing this letter from Antwerp where I came yesterday morning and I am getting ready to go to Brussels where I have an invitation.' The address mentioned in his letter was: 42, Lge Gasthuisstraat, Antwerp.

Two to three days before leaving Holland for Belgium, he had received a letter from a certain Mr. Dubois requesting him to come to Brussels and speak at the 'Palais de Justice' on 4 October. 'If Dr. Tagore delivers his lecture in Antwerp on Saturday (2 Oct), will he be able to be in Brussels on Sunday? In that case, I hope he will accept some friends at my house, and hear some music.' Mr. Dubois wrote. The 'Modern Review' of Jan 1921 brought out the translation of the commentary made by Gaston Perier, a French critic, about the event: The auditorium of the 'Palais de Justice' had been filled up well before the scheduled time by a throng of spectators. The audience had been touched by the poet's lecture on 'The Meeting of the East and the West' and his appeal for building a world of understanding, friendliness and tolerance. Rabindranath received a tremendous success. It is said that Rabindranath chanted at times some of his own Bengali songs and that the next day someone had written to him: 'For those who suffered with you when the savage clapping of hands interrupted the vibrations of your voice; for those who wish heartily the real meeting of East and West, in purity of Love and Peace...Thank you Brother.'[2]

We know that Rabindranath remained in Brussels till 6 October and was on that day at the Grand Hotel Britannique from where he wrote to Dr. Rothenstein: '...The welcome which has been accorded to me in all the countries that I have travelled in Europe has been deeply genuine and generous to the extreme. This makes it delightfully easy for me to give out the best that I have in me in an easy flow of communication. I feel it clear that my relationship with the Continental Europe is natural and unobstructed.'

Although Rabindranath received a lot of deference from the common people in Belgium there was no comparison with the warm welcome he had felt in Holland. This is clearly mentioned in the letter W.W. Pearson wrote to C.F. Andrews on 8 October from Paris.

'Today, the admiration and unbounded enthusiasm that all Europeans countries had for the first Asian who was conferred the Nobel Prize in Literature in 1913 have faded. In Belgium, there is little interest in Rabindranath Tagore. He is known mostly by some literary-minded people and those interested in the religion of the East.'

Strangely enough, however, Belgians had indirectly influenced Rabindranath's thought which reciprocally had also inspired Belgian missionaries living in West Bengal.

Rabindranath Tagore was 14 years old when he was admitted to Class 5 (equivalent to Class 9 today) at St. Xavier's College in Kolkata. It was in February 1875. The College was run by Belgian Jesuits. Father Eugène Lafont was the Principal, Father L. De Vos, in charge of discipline, and Father J. Henry in charge of studies. J. Muilender, a scholastic, and two lay brothers J. Bierme and A. Declercx were there to maintain discipline. Both Father A. de Peneranda and Father L. De Vos were in charge of Class 5.

At that time, the lessons were held from 9 am to 3 pm at the Théatre Sans-souci in Park Street.[3] The daily routine included:

9 am – 10 am: Latin

10 am – 11 am: English

11 am –12.30 pm: Choice between Bengali, Drawing, French and Sanskrit

12.30 pm – 1 pm: Lunch

1 pm – 2 pm: History or Geography

2 pm – 3 pm: Maths

The question we ask ourselves is: What made Maharshi Debendranath Tagore, the founder of the Brahmo religion,[4] enrol the youngest member of his family at St. Xavier's?

Before letting Rabindranath study at St. Xavier's, Debendranath had certainly weighed the pros and cons of such a decision.

Indeed, Debendranath felt no empathy at all for the missionaries. The first and greatest danger consisted, therefore, of putting his son in a Christian institution run by the Jesuits. Were they not known as strategists in the art of swaying the elite in their religious beliefs? By putting Rabindranath at St. Xavier's, was he not simply inviting the demon to realise his work? The teenager Rabindranath could very well be tempted by the jesuistical charm and betray his Brahmo religion! That risk was even more relevant as one of his uncles, Gnanendramohan Tagore, had already been converted to the Christian faith and married the daughter of a pastor.

Debendranath was angered and disgusted by the proselytizing of the missionaries. That is why, when he had to address a congregation as President of the Brahmo Samaj, he wouldn't hesitate to caution its members against their activities. He also publicly confessed his allergy to Christ. At a talk on 'Love' delivered in a Brahmo temple at Mechua Bazar on 22 January 1872, he explicitly urged the faithful not to be influenced by the Christian teachings. He declared: 'In the name of Christ, Europe has undergone blood-bath. If Christian influence is allowed to enter India, the very bones of our country will be crushed. Christianity stands for all that oppose freedom.' He considered that the Christian message conveyed by the missionaries was contrary to the one of tolerance most treasured by Hindus.

Till the middle of the 20th century, few were the missionaries living in India who had ever thought of adapting their style of living to the Hindu surroundings. They had joined the European colonizers and also wanted, in a way, to conquer India. Most of them had devoted their life only for the expansion of Christianity. In other words, they had tried to find ways to convert and baptize the pagans, even if it was at the cost of uprooting them from their tradition and culture. They had literally followed Christ's words: 'Go forth and make all nations my disciples. Baptize them everywhere...'[5]

Debendranath wrote: The Brahmo religion must take care of three dangers: Idolatry, Christianity and the Vedic thought.[6]

Once, speaking in the temple of the 'Brahmo Samaj of India', founded by Keshub Chandra Sen,[7] Debendranath objected to those who were swayed by Christian teaching: 'Although no form of idolatry entered the Brahmo temple from outside, many are worried at the spectrum of Christ ...The Brahmo religion is controlled by Christ. I am very sad about that. We want only God, God alone. There should be no place for the incarnation at the four corners of His Home.'

Then, why did he entrust Rabindranath's education to the Belgian Jesuits at St. Xavier's? There were other colleges of repute in Kolkata like the 'Hindu College', co-founded in 1817 by Raja Ram Mohan Roy, his mentor, and Prasanna Kumar Tagore, one of his cousins. The college was well-known as it dispensed the traditional values upheld by the Tagore family. Satyendranath, Debendranath's second son, had studied in the 'Hindu College', and later had brilliantly succeeded in the entrance exams for the 'Indian Civil Service'. Rabindranath could have also sat on the benches at 'St. James School' or 'Scottish Church College'.

St. Xavier's was definitely not the best school to learn English. The Fathers spoke badly the language of Shakespeare and their pronunciation was incorrect[8]. The fact that the majority of students belonged to the rich Anglo-Indian community should have discouraged Debendranath from choosing the College. The Anglo-Indian community couldn't fit in with the Bengalis, known for their pride in their traditional values. The Anglo-Indians eyed everything European. Some were openly contemptuous and scoffed at Indian customs and values.

They were, in other words, disciples worth of T.B. Macaulay's politics that aimed at creating 'a class of persons, Indian in blood and colour, but English in taste, in opinions, in morals, and in intellect'.[9]

While pondering the pros and cons of St. Xavier's, Debendranath noted that, upto then, nothing had prompted him to choose the College. It was risky and it presented many disadvantages! But then, what was so special as to make the scales tip in its favour?

Since the beginning of the 19th century, a succession of events had led to a political, social and cultural revival of the country as well as the emergence of Indian nationalist movements. One of these was striving for a modernization of the educational system.

Already, in 1823, Raja Ram Mohan Roy[10] had an audience with the new Governor-General, Lord Amherst, to urgently put in place, for the benefit of the Indians, a system of education in English similar to the one in Europe. Raja Ram Mohan Roy was indeed convinced that the traditional system of education was responsible for keeping his fellow countrymen in ignorance while the European system of education would encourage them to go in for scientific research and contribute to the development of the country.

The Raja's view was first misunderstood and booed by many. Even Mahatma Gandhi mocked it. However, in the Tagore family, not only was the idea accepted but Debendranath, among others, was enthusiastic about it.

From 1830 onwards, we see quite a few magazines dealing with scientific subjects. Raja Rammohun Roy's vision was obviously gaining ground, transforming the mentalities, arousing the curiosity and interest of high society. During 1847-1855, articles on zoology, astronomy, geology and physics were published in *Tattvabodhini*, a periodical started by Debendranath Tagore. During the same period, we find Iswar Chandra Vidyasagar publishing in his *Jivancarita* (1850) biographies of reputed European scientists such as Copernicus, Galileo, Newton,etc.[11]

Increasingly, there was an awareness among Indians about the necessity to acquire scientific knowledge to get the autonomy and the independence they were looking for. The scientific knowledge was included in the agenda of the nationalist movement. Bankim Chandra Chattopadhay would maintain that if the Europeans had conquered India it was only due to their technical and scientific expertise.

On 3 January 1870, an article by Dr. Mahendra Lal Sircar (1833-1904) came out in the weekly magazine *The Hindu Patriot*. The author, a Doctor in Medicine from the University of Kolkata, was calling for the establishment of an 'Indian Association for Scientific Research', urging his countrymen to take to the study of applied sciences, to master every single scientific subject: 'The Hindu mind had lost much of its original Aryan vigour on account of ignoring scientific investigations and indulging in caste restrictions and theological obscurantism.' [12]

Without freedom there was no chance of promoting the development of the country and the 'Indian Association for Scientific Research', he was yearning for, would serve that very purpose. India would then get back its national pride, honour and confidence.

Dr. Sircar's proposal met with applause and was approved by several nationalists. It, however, took six years to be ratified. Businessmen and industrialists came forth to finance the projects. Among others the list of the benefactors included the names of Jatindra Mohan Tagore and Surindra Mohan Tagore, two sons of Prasanna Kumar Tagore, as well as Debendranath Tagore.

Dr. Sircar did not know that his article would be read by Father Eugène Lafont, the Principal of St. Xavier's. On 8 January 1870, just five days after it had been published. An article was published in the *Indo* (Indo-European Correspondance), the most popular Catholic weekly in Kolkata, in which Father Lafont attested his full support to Dr. M.L. Sircar's resolution.

Father Lafont was born in 1837 in Belgium, had become a Jesuit and arrived at St. Xavier's College on 7 December 1865, two days before the end of the academic year.

When the College reopened in January 1866, Father Lafont, who was in charge of discipline and of Class 5, started to organise a series of lectures on natural sciences and laid the foundations of a laboratory. In 1867, the name of Father Lafont became popular among the citizens of Kolkata mainly on account of his meteorological observation and prediction of the terrible cyclone that had swept Kolkata on 1 November. His warnings to the local competent authorities had allowed them to take measures to limit the loss of life.

By 1875, Father Lafont's suggestions and comments regarding meteorological observations were hailed by eminent savants. He had no problem purchasing scientific equipment to build an astronomical observatory at St. Xavier's. This was erected in 1875, the year of Rabindranath's admission in the College.

St. Xavier's enjoyed a good reputation not only because the Jesuits were instilling discipline and teaching well but also for Father Lafont's lectures. He had the talent to popularise science and to explain the latest discoveries in technology with practical demonstrations. In 1868, he was giving his first popular scientific show for the general public of Kolkata. *The Hindu Patriot* wrote the following commentary: 'Not a single technical word which had not been clearly explained for the common man.' Encouraged by this success, Father Lafont went on organising other popular scientific lectures. Due to the growing attendance, he had to ask the participants for a petty sum of money so as not to face a chaotic crowd.

Father Lafont's pedagogical approach, competence and dynamism had been a source of inspiration for the Hindus seeking the creation of a scientific policy in favour of the Indian population. As for Dr. Mahendra Sircar, he had found in him not only a friend but the driving force behind the association he had been dreaming of.

Debendranath Tagore had not heard anything suspicious about the Jesuits who were running St. Xavier's. The Tagore family had even showed empathy towards the College which, within a short period of just 10 years, had become a prestigious institution in Kolkata. In 1872, Jatindra Mohan Tagore (1831-1908) of the Pathuriaghata branch of the family had even left his property located at Moghy on the bank of the Ganges at the disposal of the Fathers for a fortnight.

Although they were missionaries, the Jesuits were primarily regarded, even by the ultraconservative Hindus, as experienced teachers. Their endorsement of Dr. Sircar's scheme for the science association published in the *Indo* on 13 April 1872 had been hailed by the Indians:

> 'We do not believe that every modern discovery is a blow to Christian truth or that the propagation of science will in time eliminate all authoritative religious teaching. We are therefore glad when an intelligent and enterprising Hindu like Dr. Mahendra Sircar comes forward as the champion of physical science and advocates its propagation among his countrymen. Dr. Sircar is one of those Hindus who has proved that the adoption of western progress, properly so called, is not necessarily bound up with western mannerisms. He discerns between what is good in western progress and what is bad or simply indifferent, and he wisely refrains from attempting to convert his Bengali compatriots to science by shocking and startling them on every possible occasion.'

Debendranath Tagore was fond of astronomy and must have been captivated by Father Lafont's achievements. As for Rabindranath, he had always been attracted by natural sciences. As a child, he wrote:

> 'I felt a great desire to understand the laws of nature. I remember a certain Silanath Dutta who used to come from time to time, on Sundays, in Jorasanko. He was not a great scholar but liked to make experiments. His ability to explain these clearly was amazing. Later, when I was 12 years old, my father took me to Dalhousie, in the mountains. Our days consisted in observing nature and our evenings in observing the stars. It is in Dalhousie that my father taught me the basics of astronomy as well as how to spot the constellations. In fact, my first essay was on that subject. Later on, I collected some books on astronomy.' [13]

For Debendranath, there were now enough reasons to tip the scales in favour of St. Xavier's College: The good reputation of the institution thanks to the pedagogical competences of the fathers and to the new way of teaching they had innovated. The prospectus issued on the occasion of the opening ceremony of the College in 1960 began with the mention of the Jesuit Society as 'essentially devoted from its beginning to education'. It also underlined the fact that 'the study courses, subject to some adaptation to the needs of this country, would be similar to that pursued in the great Colleges of Europe.' In other words, the syllabus would include the study of the sciences, which, at that time, was something unheard of in the British schools. The Jesuits of St. Xavier's were therefore encouraging the Indians to acquire the knowledge that would allow them to compete with the Europeans and, one day, become independent. No doubt, the presence of Father Eugène Lafont contributed considerably to Debendranath's choice.

Remembering the College

It was in February 1875, one month after the beginning of term, that the young Rabindranath was admitted in College. He had to make up for lost time. Unfortunately, he could not cope with the studies nor could he concentrate. Only a month had gone by when his mother, Sarada Devi, passed away. That tragedy moved him deeply. He had to stay home till the end of the funeral rites. When he came back to school, the second semester was going on. Rabindranath was completely lost: 'I was deaf and dumb in the class of Latin and my exercise books had the unalterable whiteness of a widow's garment from the first to the last page.'

His teacher, Father A. de Peneranda, an astronomer Jesuit father who besides his teaching classes assisted Father Lafont in his research work, understood his student's sadness and comforted him. Rabindranath remembers well that teacher whose English pronunciation made him a laughing stock for all his classmates but who seemed so detached from worldly affairs and always in communion with God. In his book, *Reminiscences*, published in 1912, Rabindranath recalls one incident that had touched him deeply: 'Father de Peneranda came to me and seeing me confused asked me: 'Tagore, do you feel unwell?'[14] After his mother's death, Rabindranath attended his classes irregularly. Due to the lack of archives, it is not possible to know whether Rabindranath failed in 1875 but one thing is certain that he was back in the same class in 1876[15]. This was perhaps because he had not attended the required number of days to be in a position to appear in the exams. In the end, he dropped out of College in the middle of the year without getting through it.

About the Jesuits at St. Xavier's, Rabindranath wrote praiseworthy comments: 'Still today, I remember perfectly well the teachers. They were not all behaving the same way but, anyhow, I must say that they have influenced me. They radiated a divine goodness as I had never felt before. Not only had they the talent to teach but they were meek and kind.'[16]

Rabindranath never mentioned any punishment or scolding he received for having been listless or for his reveries during classes. Indeed, St. Xavier's College was fairly different from the schools he had attended before and above all the Oriental Seminary where the whacking of knuckles with sticks was common! [17]

Rabindranath never accused the Belgian Jesuit missionaries who ran the College of proselytising. They were admired for their ability to teach and for their determination to develop the personality of their students so that these students could become responsible citizens. The missionaries were also appreciated for the discipline that prevailed in their institution.

Preaching on Purity at the famous glass-casing temple in Santiniketan, Rabindranath had again spoken highly of his teacher who 'was not physically regarded as a beautiful man but who shone with an extraordinary inner beauty that was, it seemed, the result of his constant union with God more than simply being due to his renouncement of worldly affairs and honours.'[18]

When, in 1927, there was question to take up anew the Old Boys Association[19] of St. Xavier's, it was decided to contact Rabindranath and request his patronage. It was felt that once his name appearred on top of the list of the association, it would be easy to enlist new members. The alumni of the College would be honoured to have their names mentioned in the same list as the one with the name of the Nobel Laureate whom the Bengalis considered equal to Shakespeare and call 'Gurudev' (Spiritual Master).

Without any hesitation, Rabindranath accepted the request. The poet gave the secretary of the association a signed photo and, with tears in his eyes, broke into the words he had dedicated in his Memoirs to Father de Peneranda, his ideal teacher[20]. He presented a bust of Jesus in marble to St. Xavier's. Today, the bust can be found in the Principal's room of the College. It is interesting to know that during the same year, in 1927, the Maharajah Sir Prodyot Kumar Tagore offered a 'Madonna in Prayer,' an original painting by the 17th century Italian painter Guido Reni to St.Xavier's.[21]

The relations between St. Xavier's and Rabindranath grew deeper in course of time. Rabindranath became the Patron and Vice-President of the Old-Boys Association founded in 1932.[22]

St. Xavier's celebrated Rabindranath's 70th birthday anniversary. A delegation was sent to his residence at Jorasanko to congratulate the poet. The July 1932 issue of *The Xaverian*, gives a detailed description of that happening: 'The deputation from St. Xavier's headed by Raja Kshitindra Deb, left the College in the white bus and made its way to the poet's house in 6, Dwarkanath Tagore Lane. They were ushered into the reception room – small, neatly furnished and immaculately clean. Punctually, at the appointed time, Dr. Tagore entered, wearing a long flowing robe of dark brown that clung loosely to his tall frame which carried his 70 years with dignity and ease. Each member of the party was introduced to the poet who then sat down with just a suspicion of weariness. The Raja read the address, composed by himself as the representative of 'St. Xavier's Old Boys Association.' The poet listened with great interest and recollection and was plainly moved by the pledges of affection and appreciation that the address embodied. In a voice full of emotion, Tagore thanked the deputation and in a few simple words alluded to the College that had always been for him an inspiration to his youth.[23]

On another occasion, responding to a request from the College, Rabindranath participated in a fund raising for the flood victims in 1931. He accepted to preside over a cultural evening organised for that purpose. His presence naturally contributed to the success of the event. Once again, he had expressed his feelings of gratitude for his old college before the beginning of the programme. It included music, an English play and one of his comedies, which was particularly enjoyed by the audience.

Rabindranath joined other cultural activities organised at St. Xavier's. He never accused the St. Xavier's Belgian Jesuit missionaries of proselytizing. Yet, they seem to have exerted a deep and spiritual influence on him. For, if he had been put into contact with first rate pedagogues as well as with scientists, he had also met religious members of the Society of Jesus who passed onto him the devotion to Christ.

In his publication entitled 'Christ,' Rabindranath shares the fruits of his reflections and meditations on the life, teachings, birth and actions of Jesus. He writes: 'Jesus showed how great man is. In calling God, Father, he lifts up and confers on all men the title of sons of God. He unites them. Jesus preaches brotherhood. He came to explain who the son of God is. He has no ambition to establish a kingdom on earth. His strength is in himself and not in a foreign power. Nor shame or wickedness affects him. Today, his death is a symbol of life for all mankind. Jesus speaks of the kingdom of God which is in man, a kingdom for the humble, the meek. If he had said that man's greatness was somewhere else, the true greatness of man would have been affected. He said that he is son of God. He showed us that humanity is not to be found in wealth nor in good manners but within us. It is not by offering flowers, food, prayers that we love God, but by serving mankind. Jesus is a man of suffering. He sees his Father by giving his life for the others. Service is a commandment of love.'[24]

Rabindranath coined the word 'Borodin' (The Big Day) to mean Christmas in Bengali.

For him 'the Christian religion was different from the others in that it respects and makes everybody worthwhile. The Christians who give their lives for mankind are witnesses of the most precious value religion can offer.' (written on 25 December 1940).[25]

Rabindranath remained Brahmo his whole life but he discovered and sang hymns to a God who is close, good, who consoles and who is definitely different from the far away, impersonal and nameless God sang by the founders of his religion.

The vast majority of Hindus portray God as a Mother. However in Rabindranath's devotional songs, we hear him praying and calling God 'Father' or 'King of Kings'. In one of his songs, Rabindranath implores God, his loving Father: 'Are you not our Father! We see your face with full of affection.'[26] He wanted that the whole day, while eating or resting or even sleeping the mind could repeat the 'Om Pita No'si' mantra that means: You are our Father. I am not alone, I have a Father.[27]

At St. Xavier's, the classes began and ended with the well known Christian prayer, 'Our Father'. Rabindranath must have recited it at least twice a day during the three or four months he attended classes there. Later on, he composed a song whose words remind us of those expressed in 'Our Father': 'You are our Father. I want to know you as a Father. I want to receive you. Do not be angry on me. Oh Lord, forgive us our sins. Give us what is good, what makes us happy. You are the source of joy, oh Father, you are the source of goodness. You are good, you only are good, you are the best. I praise you, oh Father, I praise you.'[28]

St. Xavier's could well have been at the origin of Rabindranath's devotion for Jesus and for God as Father.

Rabindranath did not know that after his death, some Belgian Jesuits specialising in the Bengali language and literature would read his works in their original form, and being fascinated by them, would better appreciate the Bengalis and their culture. They would do credit to his words and thoughts in their homilies, lectures and articles. Besides, several of his compositions would be sang in the churches in Bengal. The first Rabindra Sangeet introduced in the liturgy was 'Anandoloke', a Brahmosangeet glorifying the Almighty: 'Your glory fills the earth and sky, the world is but a jewel of your feet, In the realms of joy and gladness You dwell, O Lord of Truth and Beauty.' Today, sixty-eight Rabindra Sangeet are included in the singing handbook used by the Bengali Christian congregations.

In 2010, the Government of West Bengal conferred its highest literary award, the 'Tagore Prize', on Father Paul Detienne, a Belgian Jesuit, who lived more than 30 years in Kolkata and who is presently in Belgium. Father Detienne is credited with having created a new literary style in the Bengali language.

The Belgian Orchestra 'Rajhans,' directed by Hans Vermeersh, execute Rabindranath's compositions from time to time. They did it recently in Portugal, the United States and Great Britain. Moreover, every year, on 11 November, the Orchestra renders a Tagore composition at Knokke-Heist, to commemorate the Indian soldiers who gave their lives on Belgian soil during the First World War.

Notes

1 Prasantha Kumar Pal. 'Rabijibani' Ananda Publishers, vol 8, p 41

2 Ibid.

3 *St. Xavier's Magazine*, July 1933, p. 33-34.

4 In 1850, Debendranath Tagore had initiated the 'Brahmo' faith. By refusing to acknowledge the authority of the Vedas, it broke away from Hinduism. The Brahmos believe in one God (infinite, undefinable, invisible –immanent and transcendent, omnipotent creator). The creation is to go back to the intimate reality.

5 Matthew's Gospel, 28, 19

6 Debendranath Tagore. *Maharshi Debendranather Patrabali*, Ed. Priyanath Sastri.

7 Keshub Chandra Sen was influenced by the Christian ideas. He became a member of the Brahmo Samaj in 1856 but, influenced by Christian ideas, founded his own breakaway Brahmo Samaj of India in 1866.

8 Udayan Namboordiry. *St. Xavier's-The making of a Calcutta Institution*, (Viking, 1995), p 46.

9 Sharp, H. (ed.). *Selections from Educational Records, Part I (1781-1839)*. Superintendent, Govt. Printing, Calcutta, 1920.

10 Raja Rammohun Roy (1772-1833). Religious and social reformer, called the 'Maker of Modern India' he co-founded the Brahmo Samaj in 1828.

11 Arun Kumar Biswas. 'Father Eugène Lafont and the Contemporary Science Movement', The Asiatic Society, 2001, p. 16.

12 Ibid, p. 48.

13 *Viswa Parichay*, *Rabindra Rachanabali*, Vol 4, p. 423.

14 *Jeevansmriti, Rabindra Rachanabali*, Vol 10, p. 53.

15 *St. Xavier's magazine*, Jan. 1942, p. 109.

16 *Jeevansmriti, Rabindra Rachanabali*, Vol 10, p. 52-53.

17 Ibid, p. 8.

18 'Suchi', *Rabindra Rachanabali*, Vol 12, p. 436 and 438.

19 Rabindranath was on the list of the Alumni in 1916 vide *The Xaverian* Vol 3, number 1, 1916

20 'The Old Boys Column,' *St. Xavier's Magazine* – Centenary Edition 1960.

21 The painting 'Madonna in Prayer' had been a part of Warren Hastings personal collection. It was auctioned to an ancestor of Pathuriaghata Tagore when the governor ne.eded some money to fight some cases. It was the wish of the donor that it be hung in the chapel.

22 *St Xavier's Magazine* 1935, Vol 4 (Platinum) Jubilee Special number p. 83.

23 'Our Past', *The Xaverian*, July 1932, p 333-334

24 'Christ', *Rabindra Rachanabali*, Vol 11, p. 509 (free translation).

25 Ibid, p. 522-523.

26 *Gitabitan*, p. 83.

27 'Christ', *Rabindra Rachanabali*, Vol 11, p. 523 to 524

28 *Gitabitan*, p. 162.

Rabindranath Tagore in Berlin

Martin Kämpchen

Rabindranath and Germany

When Rabindranath Tagore was awarded the Nobel Prize for Literature in 1913, he was as unknown in Germany as in most other countries outside India. However, a combination of lucky circumstances and the natural inclination of the German people towards romanticism and idealism propelled the Indian poet to extraordinary heights of popularity within just a few years.

Rabindranath was lucky to have a publisher in Germany, Kurt Wolff, who with enormous zeal and energy published almost all books as soon as Macmillan brought them out in London. By the time the poet visited Germany for the first time, in 1921, he with his team of enthusiastic translators had brought out an eight-volume *Collected Works* of Tagore which was unprecedented in any other country. He offered each Tagore book in various formats and editions to suit the taste and financial capacity of every prospective buyer. By 1923, the combined sale of Tagore's works in German language crossed one million copies. With Hermann Keyserling, the powerful and popular philosopher, Tagore had an ideal propagator of his ideas and ideals. He was the only person of consequence in Germany who had met Rabindranath in India before the Nobel Prize and had a solid knowledge of Indian religions and Indian life. The loquacious count made himself the mouthpiece of Tagore situating him in the proper cultural and religious environment. With their combined acute business sense, Wolff and Keyserling 'managed' the tour of Germany in 1921 and helped to create Tagore's popular success.

After the First World War which left the German nation vanquished and the population in a deep spiritual and cultural crisis, the country was in search of prophetic figures that it could look up to and emulate. Tagore arrived at the right moment. He was viewed by the German public as a messiah and wise man who had come to console and guide the country. Tagore saw himself in that role as well. This combined with the German cultural inclination towards romanticism and idealism. In the early nineteenth century, the German Romantics had discovered India as the land of their dreams where pure, childlike humanity still prevailed. India was seen as an alternative to European culture. In the early twentieth century, the German cultural subconscious mind identified Rabindranath as a late icon of such romantic yearnings.

Tagore in Berlin

Rabindranath Tagore visited Berlin on all three trips to Germany – in 1921, 1926 and 1930. In fact, in 1926 on his second trip, Tagore stayed in Berlin twice, and in 1930 he even stayed there three times. Berlin was for Tagore a 'launching pad' from where he proceeded to East Europe (Prague and Moscow), in 1926 and 1930, and to Copenhagen, in 1930. In the same year Tagore rested in Berlin for three days at the home of his friends, the Mendel family. Berlin, the capital of Germany with its long tradition of governmental power, its palatial buildings, wide avenues and parks, and its cultural eminence, had become the most important city between Paris and Vienna and Moscow. Appropriately, it was in Berlin that Tagore met numerous personages of consequence with whom he could with dignity participate in global discourses. In Berlin, for instance, Tagore was in 1926 introduced to Germany's *Reichspräsident* (President of the German Reich) Paul von Hindenburg. In 1926 and 1930, he had several meetings with Albert Einstein, his Western counterpart in iconic status and general respect. It was at the University in Berlin that he delivered, in 1921, his most acclaimed and best remembered lecture which had to be repeated by popular demand. Here he launched, in 1930, his first German exhibition of pictures at the Galerie Ferdinand Möller. Consciously or subconsciously, or perhaps by force of circumstances, Tagore chose Berlin to be the focus of his activity and mission. Only Darmstadt, the home of Hermann Keyserling, figured almost as prominently as Berlin in newspapers and in memoirs.

Popular Reaction

The ten years from 1921 to 1930, from Tagore's first to his third and last visit to Berlin, embrace an exemplary period in modern German history. The country rose from its post-war misery to unexpected economic and cultural achievements and then experienced a second downfall. In German, this decade is referred to as *Die goldenen Zwanziger* – The Golden Twenties. The First World War was the first war of massive proportion in terms of the number of soldiers and amount of arms and war material employed. In the whole of Europe nine million soldiers died, ripping apart families and the very texture of society. In the Versailles Treaty of 1919 the victorious allied nations demanded from Germany a huge amount of reparation which was to be paid over several decades. This obvious injustice to a people already vanquished, plunged Germany into an economic and emotional crisis. There was widespread unemployment, people went hungry. Hordes of beggars, many of them crippled war veterans, roamed the streets. Germany had the highest infant mortality rate in Europe, and the country went through several waves of epidemics. Accordingly, the political climate was vicious with hate campaigns, mass strikes which were struck down violently, and several attempts to overthrow the government.

This was the scenario when Tagore first appeared in Berlin, in 1921. It can be well imagined that, against this background, the reaction to him was deeply divided. All those (and they were probably the majority) who genuinely yearned for social peace and for justice from the global community saw in the poet a sign of hope, and more so, a messianic promise. Tagore declared that their real opportunity to achieve that peace was not so much through political and social action on the streets or through political and social institutions. Rather, peace should be sought by strengthening the spirit in man. More emphatically, Tagore repeated that the regeneration of the German people should be achieved by returning to the source of German culture, which he saw revealed in its affinity with ancient Indian culture. Having read of German literature, as far as we know, only poems by Heinrich Heine and Goethe's play 'Faust', Tagore had a simple and probably idealized concept of German culture. That same culture, after all, had also caused the war and was to create more destruction later.

A section of the press in Berlin ridiculed the Indian poet, or viewed the public adulation with sarcasm. This section claimed that for the general population Tagore was little more than a sensation, a diversion from the misery of daily life. Proof of this insensitive reaction by the masses was seen in the rowdiness demonstrated on 2 July 1921 when Tagore spoke at the University of Berlin. Many more people had obtained entry tickets than could be accommodated. People jostled and engaged in fisticuffs to push into the overcrowded hall. The police had to be called to create order and make a path for the speaker to reach the dais.

At best, the public was unsure of how to judge the Indian poet who was so different, so conspicuously strange. This was emphasized by his robe and his solemn, dignified demeanour. The atmosphere Tagore created around him was awe-inspiring, but also somewhat unreal and unconnected to the harsh realities on the streets. This 'otherness' was often seen as inimical to the harsh situation Germans faced at that moment, namely to rebuild the nation in the face of tremendous odds. Hence, many felt that the need of the hour was not a message of peace, but an aggressive call for hard work and dynamism.

The situation in Berlin worsened until 1923, when galloping inflation played havoc with the economy, and the German currency had to be reformed (*Währungsreform*). By that time, the stringent rules of the Versailles Treaty were being relaxed, additional treaties came into force, and the USA initiated a package of economic assistance which helped Germany to get on its feet within a miraculously short time. The Golden Twenties truly began by 1924 and came to a sudden stop in 1929 when the Great Depression plunged Germany into another – global – economic crisis.

When Rabindranath Tagore visited Berlin for a second time, in 1926, that upswing was already in evidence. Both popular and high culture flowered, especially in the city. Expressionism had begun to make a mark in art and literature from the early 1920s. This movement was yearning for the New Man, turning away from big city life, industrialization, a mechanization of life, away from capitalism and militarism. Socialism and anarchic models of society became ideals. Surrealism and symbolism and other anti-naturalistic styles became prevalent.

Theatre culture was at its prime, with the best-known theatre director, the Austrian Max Reinhardt, opening two new halls. Berlin was the metropolis with the maximum number of cinema halls. Silent films were on their way out and from 1927 onwards, the exciting new world of sound films emerged from Hollywood. Popular enthusiasm for India was fuelled by the hugely successful

Indo-German films directed by Franz Osten, especially *The Light of Asia* (1925; German title *Die Leuchte Asiens*) which was shot in India in cooperation with Himansu Rai (who had been a student of Tagore), and *Shiraz* (1928; German title *Das Grabmal einer großen Liebe*) and *A Throw of Dice* (1929; German title *Schicksalswürfel*).

Indian travelogues by German writers appeared. Special mention should be made of Hermann Keyserling's *Travel Diary of a Philosopher*, Waldemar Bonsel's *Indienfahrt* and Hermann Hesse's novel *Siddhartha*; all of these appeared between 1918 and 1922. New forms of dance, like the Charleston, created a craze. The Alexanderplatz became the square were a new upbeat mood was demonstrated with bright lights, bars and night-clubs. Fashion and sports received fresh importance.

One should have thought that Tagore's reception in Berlin would have profited from the Golden Twenties mood. Surprisingly, the reverse happened. In the disaster year of 1921, Tagore's spiritual message was shining bright; but in 1926, he was, by comparison, neglected. Let us look at the production of Tagore's plays in Berlin to gauge this transformation. In 1921, *The Post-Office* was performed 17 times at the famous theatre Volksbühne, but in 1926, there was no Tagore play anywhere in Berlin. The four Tagore plays that were staged throughout Germany, had 121 performances in 1921. In 1926. it came down to a mere 19 performances.[1] Kurt Wolff, Tagore's business-minded publisher, had published Tagore's last book, a two-volume edition of *Gora*, a year earlier, in 1925. After that, he was no longer inclined to publish Tagore's works. The Tagore enthusiasm had worn thin, and Wolff must have thought that it could not be resurrected after the economy boomed again. Was then Rabindranath Tagore merely a poet suited for the years of gloom and disaster in Germany? Was the general population attracted to him only when the chips were down? And could he be conveniently forgotten when the majority was feeling better and enjoying itself? So it may seem.

Reaction of the Cultural Elite

After I have briefly reviewed the *popular* reaction to Tagore's visits to Berlin, let me look at what happened below the surface, on a level of deeper interaction. When Rabindranath first arrived in Berlin in 1921, he stayed from 29 May to 4 June. The German Ministry of Culture had asked the young Indologist Helmuth von Glasenapp to guide and accompany Rabindranath in Berlin. In his memoirs, von Glasenapp recalled that on the first day of Tagore's stay in Berlin, he (Glasenapp) was the only one to know of the poet's presence in the capital. He got to know Tagore as an 'unassuming and kind-hearted person'. Glasenapp continued:

> We mainly conversed about Indian philosophy, and he revealed to me that Ramanuja's teaching of Pantheism appeared to him the most convincing solution to the mystery of the relation between God and the world.[2]

The young Indologist wrote:

> Germany means to him not only the country of poets and thinkers, the country of Goethe and Kant, but first of all the country which has shown an unselfish interest in the study of Indian languages and religions. He quoted [Heinrich] Heine who had said that the German nation had been the only one that did not try to capture outward treasures in India but had striven for the spiritual values of India. Tagore hopes that German and Indian minds will be able to cooperate in order to bring about a better understanding between the East and the West.[3]

Here we witness two people with their mental focus directed towards India, seriously wanting to understand each other. Helmuth von Glasenapp would later rise to be a leading professor of Indology, which lends added weight to his report.

The days in Berlin were taken up by a stream of visitors who sought a meeting with the poet at his hotel. Helene Meyer-Franck, the German translator of many of Tagore's prose works, and her husband Heinrich Meyer-Benfey had arrived from Hamburg to 'manage' Tagore's daily routine. Helene Meyer-Franck later reminisced:

> On Tagore's request we were with him every morning. We could help him looking through his mail and replied to some letters ourselves; also, we were present while he received his numerous visitors. We were impressed by Tagore's great powers of concentration and his self-control.[4]

Apart from the two much-publicized lectures at the University, Tagore also attended a dinner at the home of Government Secretary Professor Carl Heinrich Becker and a musical soirée at the home of the banker Fritz Andreae in Berlin Grunewald. He also watched a performance of his play *The Post-Office* at the *Volksbühne.*

It is essential to realize that Tagore, while in Berlin, took care of serious matters; popular acclamation was not sufficient for him. He associated with personalities of the government and the social elite. Tagore saw a need to express his thoughts among sections of society who may be expected to be farthest removed from his message of interiority. And he enjoyed, perhaps even craved, the attention of the upper echelons of society, as it added to India's prestige and gave greater weight to his voice as a representative of India.

With Fritz Andreae's wife, Edith Andreae, Rabindranath maintained a correspondence until 1939. She was a highly cultured and wealthy society lady. More importantly, Edith Andreae was the sister of Walter Rathenau, the then External Affairs Minister. He was assassinated a year later, in 1922. In the soirée which Tagore attended at their home, he was for the first time treated to classical Western music on the piano. He listened to Händel, Bach and Mozart. 'At first he was astonished', according to one account, 'then he was captivated and deeply moved'.[5] Tagore wrote two significant letters to Edith Andreae from which I wish to quote. The first was written before his voyage back to India. He wrote:

> I do not know much of those of Europe who belong to the old regime and who cling to the wreckage of the past, but the little that I have seen of the young generation of this Continent gives me the hope that this age of universal suffering will arouse our faith in the creative role[?] of [the] human spirit. [...] The old will yet go back to the machines for their sustenance, but the young will reject them and will cry for the harvest of life.[6]

Having returned to Santiniketan, Tagore again picked up his pen and wrote to Edith Andreae. This time he became more specific:

> Since my return to India I have often thought of the time I spent in Germany and regretted that I was not able to stay there for a longer period. My memory is still haunted by the faces I saw at the meetings I addressed, pale and haggard and yet glowing with the intensity of some inner vision and expectancy of a great realization.

The end of his letter repeated what he had said so often:

> Let me hope that the sufferings through which your people are passing and the change of fortune that has so suddenly overtaken them may give them that great shock necessary to reveal to them another world which is Man's mission to conquer.[7]

Tagore resolutely hoped that pain and misery would, if steadfastly resisted and endured, become a reward, in other words that pain would transform into spiritual substance.

The visits to Berlin in 1926 and 1930 were – on the popular level – less sensational, but probably of more lasting significance. From 11 to 15 September 1926, Tagore gave a lecture at the Philharmonie-Hall, addressed the Hindustan Association of Central Europe, and met the President of the Reichstag (Parliament) Paul Löbe, Carl Heinrich Becker, now the Minister of Culture, and the President of the German Reich, Paul von Hindenburg.[8]

Of special historical relevance was the interaction between Tagore and Albert Einstein. They first met at a breakfast reception of the Ministry of Culture and then again, more privately, at Einstein's home. They again got together in 1930, this time at Einstein's country house near Berlin, in Caputh, and at the Villa Mendel. Two conversations between Tagore and Einstein were recorded and later published.[9]

However, in retrospect, these conversations may not have the value which is still being accorded to them. Importance was given to them by the fact that two great representative figures, one from 'the East' and the other from 'the West', repeatedly met. The photographs of Tagore next to Einstein taken in Berlin are some of the most publicised among Tagore's iconic pictures. It was a mere coincidence that the figure from the East was a poet and the one from the West a scientist, thus consolidating the cliché that 'the East' is great in the arts and 'the West' a pioneer in the sciences. This is a cliché which, properly speaking, demeans both India and Germany by its one-sidedness. The content of these conversations are not particularly significant. Are they not rather too sketchy to really formulate a position vis-à-vis another's position? To my mind, the true significance lies in the fact that these conversations demonstrate how the two men share a common vision and have an excellent chemistry with each other. Such unison Tagore enjoyed with very few persons from the West; probably with William Rothenstein and C.F. Andrews, but certainly not with any other German-speaking personality.

The Berlin visit of 1930 was dominated by the exhibition of Tagore's paintings in the Galerie Ferdinand Möller. This is an event of historical dimension. For the first time, the Indian poet was able to present himself to the German public *directly*, that is without the medium of translation. This is important to realize, considering the ungainly or misleading quality of the early translations into English and from English into German. Tagore realized the opportunity he had with his paintings when he said that his poems are meant for his countrymen and his paintings are a gift to the West.[10] Even with direct and authentic translations available today, I still believe that the exhibition mounted in Berlin in 2011 was the most momentous event of the 150th Birth Anniversary celebrations in Germany.

In 1930, the public was suitably taken aback by these paintings, judging from the spate of articles in Berlin newspapers. They grappled with the urge to put labels on Tagore's art. Was he influenced by Indian art or European art? Was he like anything they had seen before? Such searching questions emerge when a totally new form of seeing the world through art arrives. The noted art historian Heinrich Lützeler, commented:

> [...] what surprised me in Tagore's paintings, was the close connection with Europe, contrary to his literary work which has its roots entirely in Indian tradition. The form, especially colours and lines, but also the themes of Tagore's paintings are strongly influenced by symbolism and the 'Jugendstil.'[11]

Tagore himself denied any direct European influence in his paintings, but this did not stop the public from making speculative comparisons. Those most frequently made in newspaper reports were with Edvard Munch, Emil Nolde and Paul Klee.[12]

Finally, one last defining event in Berlin was the recording of a short 'sound film' in the legendary UFA Studio of Neubabelsberg. A letter from the UFA Studio dated 17 July 1930 to Tagore reads: 'Dear Sir! Thanking you again for your extreme kindness to visit our studios [,] we have the great pleasure to send you the little souvenir sound film we made yesterday and which turned out very well indeed. [...]'

This was probably the first 'sound film' made of Rabindranath Tagore. So it was in Berlin that he entered the charmed circle of the modern media.

During the time of National Socialism (1933-1945), Rabindranath was anathema in Germany. However, when East Germany turned into the communist-ruled German Democratic Republic (GDR) Rabindranath was looked upon as a brother-in-arms. Rabindranath's insistence on international understanding and the brotherhood of men resonated with the communists, and he was published again. Most books by Tagore were published by the government publishing house Volk und Welt in East-Berlin.[13] In 1961, Tagore's Centenary Year, several books were brought out, among them *Kabuliwallah, O Kabuliwallah*[14], a book containing short stories, a short version of *My Boyhood Days*, poems and aphorisms. Some of the texts were translated from English, others directly from Bengali. Further, an anthology of various prose texts[15] and an old translation of *The Home and the World*[16].

After the Centenary Year Volk und Welt managed to bring out a four-volume *Rabindranath Tagore Selected Works* with prose works. The most noteworthy volume was a translation from Bengali of *Gora*[17]. German Unification brought the efforts of this East-Berlin government publisher to an abrupt halt. Unified Germany has, though slowly, rediscovered the centrality of Berlin as the political and cultural capital. The Indian Embassy's cultural centre has been named Tagore Centre, and a High School-cum-College (*Gymnasium*) of the city is now known as Tagore Gymnasium. A Tagore Street had existed already during the time of the GDR. One of the greatest feats of the Tagore's 150th birth-anniversary celebrations in 2011-12 was the exhibition of Rabindranath's paintings (*The Last Harvest*) in a Berlin museum. My own efforts to make Rabindranath known in Germany have also shifted to 'Berlin: My last volume' of Tagore poetry translations into German has been brought out by Insel Verlag in Berlin'.[18]

Notes

1 Reeta Sanatani, *Rabindranath Tagore und das deutsche Theater der zwanziger Jahre: Eine Studie zur Übersetzungs- und Wirkungsgeschichte seiner Dramen in Deutschland.* (Frankfurt / Bern: Peter Lang 1983), pp. 66-80.

2 Helmuth von Glasenapp, *Meine Lebensreise.* Menschen' Länder und Dinge, die ich sah. (Wiesbaden: F.A. Brockhaus 1964), p.107.

3 Helmuth von Glasenapp, ‚Tagore's Arrival in Berlin'. In: *Deutsche Allgemeine Zeitung*, 30.5.1921. English translation: *Rabindranath Tagore and Germany.*(Chapter: 'Rabindranath Tagore in Germany: A Cross-Section of Contemporary Reports'. Edited and translated by Dietmar Rothermund). (New Delhi: Federation of Indo-German Societies in India, 2011), p. 41.

4 Martin Kämpchen and Prasanta Kumar Paul (ed.). *My dear Master. Correspondence of Helene Meyer-Franck and Heinrich Meyer-Benfey with Rabindranath Tagore.* (Kolkata: Visva-Bharati (2nd edition) 2010), p. 135. This is an excerpt from Helene Meyer-Franck's biography of her husband (*Heinrich Meyer-Benfey.* [Privately published] 1946, p.34). The translation into English is by the editor.

5 Ursula von Mangoldt, 'Rabindranath Tagore'. In: *Gebrochene Lebenslinien. Mein Weg zwischen den Zeiten.* (Freiburg: Verlag Herder 1981), p. 110. Translated from German by the author.

6 Letter by Rabindranath Tagore to Edith Andreae, dated 26 June 1921 from Boulogne-sur-Seine (Rabindra-Bhavan archive, Santiniketan).

7 Letter by Rabindranath Tagore to Edith Andreae, dated 27 October 1921 from Santiniketan (Rabindra-Bhavan archive, Santiniketan).

8 See Lothar Günther, Hans-Joachim Rehmer, *Inder, Indien und Berlin. 100 Jahre Begegnung Berlin und Indien.* (Berlin: Lotos Verlag Roland Beer, 1999), p. 66.

9 See *The Kenyon Review & Stand Magazine*, 13 (2001) 2, pp. 1-33.

10 Op. cit., p. 67.

11 Heinrich Luetzeler, 'Tagore's Paintings', *World Window*, vol. 2 no. 1 (May) 1962, p. 10.

12 See Martin Kämpchen, 'Rabindranath Tagore's Paintings in Germany'. In: R. Siva Kumar. (ed.)*The Last Harvest. Paintings of Rabindranath Tagore.* (Ahmedabad: Mapin Publishing, 2011), p. 30-33.

13 See Martin Kämpchen, *Rabindranath Tagore and Germany: A Bibliography.* Santiniketan: Rabindra-Bhavan, Visva-Bharati 1997 passim.

14 Karlernst Ziem (ed.) (Berlin: Alfred Holz Verlag, 1961), p. 204.

15 Karl Fischer (ed.) *Rabindranath Tagore 1861-1941.* (Berlin [publisher not mentioned] 1961), pp. 71.

16 Helene Meyer-Franck (trans). (Berlin: Verlag Volk und Welt, 1961), 304 pp.

17 Translated from Bengali by Gisela Leiste. (Berlin: Verlag Volk und Welt, 1982), 637 pp.

18 Martin Kämpchen. (trans.) *Rabindranath Tagore, Gedichte und Lieder.* (Berlin: Insel Verlag, 2011), 152 pp.

Tagore in Denmark

Annemette Karpen

'Denmark is the Country I Have Visited That Received me Most Cordially'

I have chosen this quotation by Rabindranath Tagore at his press meeting in May 1921 when he first visited Copenhagen. However, before we get to that visit, I would like to take a look at that particular historical period.

During the period from 1890 to 1913, a crude and ugly nationalism had erected walls of hatred between the various countries in Europe and of which we know the result: World War I. Christian values were not as convincing any longer and had declined to some extent. People, in general, were ready for new ideas and spiritual values. Especially people of artistic, intellectual and spiritual backgrounds were seeking these values elsewhere. World War I had made it difficult to travel between countries as people were used to do and the war itself made the world a terrible place to be in – to some it seemed like the end of the world. Tagore himself talked of the world war as the biggest threat to humankind.

The Theosophical Society was founded in New York in 1874 by Madame Blavatsky (1831-91). In the area of Adyar in Madras in South India, the Society established a big centre in 1905. It was a movement that sought to reach beyond established religions for something higher and for unifying the entire world with the cosmos. It is this kind of thought that we find in Tagore's poetical works – thoughts expressed by him in an entirely new manner of utilizing the Bengali and English languages. An expression of gratitude and love for nature, animals and human beings is the dominating theme. Hereby, he creates a universalism carrying words into higher spheres of purity and simplicity. In Denmark a branch of Theosophical Society, Teosofisk Selskab, was founded in 1918.

In Europe these thoughts and spiritual ideas had spread. Rudolf Steiner (1861-1925) had been involved with the Theosophical Society, but founded the Antroposophical Society at Christmas time in 1913. This new movement was based more on Christian thought as perceived by Johann Wolfgang von Goethe (1749-1832) and the German idealism of Georg Wilhelm Friedrich Hegel (1770-1831) and J. G. Fichte (1762-1814). Antroposophy was a manner of realization which could lead the human spirit to the World Spirit. In Denmark, the Antroposofisk Selskab was founded in 1923.

In 1920, around the time of Tagore's visit, a Danish farmer, Martinus Thomsen (1890-1981), talked of getting a spiritual revelation in which he experienced universal love and an increase of his intelligence and intuition. He created the Third Testament based on elements from both Christianity and Hinduism. The Martinus Institute was only founded 1932 in Copenhagen, but there were regular meetings round Martinus and his spiritual work prior to that.

In this manner we see that already there were three relatively-strong, spiritual movements in Copenhagen around the time of Tagore's first visit to the city.

Now it would also be interesting to take a look at who had recommended Tagore for the Nobel Prize. The Swedish national poet, Verner von Heidenstam (1859-1940), had become a member of the Swedish Academy in 1912. When the Academy held the meeting in which it was discussed as to whom the Nobel Prize in Literature should be given in 1913, Heidenstam suggested Tagore as a possibility – a poet whose works could be compared with the poetry of J.W. Goethe. His poetry was full of love and a profound spirituality saturating all thoughts and feelings with a purity of heart along with a noble naturally-elevated style of writing – works which contained nothing controversial, offensive, vain, worldly or small minded. The Academy had never given the Nobel Prize earlier to anybody outside the Western world, and hence this would be both a new concept and a big step towards the wider world. However, the academy members were captivated by Heidenstam's enthusiasm and the prize was given to Tagore for *Gitanjali* 'for the depth and high topics and the beauty and freshness, which in a brilliant way gives spirit to his poetry in his own English translation even with the Western fiction and poetry.' (My translation from Swedish from the website www.nobelprize.org)

Shortly after the Nobel Prize was given to Tagore in 1913, the breakout of World War I made it almost impossible to travel privately and that, together with the personal business of Tagore, postponed his visit to May 1921.

From 1913 onwards, a number of Tagore's works were translated into Danish. In 1913, *Gitanjali* (Sangofre) was published followed in 1914 by *Gitanjali & The Gardener* (Sangofre & Blomstrenes Bevogter), in 1917, *The Stray Birds* (Flakkende Fugle) and, in 1918, *The Crescent Moon* (Halvmånen) appeared followed by *The Hungry Stones* (Hungrige stene) in 1919, *The Home and the World* (Hjemmet og Verden) in 1920 and *The Lover's Gift* and *The Crossroads* (Elskerens & Skillevejen) and the play, *The Post Office*, (Posthuset) both in 1921. The books were published by Povl Branner at Pio's Editions or Branner & Korch.

With this selection of literature, Tagore's Danish readers had a fine opportunity to get acquainted with his work, his great mind and the many levels of his thought. We also have to take into consideration that Tagore's absence from the Award Ceremony in 1913 had created a mystic aura around him, and it was still the age when there was neither radio nor TV broadcasting.

On May 22, Tagore travelled by train and ferry from Germany to Denmark. Tagore was even more surprised with the warm greeting he received from the Danes at the various stations the train passed through before arriving at its final destination at the Copenhagen Central Station, where thousands of people had assembled to welcome their famous guest. People were pushing and pressing each other to get a closer look at Tagore while eagerly shouting 'Long Live Tagore' and 'Hurrah'. Inside the arrival hall and outside the station were still more crowds climbing onto cars, platforms, fences, etc. to get a look at the celebrated poet. The police had great difficulty in controlling the masses and finally gave up. Tagore's visit was only a short one for he was on his way to Stockholm in Sweden, but, nevertheless, it was the highlight of the 1913 spring in Copenhagen. The Danes were ready to receive not just his person, but also his art and his spiritual and political messages. Everything would be closely followed by the press with articles and photos of Tagore on the front pages.

Tagore was installed at Hotel D'Angleterre, one of the finest and oldest hotels, at The King's Square in the centre of Copenhagen. The next morning, he was taken by car and horse cart to The Eremitage Forest by his Danish editor, Povl Branner. Tagore expressed his pleasure on seeing the Danish landscape in the springtime. A landscape of ancient oak and beech forest and open areas of lush green meadows with white-flowered fruit trees and populated by flocks of deer. 'No place in the world can surpass this in beauty', he exclaimed (Politiken May 23, 1921).

Tagore then visited his publisher, Mr. Branner, for lunch in his home. Also invited were Prof. Ammundsen and Prof. Wilhelm Andersen as well as Prof. Edward Lehmann, who had written a book on Buddhism and whom Tagore had especially asked to meet (Politiken May 24, 1921). During this lunch, there was an intimate and easygoing atmosphere between all the invited guests. There were also a few writers present such as Preben BilleBrahe, Kai Hoffmann, Helge Rode and Otto Rung. In the afternoon, Tagore had asked to meet the Danish philosopher, Prof. Harald Høffding, at his honorary residence at Carlsberg. During this meeting, Tagore expressed the wish for greater collaboration between India and Europe and the possibilities for Indian students to come to European universities. After this visit, a big press meeting took place at the hotel. Tagore was asked many questions on art, education, politics and the caste system. Tagore mentioned that he knew about Danish culture and admired the works of Hans Christian Andersen, Grundtvig's ideas on the people's high school, and that he was well aware of Danish agriculture and the various farmers' and workers' cooperatives (Berlingske Tidende, May 22, 1921).

We can see from this that Tagore during his studies in England had also acquainted himself with Danish farmers' cooperatives, agricultural planning and workers' cooperatives for better housing, banking and insurance. He had also studied N.F.S. Grundvig's (1783-1879) ideas about education for life leading to the people's high school movement developing in Denmark during the last part of 1800s. Tagore also became very agitated when he talked about Western imperialism and the suppression of the Indian population (Politiken, May 23, 1921).

After the press meeting, an informal dinner at Restaurant Nimb in Tivoli Gardens was offered to a select group of writers, poets and scientists. Between the dessert and coffee, Tagore was invited to The Student's Society to be honoured, and, at midnight, he was greeted by several thousand people carrying lighted torches across Copenhagen all the way to Hotel D'Angleterre where Tagore then stood at the balcony to greet them. The crowd sang the National Anthems and Tagore was very touched by this expression of cordiality. With a 'Viva' for The World Society of the Hearts, he said that in that very moment a link between Denmark and India had been established.

The following day, Tagore gave a lecture on Indian culture at The University of Copenhagen for a full auditorium. Outside the university, at Our Lady's Square, a huge group of people had gathered asking for a second lecture. Tagore immediately delivered an extra lecture addressing the crowd standing at the window. Everything was recorded in detail and all the major newspapers published articles and photos of Tagore on their front pages. In *Politiken*, one read: It seems that Tagore created such an enthusiasm in Copenhagen by his mild wisdom of love and in a few days he has won a huge congregation and The Society of Hearts seems to have gotten an affiliation in Denmark.

The next day, Tagore sailed by ferry to Stockholm in Sweden to receive his Nobel Prize.

No other writer or poet had been received like this any time before or after in Copenhagen. What was the reason for that?

Gitanjali was translated in 1913 by Louis von Kohl (1882-1929) from the English prose version by Tagore and published by Povl Branner at Pio Edition. Louis von Kohl decided, as soon as he had read *Gitanjali*, to make the Danish translation during the summer of 1913. It was hence possible to publish the work as quickly as possible after the Nobel Prize Award. This would also be the first translation of a work by Tagore into a European language outside his own English versions of his works. Louis von Kohl aimed at keeping the free prose of Tagore's own English versions in order to maintain the simple style and form peculiar for him. In his foreword to the 1913 edition of *Gitanjali,* he writes that he hopes he is giving his readers a clear picture of how great a poet and how elevated a personality Tagore was – and how deep and human thoughts are to be found in his poetry although the style could seem very different from the style of European poetry. Louis von Kohl also writes that his translation may only give a weak impression of the beauty of the original text. In the prelude (p. XX) he writes:'...The personality he reveals in his poems is so humanly valuable that only a few other poets or authors in world literature could reach his level of aspiring idealism, unselfish activity and love for all living beings and plants around him'. The following works by Tagore were translated from English into Danish by other translators as well. The works would, of course, attract emancipated female and male readers by their poetry and the beauty of language and content. Tagore also had a deep understanding of the female mind. It is seen in his poems and female qualities are often celebrated. In his novels, he deals with all the injustice faced by the Hindu woman: the tediousness of married life with frequent pregnancies, house and religious duties, family honour and the seclusion of the women of the upper classes. He pities the sacrifices most women had to make in order to follow traditional Hindu customs. Tagore was convinced that women should also have the possibilities of education, freedom, artistic expression, and work as men did. These viewpoints appealed to the Danish readers as women in Denmark were quite emancipated in the early 1900s. They had obtained access to all types of education as well as the right to vote in 1915.

In the work, *The Home and the World*, we read about the development of a very young and shy married woman, Bimala, living in aristocratic purdah and secluded from the outside world. With the effort of her husband she becomes an educated girl and by her love for a friend of her husband's, she becomes aware of the outside world and gets entangled in the politics of the Swadeshi movement of the time and the moral dilemma of the lover. The story is seen from the viewpoint Bimala, of the husband Nikhil and the lover Sandeep as the intrigue unfolds.

Further, theatre-goers would benefit from Tagore's dramatic work as the prestigious private Dagmar Theatre staged *The Post Office* (*Posthuset*) translated and directed by Johannes Anker Larsen (1874-1957) He was a very interesting personality, who, after his unfinished university studies, became an actor, stage director, translator and novel writer. Larsen also had another side to his personality – that of a mystic, not the type who gathered a crowd around him, but more of a private kind of personality. He had written a number of books all with some religious content inspired by theosophical thought, and, from 1910 onwards, his books had a direction that was decidedly veered towards mystical experiences. In 1923, he won the prestigious book prize of 70.000 kr. by Gyldendal Edition for his novel, De Vises Sten, a literary tour-de-force dealing with a person's spiritual development, but full of esoteric and mystical incidents. The book was translated into various languages and he had many followers in Denmark and abroad.

It is obvious that Larsen had read Tagore's books published in Denmark from 1913 until 1921. It is also obvious that he had met with Tagore during the latter's stay in Copenhagen in May 1921 in order to get permission for translating *The Post Office* (*Posthuset*) into Danish and to stage and direct it at Dagmar Theatre. Like Tagore, Larsen also developed a personality in which the worldly, vain and selfish evaporate and get absorbed in the universal spirit instead.

Posthuset was played seventeen times and presented in conjunction with a Danish theatre piece, En sjæl efter døden. In the theatre catalogue, we read that some of the best actors of the time, such as Poul Reumert, Holger Reenberg and Elith Pio, were selected for the various roles. There are also some photos from the play displaying a simple but effective stage décor together with realistic Indian costumes. In the catalogue, Robert Neiiendam has written that those who had read *The Crescent Moon* know that he has the profound knowledge of a child's mind and way of thinking. This knowledge is used in *Posthuset*, which is about a sick child's fantasies and longing.'

A critic wrote in the magazine, *Teatret*, in 1919: '... It is impossible to see where the experiences of the poet end and where his fantasy begins, it is impossible to penetrate the work: it does not open as much as the smallest scratch for the spectator...' Further in the review the critic writes that at Dagmar Theatre the actors played their parts with discretion, subdued movements, low speech and calm mimique. The dialogue did not reach out to the audience. Only Mr. Elith Pio in his small part as the messenger acted with vivacity and made that part theatrically interesting.

It is obvious that the symbolic inner life of *Posthuset* was not understood very well by this critic, who saw it more as a realistic type of story. He did not see that the stuffy confines the child was trying to escape from by his physical death was the symbol of spiritual freedom from the physical world.

However, *Posthuset* was staged as a radio drama directed by Sigurd Vantzin and was broadcast on April 29, 1932, April 13, and September 26, 1951 and on October 6, 1961.

This visit to Copenhagen was to be the focal point for Tagore's influence on Danish culture since Tagore and his large group of enthusiastic followers were ready for understanding each other. His thoughts would influence various esoteric, philosophical and grassroots political organizations over the next decades through their trans-nationalism and universalism.

During the next years, a few other works of Tagore were published. *My Reminiscences* (*Mine Erindringer*) was published in 1922, *Fruit Gathering* (*Frugthøsten*) in 1923 and *Fireflies* (*Ildfluer*) in 1926.

In 1926, Tagore visited Copenhagen for the second time. However, this time there was not as much celebration as during his first visit. He was there for two days, September 6 and 7, and gave a reading at the palace of The Odd Fellow Lodge. The evening was introduced by the Sanskrit professor, Prof. Tuxen; Prof. Harald Høffding, a friend from the previous visit, was among the invited audience. Tagore read from his own works in Bengali and English and the reading was well received by the audience.

During one of his big lecture tours in England, Tagore, by accident, met a Danish man with whom he developed a lifelong bond. . It is the gardener, Emmanuel Sørensen (1890-1984), who was working at the Darlington Hall where Tagore stayed for three months during his visit to England. Sørensen was a mystic and practised silence as a tool for obtaining spiritual revelation. He was educated as a gardener in Århus, Denmark, and went first to Italy, and then to France to learn more about different types of gardening. After a few years, he went to England and obtained work at the gardens and parks of the manor house, Darlington Hall. Tagore found Sørensen very interesting and saw great possibilities in him. The consequence was that Tagore invited him to accompany him on the next big lecture tour in 1930 that included countries such as France, Italy, Greece, Egypt, Palestine and Syria. Like Tagore, Sørensen did not believe in traditional education, which he called 'headucation'. After this extensive tour, Tagore asked Sørensen to follow him back to India and teach Indians about silence, as he expressed it. In India, Tagore presented Sørensen to Mahatma Gandhi, Nehru, Lama Govinda, Jagdish Bose and Ramana Maharshi, the stalwarts of the time. Sørensen felt himself drawn very close to Ramana Maharshi and got the name Sunyata. He gave up his work as gardener and became a holy man at Mount Kalimath. He was also to become a very close friend of the Nehru family. Sunyata stayed on in Tagore's homeland for 30 years until he was asked to go to California to teach the Americans about silence. I include this story to show how two great minds feel drawn towards each other by intuition, when they meet by accident.

In 1930, an exhibition of Tagore's works was presented first in Paris in May and from there it arrived in Copenhagen and was presented at The Royal Academy, Charlottenborg, with the vernisage held on August 8. From there, the exhibition had to go further to Birmingham, London, Berlin, München, Dresden, Geneva, Moscow, Boston, New York ending in Philadelphia in May 1931. The exhibition displayed watercolours and drawings in three big halls. At the opening, Tagore was present and met with artists, book publishers and scientists.

The art critics were not very impressed with his works which often had an abstract or symbolic form and character. However, that was a typical response for the time when many Danish artists had begun to express themselves in abstract and nonfigurative works, but were still not really understood or appreciated by the general public and critics. Tagore mentions in the catalogue created for the exhibition that his paintings are visual poems of lines which did not contain an idea or the visualization of something factual, but were lines where the forms gave rhythmical meaning. This type of artistic expression definitely met with sympathy by the Danish avant-garde artists and made a great and inspiring impact on their minds. However, it would take about twenty years before abstract and nonfigurative art was understood and accepted by the general art public in Denmark.

After the opening, Tagore went straight to Denmark's Radio House and read from *Gitanjali* and *The Gardener*. The poems were first read in Danish by Mr. Manniche from The International People's High School, and thereafter Tagore read the poems both in English and Bengali. It was a direct broadcasting to the Danish radio audience. The critics called his reading style and the rendition of his works an aesthetical pleasure. They also emphasized that this was the first poet to unite West and East in harmony (Berlingske Tidende August 10, 1930).

In the evening, Tagore gave a lecture on his own art and art in general at Borup's People's High School. He spoke about art as we experience it in reality and as we do in art works. He also wrote in the art catalogue of his exhibition at Charlottenborg that some people see art in daily things such as a donkey or a bunch of flowers, things overlooked by others because they know them so well. The lines and harmony of the animal or flowers in nature arouse a mood or feeling in some people, while others get the same mood or feeling when they watch a painting. All art is about feelings, he says. The great master's art works are experiments of reshaping what they felt in the moment of inspiration. These feelings are given to all people, rich as well as poor. After the lecture, the director of Borup, Mr. Rosenkjær mentioned that Tagore, by his lecture, had built a bridge between East and West bringing India closer to Denmark.

On August 10th, Tagore gave a lecture at Kronborg Castle in the majestic Hall of Knights. The lecture was called 'My Religion' and described his childhood and upbringing in a Brahmo milieu and the religious crisis that Tagore experienced at the age of eighteen, a crisis that had had a deep impact on him and made him understand the reality of religious values. From the above we can see that Tagore made a huge impression on artists, intellectuals, academics, scientists and on people who are drawn by the esoteric--- impressions that have influenced many societies and organizations towards world peace. People may not have been ready for his thoughts of trans-nationalism, as we can see that in the period from 1921- 2013 many wars, uprisings and conflicts occurred. Tagore's thoughts and ideas about caring for nature and humankind, friendly relations between countries, the sharing of scientific results, and thus creating his trans-nationalism are more important than ever. Tagore's intelligence and sensitivity made it possible for him to look into the future and foresee some of the problems that we are dealing with in contemporary times. His thoughts are relevant now, more than ever, for the modern world community.

In 2011, in the memory of Tagore, there was a conference in Denmark at the University of Copenhagen entitled 'Tagore (1861-1941): Music, Art, Nation'. Organized on September 19, 2011 by the Centre for Global South Asian Studies it had prominent speakers. H.E. Ashok Kumar Attri, Ambassador of India to Denmark, gave the introductory address on Tagore; followed by Dr. Reba Som, Director of ICCR's Rabindranath Tagore Centre in Calcutta, who spoke on 'Pictures and Songs' – the Tagorean connection between the first exhibition in Copenhagen in 1930 and the exhibition that was being held at the university in conjunction with the current conference. Prof. Peter B. Andersen spoke on 'Tagore in Denmark'; Prof. William Radice, SOAS, London University spoke on 'Tagore, *Gitanjali* and the Nobel Prize'; Annemette Karpen, M.A., Film Director, spoke on Tagore's works utilized by Satyajit Ray in his films, and Dr. Ashok Bhattacharya from Calcutta spoke about 'Tagore's Pedagogical Ideals as Manifested in Santiniketan'.

The travelling exhibitions 'Tagore and Katha' and 'Tagore in Advertisement' were shown at the University during the following three weeks. The Katha embroideries gave the spectators a glimpse of an unknown side of Indian crafts through their illustration of selected songs, poems and novels by Tagore. Tagore's featuring in various advertisements, endorsing a variety of products between 1889 and 1941, was a little known aspect of Tagore for many people. This, however, was not done for a commercial reason, but in order to urge his countrymen to buy Indian-made products -- a very wise way of utilizing his popularity for the Swadeshi cause.

The last question before we conclude this essay is how do we spread the knowledge about Tagore, his works and thoughts to a modern Danish audience? My suggestion is to let children hear about Tagore, for instance, by enacting some of his dramatic works. *Chitrangada*, for example, has great poetry, as well as valour, romantic love and unselfishness on the part of both Arjuna and Chitrangada, and it would inspire children and young people. At a young age, the human mind is very impressionable and ready to absorb beauty, courage and great deeds. I am, at the moment, working on some projects of that nature and I am looking forward to share this with the younger generations – presenting Tagore, an Indian, a new type of man encompassing all of mankind in his heart, mind and soul.

References:

Translated works of Tagore.

Gitanjali (1913), *Gitanjali & Blomstrenes bevogter* (1914), *Flakkende fugle* (1917), *Halvmånen* (1918), *Hungrige stene* (1919) *Hjemmet og verden* (1920), *Elskerens gave & skillevejen* (1921), *Posthuset* (1921), *Mine erindringer* (1922), *Frugthøsten* (1923), *Ildfluer* (1926). www.nobelprize.org

Newspaper articles:

'Nobelpræmierne', *Kristeligt Dagblad,* November 14, 1913.

'Rabindranath Tagore', *A Chronique in Politiken*, November 20, 1913.

'Den indiske digters søndag i København', *Berlingske Tidende,* May 22, 1921.

Berlingske Tidende, May 23, 1921.

'Nobelpristagerens ankomst til København i aftes.' *Politiken*, May 22, 1921.

'Tagores første dag i København', *Politiken*, May 23, 1921.

'Tagore i København', *Kristeligt Dagblad*, September 7, 1926.

Berlingske Tidende, August 10, 1930.

Torben Brostrøm. 'Tagore', *Information*, February 2, 2009.

Daniel Henscher. 'Indisk forfatter tog danskerne med storm', *Kristeligt Dagblad*, November 14, 2011.

Study material on Tagore in Denmark by Daniel Henscher Catalogue from Rabindranath Tagore Exhibition 1930 Charlottenborg, Danmarks Kunstbibliotek s 22.975

Theatre Catalogue, Dagmar Teatret 1921-22 Teatret 1921 (a Danish theatre magazine) Catalogue 'Tagore in Advertisement' ICCR.

Catalogue 'Tagore & Katha' ICCR.

'Mystiker og digter – Anker Larsens liv og forfatterskab af Knud Wentzel, Syddansk Universitetsforlag 2011.

Karpen Annemette. 'Albatrossen på däcket' af Kjell Epsmark Stockholm 2008 'The Genius of Tagore', a lecture held at The Tagore Conference, University of Copenhagen, September 19, 2011.

Karpen Annemette. 'Two Geniuses Meet', Text used in Satyajit Rays Films such as Charulata & Teen Kanya, a lecture held at Lund University & Skissernes Museum in Lund, Sweden during the Tagore Celebrations, 20-24. March 2011.

Karpen Annemette. 'Tagore', Gyldendals Teaterleksikon.

Turning 'East': Orientalist Variations on Tagore from the 'Eastern Corner of Europe'

Ana Jelnikar

I recall from my primary school language and literature classes the wide range of international writers we pupils were exposed to. Flicking through one of the textbooks, amidst the set readings from the Slovenian and other literatures of the former Yugoslavia, and alongside excerpts from Tolstoy, Boccacio, Ovid, Rolland, Gogol, Hemingway and others, there was also Tagore's poem 'Authorship'. With our family moving to England before we got to Tagore's poem in class, I only came to 'discover' the great poet as part of the British Open University MA postcolonial literature course, while my Slovenian friends had all read him in both primary and secondary school.

When exploring W. B. Yeats's reading of Tagore, whose influential introduction to the English *Gitanjali* (1912) had largely shaped Western perceptions of the Indian poet, I became interested in understanding what made Slovenes, who then had no direct historical links with India, respond so keenly to Tagore (cf. Jelnikar 2008). The dominant Saidian Orientalist model of Western appropriations of the 'Orient', the limited applicability of which was already apparent with respect to the Irish poet (though Yeats was not always above the imperialist mindset when it came to India), was even less relevant for a people who had no imperialist interests as regards 'the Orient' but were themselves then part of a multi-national Empire and profoundly concerned with issues of emancipation against various denationalizing forces in the latter half of 19th and the first half of 20th century. A more open and reciprocal model of 'Otherness' was needed, underpinned not by the desire to rule but by a strong sense of identification, if I was to make sense of the Slovenian response to Tagore. The notion of *situational identification*, whereby sympathies are forged between individuals and inspirations derived from a sense of shared predicaments, or as Patrick Colm Hogan puts it, 'we develop an immediate sense of intimacy with someone as we intuit shared felings, ideas, references, [and] expectations' (Hogan 2004: 26), seemed a particularly useful notion with which to frame some of the lesser-known responses to Tagore from the margins of Europe.

I soon discovered there was one particular poet, Srečko Kosovel (1904-1926) – today regarded as Slovenia's foremost avant-garde voice of the interwar period – who was especially drawn to Tagore. Kosovel was convinced of Tagore's relevance in showing a new direction out of the crisis Europe in general and the Slovenian people in particular in the post-WWI era. When in 1925, aged twenty-one and within months of his untimely death, he was getting his first manuscript ready for publication, he decided to title it *Zlati čoln* (*The Golden Boat*), alluding to his Indian mentor and contemporary.

Our understanding of Tagore's complex reputation in the West requires therefore a more differentiated view of Europe. By considering the particulars of the Slovenian response with special attention to one voice, I intend to problematize what is still too often seen in monolithic terms as Tagore's 'Western reputation'. What of the fellow poets and like-minded individuals in the West who endorsed Tagore's poetry and philosophy outside the strictures of an imposed – or adopted – mystic identity?

The Nobel Prize and After

In November 1913, through the efforts of influential supporters, most notably Yeats, Pound and Sturge Moore, Tagore became the first non-European to receive the Nobel Prize for Literature. What followed was an unprecedented response to any poet in the history of letters. Many interrelated factors came into play as various countries, groups and individuals responded to Tagore, each in their own way, even as they drew on the common stock of perceptions that guided Western imagination as regards the East in the early decades of the twentieth century. Often Tagore's reputation could not be sustained, complicated as it was by false and narrow expectations, changing literary trends, and weak translations. Particularly in the Anglo-American world, the tremendous enthusiasm with which the 'mystic from the East' was initially received soon deteriorated into disappointment and rejection. Amit Chaudhuri, for example, finds it shocking that 'respectable people like Yeats and Pound' lost their regard for Tagore's work so quickly, and that as a result Tagore's reputation was seriously damaged in the English-speaking West (2002: xviii).

If Tagore's fame in the Anglo-American world had practically run the length of its course by the end of the second decade of 20th century, on the continent, it was positively on the rise. Tagore's 'most astonishing successes' were his European tours of 1921, 1926 and 1930 (Radice 1994: 27). It was indeed in the 1920s that Tagore's popularity reached unprecedented heights across Central and Eastern Europe, and it is to this wave of his popularity that Kosovel's response belongs to.

The extraordinary impact Tagore had on the minds of literary men and women as well as the sheer sensationalism surrounding the poet can be partly explained away circumstantially. Broadly speaking, the West in the early 20th century was once again in need of a spiritual injection from the East. The following lines from Kosovel capture the atmosphere of disillusionment and spiritual bankruptcy after the events of WWI:

> A tired European
> stares sadly into a golden evening
> even sadder
> than his soul.
> […]

A civilization without heart.

A heart without civilization.

An exhausting struggle.

An evacuation of souls.

[...]

("Kons", Kosovel 2008: 100)

Exhausted from what another European poet would call a 'filthy modern tide' (Yeats 1985: 375), people in the West were ready for 'a book of the soul' in which 'life is the visible expression of the eternal', to quote one enthusiastic reviewer of *Gitanjali* (*The Daily News and Leader*, 21/01/1913, in Kundu et all eds. 2000: 18-19). Branded, predictably, a Wise Man from the East, its author was seen to be representative of Indian civilization itself; his was the quintessentially Indian voice. It must be conceded that with his white beard and long-flowing robes, he fitted the part well.

Part of the disaffection with Tagore that some Western readers experienced also arose from their perceptions that he was somehow not 'Indian' or 'Eastern' enough. He did not sustain their expectations derived from the exotic diet of Omar Khayyam or even Kipling (Aronson 1978: 11-20). Prefabricated conceptions of 'East' and 'West' also presented a major drawback to a critical appraisal of the writer's literary sensibility. Tagore understood the logic at work. He is on record to have said to his friend Edward Thompson: 'They have labelled me a mystic and when I produce something that is not mystical they are offended' (in Thompson 1998: 116).

The problem of the narrowness of such Eurocentric readings presented itself forcefully to Aronson, as he considered the vast body of responses to Tagore from Europe and the United States between 1920 and 1940: 'Again and again literary critics refuse to discuss the East in terms of human beings and human experiences' (Aronson 1978: 15). This attitude has since been theoretically instituted as 'Orientalism' and the Westerners who look at the Orient and 'conceive of humanity either in large collective terms or in abstract generalities,' giving little scope to 'existential human identities' as 'Orientalists'(Said 1995: 154-5).

True, the Saidian cultural critique can provide a relevant template for analyzing the way Tagore was imagined in the West and how his constructed image served to reinforce the colonial point-of-view, but – aside from reservations we might have against reducing human curiosity about, and desire to know, 'the other' to conscious or unconscious motivations of power – his analysis proves much more limited once we move away from western imperialist nations to the margins of the European world.

Germany is the most often given example to demonstrate that the association of Western Orientalism with colonizing power is at best only partly true. The fact that German scholars were the key translators and commentators of ancient Indian texts at the beginning of the nineteenth century, and that Germany then had virtually no interests in India or China, disproves any necessary direct link between Orientalism and the project of colonial subjugation (Clarke 1998: 27). This is of course not to say that there were no continuities between the German view of an idealized East and those of the colonizing western powers. As Imre Bangha notes, German scholars still perceived Asians as 'the other', but having themselves been 'for long subjects of a fragmented nation and dominated by nearby foreign powers', their views of Indian culture threatened by European Imperialism was often driven by sympathy. Their cross-cultural response was framed by their perceived sense of commonality and joint purpose with the Indian poet (2008: 14). Their sympathies often lay with the Indian freedom movement, and they genuinely looked to Tagore (and/or Gandhi) for moral sustenance as well as alternatives to imperialist ideologies, seeking 'to substitute a more holistic paradigm for old mechanistic and dualistic ways of thinking' (Clarke 1998: 105). Their ideas, as J. J. Clarke would have it, served as a 'corrective mirror' to Europe. The perceived otherness of the Orient need not always be one of mutual antipathy, a way of affirming Europe's triumphant superiority, but it 'also provides a conceptual framework that allows much fertile cross-referencing, the discovery of similarities, analogies, and models' (*Ibid.*: 27).

Clearly then we must allow for various motivations and responses, even within the same parameters of othering, when discussing how 'the West' engaged with and related to 'the East'. A more open and reciprocal model of otherness is needed if we are to make sense of how Europeans responded to Tagore more fully.

Identification Paradigm

Many individuals and groups from within Europe celebrated Tagore from their own real or imagined position of 'otherness'. We only need to look at the letters Tagore received from an average European reader to appreciate this point. Many of these show that apart from identification along personal lines, there were numerous marginalized groups and organizations who wrote to the poet for endorsement of their cause. Alongside appeals by various women's organizations from across Europe, the letters included, for example, an appeal from a *Hilfskomitee* for German emigrants in what was then Czechoslovakia, or a submission from the editors of a Roma journal describing the plight of the Roma under Tzarist Russia and contrasting it with the more favorable circumstances emerging with the October revolution. A letter from the Czech 'Union of German societies for the prevention of cruelty to animals' asked Tagore for moral support to set up an anti-vivisection hospital (Rabindra Bhavana archives, cf. cited works).

I see the value of these letters primarily in showcasing the broad range of concerns that drew people to Tagore, as they intuited shared predicaments and concerns with him. Tagore in turn took these gestures seriously, giving support where he could. In September 1940, to give one last example, he received a letter from a Polish woman by the name of Marya Falk, a Lecturer in Slavonic Languages at Calcutta University and initiator of the newly-established Indo-Polish Association there, to which Tagore accepted the honorary chairmanship (cf. Pobozniak 1961: 354). Her letter conveys sympathy derived from identifying Poland's own relation to her struggle for independence with India's historical predicament, from which, the author hopes, echoing Tagore's philosophy, the 'call to true humanity might again be heard in the world'. Tagore in turn sent the message: 'I warmly associate myself with the 150th Anniversary of the May Constitution of Poland – may justice and humanity prevail in a peaceful reconstruction of civilization'.[1]

Once again, letters such as this make us appreciate the scope and diversity of responses Tagore occasioned *within* Europe. It is also here that the lines between colonial, nationalist, class and gender struggles get blurred and merge into a broader emancipatory discourse. Freethinking individuals and groups would often forge trans-national links and solidarities, seeking solutions to oppression in international fraternalism (cf. Boehmer 2002: 17). Personal relationships and networks, realized through travel and correspondence, were crucial for creating worldwide platforms of solidarity. When in 1919 Romain Rolland sent Tagore, along with his first letter to him, a document entitled 'Declaration of Independence of the spirit', this was an explicit attempt to forge closer ties between 'East' and 'West'. It sought to enlist Tagore's support for a protest mounted in the name of the ideal of spiritual oneness of humanity that would obliterate the frontiers of nations. Tagore was only too glad to put his name down to a document, a part of which read as follows:

> We serve Truth alone which is free, with no frontiers, with no limits, with no prejudices of race and caste [...] We do not recognize nations. We recognize the people one and universal, – the people who suffer, who struggle, who fall and rise again, and who ever march forward on the rough road [...] – the people comprising all men, all equally our brothers. And it is in order to make them, like ourselves aware of this fraternity, that we raise above their blind battles the Arch of Alliance, of the Free Spirit, one and manifold, eternal (in Aronson and Kripalani eds. 1945: 22).

Eager for signs of common purpose with the West, Tagore gladly accepted the invitation to 'join the ranks of those free souls' as he wrote back to Rolland, encouraged that 'the higher conscience of Europe had been able to assert itself' (*Ibid.*: 106). How internationally resonant such documents were is borne out also by the fact that Kosovel translated Rolland's *Déclaration d'independence de l'esprit* into Slovenian in 1926, adapting it somewhat for the Slovenian context.

The utopian thrust and revolutionary fervor of such documents certainly found echoes in some of Tagore's own grander proclamations: 'Our fight is a spiritual fight, it is for Man. We are to emancipate Man from the meshes that he himself has woven round him – these organizations of National Egoism' (in Bhattacharya 2005: 55). While standing in contrast to the poet's more realistic project of building an educational institution in India for the encouragement of hermeneutical dialogue between 'East' and 'West', the proclamation nonetheless anticipates Tagore's active involvement in the Orientalist enterprise (of Clarke's kind), to which the winning of the Nobel Prize had opened a door.

Having established the diversity of responses that Tagore elicited within Europe, particularly with respect to the more marginalized individuals and groups, and the various networks of trans-national solidarity this gave rise to, it is worth noting that Tagore, in turn, felt himself drawn to the margins of Europe, sensing, as he wrote in 1926 to Leonard Elmhirst, 'a mysterious feeling of kinship' with the less industrialized part of the continent as opposed to the powerful countries of the West (in Dutta and Robinson eds. 2005: 339). Recovering from severe exhaustion at the sanatorium at Lake Balaton in Hungary on what was his fifth and longest European tour, he wrote in the same letter:

> Doctors advise me to take the shorter eastern route to India through Yugoslavia, Serbia [*sic*], Constantinople, Greece and Egypt. The prescription is very much like the French wine ordered for me in Milan; it is tempting. The people [in] this eastern corner of Europe are perfectly charming – their personality unshrouded by the grey monotony of a uniform civilization that has overspread the western world. It is mixed with something primitive and therefore is fresh and vital and warmly human (*Ibid.*: 340).

His return journey to India took him on a whistle-stop tour of the Balkans between 13 and 17 November, with two days spent in Zagreb (Croatia) and two in Belgrade (Serbia), before going onto Bulgaria, Romania, Greece and Egypt to finally return to India. Srečko Kosovel's life was cut short just months before Tagore came to Yugoslavia, so he never had the chance to partake in the Tagoreana surrounding the poet's visit in the Balkans. What he would have made of the sensationalism of it is difficult to say.[2] What we know of his response, however, belongs to those whom Aronson refers to as 'the lonely and the outcast' who could and indeed *did* adequately respond to what, as far as the mass-media were concerned, was merely one sensation out of many' (Aronson 1978: xv, original emphasis). The mass outfit of Tagore's foreign addresses notwithstanding, his message was, to be sure, intended for the individual.

Claiming the Indian Laureate

When Tagore's English *Gitanjali* first came out in 1912, edited and famously introduced by W. B. Yeats, the Irish poet's eulogy to Tagore travelled far beyond the English-speaking world. In the first article to be written on Tagore in Slovenia, Oton Župančič (1878-1949), the leading modernist poet of the pre-war generation, based his piece largely on Yeats's preface. If Tagore in England was made famous through the efforts of the Anglo-American-Irish literary elite, also in Slovenia, it was the enthusiasm of some of the country's foremost writers that introduced Tagore to the reading public and generated an unprecedented response to any literary figure of international stature. Following some of the early translations done by Miran Jarc (1900-1942) and France Bevk (1890-1970), it was the talented poet Alojz Gradnik (1882-1967) who devoted himself to translating Tagore's works.

One after another, the following titles came out in Gradnik's translation: *Rastoči mesec / The Crescent Moon* (1917; sold out within months and republished in 1921), *Ptice Selivke / Stray Birds* (1921), *Vrtnar / The Gardener* (1922), *Žetev / Fruit Gathering* (1922) and *Gitandžali ali žrtveni spevi / The Gitanjali: Song Offerings* (1924). These collections are being reprinted to this day.[3] Alongside many newspaper and journal articles about the poet, as well as translations of his novels (*The Home and the World, The Wreck, Gora*), essayistic writings (*Sadhana*, excerpts from *Nationalism*, and *The Religion of Man*) and the staging of the plays *The Post Office* and *Chitra* at the Ljubljana City Theatre, Tagore can be said to have found a permanent place in the Slovenian letters.

Tagore's fame in Slovenia peaked around the time of these publications. Kosovel's response to Tagore's poetry and philosophy also belongs to this particular wave of his popularity, in which the admiration for the creative writer is beginning to take precedence over the earlier, more politically-motivated appraisals. Slovenes' initial response to Tagore, though largely dominated by extra-literary factors rather than any real appreciation of the poet's artistic credo, nonetheless marks an important stage in the building of his reputation. Moreover, it was a sense of shared concerns which drew Slovenes to Tagore and what he stood for. In fact, from its beginnings, Tagore's popularity in Slovenia was connected less with the romantic side of Orientalism that looked towards India for a redemptive spiritual injection and saw in Tagore above all 'the exotic and bearded Oriental prophet' (Petrović 1970: 13), than with a sense of identification with the poet and his people, derived from a perceived common goal of striving for political and cultural independence.

In the first substantial article entitled 'Last year's rivals for the Nobel Prize' (1914), Tagore's winning of the Nobel Prize is juxtaposed to the defeat of the rival nominee, the Austrian poet Peter Rossegger. This is because, amongst the Slovenes Peter Rossegger was known less for his literary credentials than for his aggressive Germanization policy pursued against Slovenes in Southern Carinthia and Southern Styria. Against this background, Tagore is referred to as 'a spiritual giant of enormous horizons' and set in opposition to a parochial writer who 'fans the flames of nationalist hatred'. He is celebrated for his love of humanity as opposed to love of nation. His patriotic songs are not 'boisterous fighting hymns', but perfect expressions of 'his universalism'. Here, in short, is 'a patriot' whose voice is tuned to the deepest harmonies of humanity, refusing to surrender the struggle for liberation to a nationalist agenda (Lokar 1914: 246). Whatever the motivations behind the article, its analysis of Tagore's distinct anti-imperialist position was not without insight.

For indeed, through a critique of both imperialism and its anti-colonial nationalist derivation, Tagore gave his anti-colonialism a significantly broader base, grounding it in a universalist ethos. It was precisely this high ideal underscored by the article that was to resonate so strongly with Kosovel. It was against the political circumstances of the early decades of the twentieth century, as Slovenes were caught in the cross-fire of a number of aggressive nationalisms, that Kosovel strove for a definition of Slovenianness that – even as it remained sensitive to the particular needs of the Slovenian people, espousing their right to self-determination – refused to yield to an inward-looking or separatist stance. Both Tagore and Kosovel, writing from a profound awareness of their region's subjugated status, endorsed an anti-imperialist stance that rejected nationalism as a viable means of liberation, embracing instead a creative universalist ideal.

Srečko Kosovel Turns 'East'

Imre Bangha has observed in relation to Hungary that Tagore's greatest supporters were to be found among the readers and writers who were born or lived in regions 'lost' after WWI. These writers, he claims, 'had an additional motivation to perceive the irrationality of western thought that led to a war and then to a peace that they considered unjust. Their disillusionment urged them to examine whether Rabindranath would offer an alternative to western thinking' (2008: 89-90). Something similar can be said of Srečko Kosovel whose hometown was unjustly "lost" to Italy following the break-up of the Austro-Hungarian Empire.

Born in 1904 as the youngest of five children in the town of Sežana not far from Trieste Srečko Kosovel was growing up in what were for many Slovenes historically trying times. After the dissolution of the Austro-Hungarian Empire in 1918, Slovenes joined the newly-founded nation state of South Slavic peoples: the Kingdom of Serbs, Croats and Slovenes (in 1927 officially renamed Yugoslavia). Enthusiasm for the creation of the new state, which offered guarantees against Italy and Austria and possibilities of national emancipation was, however, mitigated by the fact that a large number of Slovenes remained outside the borders of the newly established state. The Treaty of Rapallo (1920), fulfilling some of Italy's territorial claims conceded by the secret Treaty of London in 1915 (when Italy joined the Allies), allocated large swaths of ethnically Slovene territory, including Kosovel's native region of Primorska, to Italy.

What made the resolution of these territorial issues particularly devastating were the assimilation measures introduced by the Italians towards the Slovene and Croat population living within Italy's borders. In 1920, the seat of Slav cultural life, the *Narodni Dom* (National House) in Trieste, was torched by a mob with the consent of the Triestine police and authorities. After the fascist take-over in 1922, the new assimilation policies enforced between 1924 and 1927 'transformed five hundred Slovene and Croatian primary schools into Italian-language schools, deported one thousand Slavic teachers (personified as "the resistance of a foreign race") to other parts of Italy, and closed around five hundred Slav societies and a slightly smaller number of libraries' (Sluga 2001: 48). Kosovel's father was forced to retire for refusing to abide by the Italian-only language policy. This brought the family severe financial difficulties. They even lost the roof over their heads, since their accommodation was tied to Anton's teaching post. By 1926 non-Italian names had to be Italianized (Srečko, meaning 'lucky,' became Felice), and by 1927, shortly after Kosovel's death, the use of Slovene was prohibited in public.

The post-war situation, further aggravated by the centralizing tendencies of the new South-Slavic state and the pressures of Germanization to the north, alerted Kosovel powerfully to the pathological forms of nationalism where identifying as Italian, Austrian, or Slovene overrode notions of shared human identity or precluded the possibility of hybrid or multiple identities. He launched an uncompromising attack on nationalism in his writing, with lines such as 'nationalism is a lie' also entering his poetry. At the same time, he strove to salvage the concept of *narod* (a people) from hijacking by nationalism: 'A *narod* for us can only ever mean a nation which has freed itself from nationalism' (CW 3: 624, emphasis original). Driving a wedge between nationhood and nationalism meant for Kosovel demarcating the important sense of national selfhood from a self-indulgent celebration of one's own identity. Nationhood requires a measure of selflessness, lest it should lead down 'the wide road of national egoism' (*Ibid.*: 67). Like Tagore, Kosovel became the severest critic of his country's social and political realities. The 'patriot' is as short-sighted as the 'underling,' and the two representatives of what Kosovel saw as the dominant tendencies existing within the Slovenian culture came under constant attack in his writing (CW 3: 14; 41; 61; 700; 701). The Slovenian people needed to progress beyond the mere defense of a people, and rise to the real, pressing task of liberating them (CW 3: 60). From Tagore's book *Nationalism* (1917) in response to which he wrote an article titled 'Nationhood and Education', Kosovel got vital inputs for thinking through these pressing issues. His clarion call was high-sounding indeed: 'We need to raise our country to the heights of the countries of the world, to the breadth of human rights, to the depths of ethical problems. That for us is the cultural mission of Slovenianness' (CW 3: 60).

Tagore too stressed the need to understand local problems in a global perspective, and seek solutions in world-wide cooperation. These, he would reiterate, were no longer to be found 'in the seclusion of our own national workshops' but had to be found through intercultural negotiations (2002: 76). It would be easy to put Tagore and Kosovel in conversation with each other, but what I want to suggest here is merely that it was from the above-sketched historical positioning that Kosovel gained his sense of intimacy and shared concerns with Tagore, responding readily to Tagore's message of anti-nationalism and universalism.

For Kosovel indeed saw his native region affected by imperialist forces similar to those that subjugated India. In his writings he would align the plight of Primorska with the 'unnatural act' he saw in the 'colonization of non-European lands' (CW 3: 65-6). Moreover, he would see it in the larger context of the plight of all who are – in his own vocabulary – 'beaten', 'downtrodden', 'subjugated'. 'Injustice is injustice, whether suffered by one, thousands or millions' (*Ibid.*: 48).

Painfully aware of the historical realities of their time, both poets deplored the fact that the meeting of cultures had come on the back of conquest and colonization, rather than in a spirit of free exchange, but argued, against the odds, for a non-hierarchical dialogue between cultures. How to resist foreign impositions and yet not bar oneself from the discoveries of the modern age, whether in science, technology, economics, politics, art, or literature; how to adjust creatively and retain agency as opposed to imitate slavishly or conform unthinkingly, and what are the implications of global expansion for cultural identities? These were questions that preoccupied both thinkers.

Another important aspect to Kosovel's situational identification with Tagore stems from the fact that both writers were perceived as occupants of the large ideological constructs of the 'East'. In the context of Kosovel, I am referring to the tradition of representation that predates fascism and goes back to the Enlightenment, in which 'Eastern Europe' or 'the Balkan East' is imagined as the Western half's lesser other. In this representational framework, Germans and Italians were seen as cultural equals: bourgeois, modern, nationally evolved, and essentially 'Western', while Slavs were backward peasants, lacking national consciousness and 'Eastern' (Sluga 2001: 2). Such mental geography was instrumental in influencing political decisions and historic events, for what helped justify and consolidate the Italian claim to authority over the disputed Adriatic border region was their alleged racial, cultural and linguistic superiority.

In that sense both Tagore and Kosovel were projected as belonging to an inferior and governable race, Indian and (Balkan) Slav respectively. Both were at the receiving end of what Raymond F. Betts has termed 'the peculiar geography of imperialism', whereby Western Europe was the centre of the world, 'radiat[ing] outward' from its core 'those attributes we describe today as "modern" ' (Betts 1998:7). Not wanting to oversimplify what is more complex, I wish to simply reiterate that it is from the particular historical positioning in which the Slovenes under Italian occupation were culturally and politically oppressed (and ideologically othered) that Kosovel saw himself as occupying the same space vis-à-vis the imperial West as Tagore.

> At a time when we are being lashed by European imperialisms, we are down on our knees, praying to God to grant us our rights and give us righteous masters. And these masters let us have our God but take away all the rights God has given to man (CW 3: 35).

But if Kosovel could understand the violence of a colonial encounter based on the binaries of imperial imagination, he was also aware of the opportunities offered by cross-cultural contact. With energy worthy of Tagore, his artistic temperament in the final instance celebrated the meeting of 'East' and 'West', as he extended the notion of 'East' to encompass Asia:

> We happen to be living at the crossroads of Western and Eastern Europe, on the battlefront of Eastern culture with Western, in an age which is the most exciting and the most interesting in its multiplicity of idioms and movements in politics, economy and art, because our age carries within itself all the idioms of the cultural and political past of Europe and possibly the future of Asia (CW 3: 178).

The reference to Asia is no doubt an allusion to Tagore's own understanding of Asia's future relationship with the world, which Kosovel was familiar with from reading *Nationalism*. And the fact that Kosovel saw his own position defined in terms of an 'East-West' juncture – at once a point of division and contact – enabled him to relate to Tagore's own project of exploiting the divide for a creative encounter: the forging of a new emancipated individual – 'new man' in Kosovel's parlance; 'universal man' in Tagore's – who would somehow be free of these divisions.

The comparison between Tagore and Kosovel certainly brings to the fore common questions of civilizational identity, particularly in the new climate of self-questioning in the West after the war. Imperialism was past its prime and challenges to colonialism could no longer be ignored. Imperialist wars had collapsed the world, as Tagore put it, into 'the biggest orgy of evil', and Europe, as Kosovel kept reiterating, was 'in crisis'. Many intellectuals and avant-garde artists of the 1920s voiced their protest against aspects of European civilization and some of its certainties. And it would seem that their objections were not unrelated to the emergence of new colonial writers and their presence in avant-garde circles in the West, whose own concerns posed a challenge to European cultural authority and contributed to the 'volatile new cosmopolitan climate' (Boehmer 1995: 123).

If the suffering of his own people was a symptom of global social forces – those of capitalist Europe with her onslaught on the rest of the world and an outlook promoting sharp distinctions between races and civilizations – then, Kosovel felt, the solution too had to be sought across the world in a new social order (CW 3: 31). Certainly for those writers who resisted the civilizational crisis in anti-bourgeois and anti-capitalist terms, the Russian revolution of 1917 offered a realistic hope, however short-lived, for the ideal of a new, non-exploitative, classless society. Moreover, it unleashed what Timothy Brennan has argued was 'a full-blown *culture* of anti-imperialism for the first time' (2002: 19, emphasis original). This last point is crucial if we are to understand the final aspect to Kosovel's sense of situational identification with Tagore, in which the Indian poet is aligned with the proletarian movement, the connection Kosovel made in a lecture he delivered to the miners in Zagorje shortly before he died. Indeed, seeing in Tagore a spiritual and intellectual kin, Kosovel co-opted him into the ranks of those 'intellectuals, famous artists and scientists' who had taken up 'a relentless fight against injustice and violence' and who had 'joined the proletarian movement' (CW 3: 27).

Stressing the links between the inter-war avant-gardes, the colonies and anti-imperialist consciousness, Brennan submits that 'the Russian Revolution [...] was an anticolonial revolution'. This he takes to mean 'its sponsorship of anticolonial rhetoric' which 'thrived in the art columns of left newspapers, cabarets or the political underground, mainstream radio, the cultural groups of the Popular Front, Bolshevik theater troupes', meeting with responses and contributions from 'the various avant-garde arts'. Brennan cannot overstate the implications of the revolution for the 'the idea of the West'. It 'delivered Europe', he says, 'into a radical non-Western curiosity and sympathy that had not existed in quite this way before'. It 'altered European agendas and tastes by situating the European in a global relationship that was previously unimaginable' (2002: 192-3).

The idea of social revolution was now combined with anti-imperialist thought. This was because an analogy was being made between the capitalist's exploitation of the worker and imperialist's exploitation of the colonized. The notion of imperialism as rapacious capitalism expanded overseas in search of new markets, resources and people to exploit was theorized by Lenin in his key text *Imperialism, the Highest Stage of Capitalism* (1920). It is therefore not surprising to see Kosovel, who subscribed to this view, assume, with Lenin, 'the social standpoint of those whom modern capitalism as a world system most exploits and oppresses, even when they are not 'proletarian' in a conventional sense' (Larsen 2000: 29).

When Kosovel turned towards the 'East' for inspiration, anticipating a 'new morning', this morning, he said, would come 'in a red mantle', hence its irradiating core was Russia and not primarily 'the Orient' of Tagore (CW 3: 93). And yet, of course, the two were closely related. In an important aspect of Kosovel's identification with Tagore, therefore, the anti-capitalist and anti-colonial struggles converged, so the 'East' became as much the promise of a new world order associated with the Bolshevik Revolution as it was evocative of the old romantic 'Orient' that would help heal the deep spiritual 'crisis' of the post-War European generation. Even though Kosovel's enchantment with Tagore drew on some aspects of the same language that guided the imaginations of Europeans when they turned towards the East in the early decades of 20th century – notions such as 'simplicity', 'naturalness' and 'child-likeness' do crop up also in Kosovel's writing (CW 3: 509, 558, 561) – to stop here would be to miss why Tagore was so important to Kosovel or how his ideas energized the project of self-emancipation that both poets shared.

For all the enthusiasm the young poet felt towards his older Indian contemporary, there was nothing of blind veneration in the way he perceived him. Rather Kosovel studied his poetry and his philosophical writings seriously, taking 'lessons' from him when they struck a chord, and urging others to do the same. Significantly, when works were not yet available in the Slovenian translation, as was the case with *Nationalism*, *Sadhana* and *Personality*, he got hold of them in German or Serbo-Croatian (the languages he could read alongside French, Italian, and Russian). It is therefore more in the spirit of parity that Kosovel approaches Tagore, as opposed to an Eastern guru at whose feet one should sit, or, following the colonial mindset, 'an Oriental' who ought to be patronized. Reading Tagore for Kosovel meant above all encountering a voice that shared some of the age's deepest cultural and intellectual concerns, spanning nationalism, scientific and technological revolutions, environmentalism and feminism alike, and that helped him both articulate a critique of Europe and portend a solution to it.

Notes

1 Rabindra Bhavana archives, Santiniketan, Serial no. 297, Poland File, letter, April 12, 1941

2 For Tagore's reception in Yugoslavia and the tumultuous events surrounding his visit there in November 1926, cf. Petrović 1970.

3 The latest addition to Tagore works in Slovenian translation is a selected book of short stories, *Izbrane zgodbe*, tr. by Dušanka Zabukovec, selected and introduced by Ana Jelnikar, Ljubljana: Cankarjeva založba, 2010.

References

Aronson, Alex. *Rabindranath Tagore through Western Eyes*, (Calcutta: Rddhi, 1978).

Aronson, Alex and Krishna Kripalani eds. *Rolland and Tagore*, (Kolkata: Visva-Bharati, 1945).

Bangha, Imre. *Hungry Tiger: Encounters between Hungarian and Bengali Literary Cultures*, (New Delhi: Sahitya Akademi, 2008).

Betts, F. Raymond. *Decolonization*, (New York: Routledge, 1998).

Bhattacharya, Sabyasachi. *The Mahatma and the Poet: Letters and Debates between Gandhi and Tagore 1915-1941*, (New Delhi: National Book Trust, 2005).

Boehmer, Elleke. *Empire, the National, and the Postcolonial 1890-1920*, (Oxford: Oxford University Press, 2002).

Brennan Timothy. 'Postcolonial Studies between the European Wars: An Intellectual History', in *Marxism, Modernity and Postcolonial Studies*, eds. Crystal Bartolovich and Neil Lazarus, (Cambridge: Cambridge University Press, 2002), pp. 185-203.

Chaudhuri, Amit. 'Introduction', in *The Picador Book of Modern Indian Literature*, ed. Chaudhuri, (London: Picador, 2001), pp. xvii-xxxiv.

Clarke, J., J. *Oriental Enlightenment; The Encounter Between Asian and Western Thought*, (London: Routledge, 1997).

Dutta, Krishna. Andrew Robinson, eds. *Selected Letters of Rabindranath Tagore*, (Cambridge: Cambridge University Press, 2005).

Hogan, Patrick Colm. *Empire and Poetic Voice: Cognitive and Cultural Studies of Literary Tradition and Colonialism*, (Albany: State University of New York Press, 2004).

Jelnikar, Ana. 'Yeats's (Mis)Reading of Tagore: Interpreting and Alien Culture', *University of Toronto Quarterly*, 77:4, 2008, pp. 1005-24.

Kosovel, Srečko. *Zbrano Delo 3*, prvi del [Collected works 3, part one, CW 3], ed. by Anton Ocvirk, (Ljubljana: DZS, 1977).

Kosovel, Srečko. *The Golden Boat* [Translated from original Slovene into English], tr. and ed. by Bert Pribac and David Brooks, assisted by Teja Brooks Pribac, (Cambridge: Salt Publishing, 2008).

Kundu, Kalyan, and others, eds. *Imagining Tagore: Rabindranath and the British Press (1912 – 1941)*, (Calcutta: Sahitya Samsad, 2000).

Larsen, Neil. 'Imperialism, Colonialism, Postcolonialism', in *A Companion to Postcolonial Studies*, eds. Henry Schwarz and Sangeeta Ray, (London: Blackwell Publishers, 2000) pp. 28-52.

Lokar, Janko. 'Lanska tekmeca za Noblovo književno nagrado' [Last Year's rivals for the Nobel Prize], *Slovan*, 12: 6, 1914, pp. 242-247.

Petrović, Svetozar. 'Tagore in Yugoslavia', *Indian Literature*, 13: 2, 1970, pp. 5-29.

Pobozniak, Tadeusz. 'Tagore in Poland' in *Rabindranath Tagore: A Centenary Volume 1861-1961*, (New Delhi: Sahitya Akademi, 1961), pp. 348-55.

Radice, William. 'Introduction', in *Tagore: Selected Short Stories*, tr. by William Radice, (London: Penguin Books, 1994), pp. 1-28.

Said, E., W. *Orientalism*, (Harmondsworth: Penguin, 1995).

Sluga, Glenda. *Difference, Identity, and Sovereignty in Twentieth-Century Europe: The Problem of Trieste and the Italo-Yugoslav Border*, (New York: State University of New York Press, 2001).

Tagore, Rabindranath. 'Thoughts from Tagore' [1920?], in *The English Writings of Rabindranath Tagore: Volume Three, sA Miscellany*, ed. by Sisir Kumar Das, (New Delhi: Sahitya Akademi, 1996), pp. 29-82.

Thompson, E. P. *Alien Homage: Edward Thompson and Rabindranath Tagore*, (Delhi: Oxford University Press, 1998).

Župančič, Oton. 'Rabindranath Tagore', *Slovan*, 12: 6, 1914, pp. 31-2.

Rabindra Bhavana archives, Santiniketan:

- Serial no. 74, Czechoslovakia File, letter, December 13, 1937.
- Serial no. 333, Russia File, letter, September 24, 1930.
- Serial no. 74, Czechoslovakia File, letter, January 30, 1935.
- Serial no. 297, Poland File, letter, April 12, 1941.

Rabindranath Tagore's Czechoslovak Connection

Martin Hirbek

At a moment when the westernmost and easternmost regions of our shared continent are rumbling with artillery fire, the weak voice of western democracy calls on you at the year's close: long live the world, but a world of equal and free peoples.

– Karel Čapek's address to Rabindranath Tagore transmitted on the airwaves of the Czechoslovak Radio on the Christmas Eve of 1937.

Friends in Czechoslovakia, in the terrible storm of hatred and violence raging over humanity, accept the goodwill of an old idealist who clings to his faith in the common destiny of the East and West and all people on the Earth.

– Rabindranath Tagore's cable in response to the message by Karel Čapek.

The great Bengali litterateur and versatile artist paid two visits to Prague in his lifetime. Those short encounters with Prague intelligentsia and the art-world left lasting impressions, invited numerous responses, and further popularised his works among the general public.

The popularity of Tagore in Czechoslovakia cannot be explained just by the welcome he generally received in Europe after he was awarded the Nobel Prize in 1913. Indeed, his search for individual divinity in *Gitanjali* resonated in minds of Europeans, shocked by the atrocities of the First World War, and his critique of technocratic aspects of civilisation in the three lectures on nationalism was well in tune with the ideas of some Western authors in the 1920's. But there is a special thread to follow in order to unravel the role of Tagore in the tapestry of Czech-Indian relations of that time.

The stage for his reception in Prague in 1921 and again in 1926 was, in a way, being set from the first half of the 19th century when the Czech national revival started gaining momentum. That is, interestingly, around the same time when we can locate the beginnings of the Bengal Renaissance. The national revival culminated after the First World War in 1918 when the Czechoslovak republic was formed out of the Slav dominated northern territories of the Austro-Hungarian Empire.

The question of what is particular about Tagore's Prague connection has to be framed by a search into what role did the imagination of and knowledge about India have to play in the Czech national revival. Since the dominant Austrian discourse held minority languages and populations to be culturally and socially inferior, the first task of our revivalists was to establish that the Czech language was equal, if not superior, to German. One of their major arguments was the newly-discovered similarity of Slav languages to Sanskrit. The revivalists argued that Czech with its complex system of inflection was, indeed, a language with very antique roots that predate those of German. This argument later developed into a specific sort of India-as-a-cradle-of-civilisation theory which explored and construed close links between Slav and Indian cultures in the field of language, mythology, and religion. This revivalist construction even stimulated such curious experiments as Czech poetry in old Indian metres.

This endeavour reflected that of German authors who also emphasised an affinity between Germans and ancient India in order to support the claim of Germany, then in the process of unification, for a future role of leadership in Europe. But our revivalists were not interested exclusively in India's past. An unknown author published in the Review *Česká včela* (*The Czech Bee*) in 1845 a fairly knowledgeable account of the Bengal Renaissance and the reformist effort of the Brahmo Samaj and related it directly to the struggle of Czech revivalists. His article ends with a proclamation: 'Blessed is a nation who has men like Dwarkanath', referring to Rabindranath's grandfather, Dwarkanath Tagore.

The Czechs, who built their national consciousness on the opposition to German cultural dominance and the Austro-Hungarian state, thus not only fought to prove that the Slavs' affinity with India was much closer than that of the Germans but also established a sense of equality and common destiny among Slavs and Indians on the basis of parallel and, more importantly, coeval struggles of two peoples against foreign oppression. A positive image of India and her movement for self-determination in Czech discourse remained largely intact till India's independence, despite the fact that it was based partly on many misguided presumptions. Tagore was thus received by Czechs not only as an Eastern seer and a celebrated writer but also as a cultural ambassador of a brotherly nation fighting for its independence.

No less important for the later reception of Tagore was the early establishment of Indology in Prague. Regular courses in Sanskrit started as early as 1850 and by the end of the century, fully-fledged classical Indology departments were already firmly established at both Czech and German Universities. Around the time of the creation of Czechoslovakia, research in modern Indian languages and literatures broadened the scope of Prague Indology.

Tagore first visited Prague at the invitation of our most prominent Indologists of that time, Prof. Vincenc Lesný (1882-1953) and Prof. Moriz Winternitz (1863-1937), who held chairs at the Czech and German universities respectively. While Winternitz is perhaps more internationally renowned for his opus magnum *The History of Indian Literature* (*Geschichte der indischen Literatur*), Lesný was the founder of Bengali studies in Prague. He studied Bengali by himself and became acquainted with Tagore's works before the First World War. There is evidence that they exchanged correspondence as early as 1913, the year Tagore was awarded the Nobel Prize. Lesný has to his credit that he was the first European translator of Tagore directly from Bengali originals without the mediation of the English versions. He translated from Bengali and published *Examples of Poetry and Prose* by Tagore as early as 1914. The same year, a Czech translation of Tagore's *Gitanjali* from English by František Balej appeared. The third Czech translator of Tagore, albeit again from English, was Věra Stephan, an unknown 'girl 20 years of age' of Russian origin, who asked Tagore to grant her permission to translate his works into 'Russian and Bohemian' (i.e. Czech) in a passionate letter dated December 2, 1913. Despite the fact that Věra, not knowing the location of Tagore's residence, sent the letter 'with a lacking address and firm hope that it will reach its aim' it did, indeed, reach Tagore (see Rabindranath Tagore Correspondence Archive, Rabindra Bhavan, Vishva-Bharati University, Santiniketan, File 74, No. 15-17). Věra Stephan's Czech translation of the *Crescent Moon* appeared in 1915 and that of *Gardener* in 1916. Besides that, the archives bear evidence of other amateur translations of individual poems by Tagore's Czech admirers from the same period.

Lesný naturally became the main personality to introduce Tagore to the Czech audience. The two men met in October 1920 in London, having been introduced by Prabhu Dutt Shastri of the Calcutta University, and Lesný invited Tagore to Prague. His works were already well known to Czech readers. Besides the above mentioned ones, translations of *Sadhana*, *Nationalism*, selected short stories, *The King of the Dark Chamber*, and *The Home and the World* appeared by the time of his first visit in 1921. During those three days, Tagore delivered three lectures, among them one at the convocation hall of the Charles University and one for the broad public in the largest auditorium in Prague – the Lucerna hall.

The latter was attended by the famous musical composer, Leoš Janáček (1854-1928), who wrote in a passionate report to the Lidové noviny newspaper (June 22, 1921):

> It was not a speech - it sounded like a song of a nightingale, smooth, simple, void of any harshness of the diphthongs. It occurred to me to fall in with a gay chord with the initial sounds of the poem he read out. I heard soft harmonious voices or sounds, but it was incoherent to me. The melody kept on falling down in a torrent of tones.

Janáček was known for noting down the speech-melodies of people he encountered and Tagore, who recited his poems both in English and Bengali, provided him with a unique opportunity in this sense. The lecture inspired Janáček to turn one of the poems from the Gardener, the Wandering Madman, into a five-minute long choral composition. Janáček himself apparently identified with the character of the Wandering Madman and the last lines of the poem are even engraved on his tombstone.

The second visit by Tagore was on the invitation of the Prague P.E.N. Club and its head, the world-known writer, Karel Čapek. Besides lectures and recitals, Tagore also attended performances of two of his plays, Chitra and The Post Office, which were staged at the National Theatre.

Tagore's popularity in Czechoslovakia, as we noted at the beginning, was partly due to the role that the imagination of and knowledge about India had in the process of our national revival. The notion of the affinity between Czechs and Indians and the opposition between Czechs and Germans again resurfaced during his visits and more profoundly still, with the advent of the Second World War. During his second visit, Tagore was complaining of 'cyclonic fury of engagements which still shows no signs of abatement.' This is further explained by his secretary, Prashanta Mahalanobis: 'We have been going from place to place like gypsies. Poet had a nervous breakdown in Prague…there are two different sets in Prague, Czechs and Germans, and everything had to be repeated twice.' (*Selected letters of Rabindranath Tagore* by Dutta, K. and Robinson A., Cambridge University Press, Cambridge, 1997, p. 339).

In 1937, on the last Christmas Eve before Czechoslovakia was dismembered by the Munich agreement of 1938, an unusual radio message was broadcast from Prague. It was an appeal for moral support and peace in the face of approaching occupation, addressed to Albert Einstein, representing the West, and to Rabindranath Tagore, the voice of the East. Karel Čapek, our most recognised and pacifist writer of that time, read out the message in Czech and Prof. Lesný in English with a few Bengali words at the end. Although Tagore was unable to reply instantly on airwaves of the Czechoslovak Radio from Santiniketan, which was the original plan and had to be stalled due to technical reasons and lapses in organisation, he wrote back a telegram of goodwill and a poem composed in response that night.

Reciprocating Tagore's visits to Prague, several Czech personalities stayed with him in Santiniketan. Both Lesný and Winternitz held tenure there as visiting professors in 1920's. Winternitz wrote a short book in German entitled *Rabindranath Tagore: Religion and World-View of the Poet* (*Rabindranath Tagore: Religion und Weltanschauung des Dichters*) and Lesný published a Czech monograph which was later translated into English as *Rabindranath Tagore: His Personality and Work* (*Rabíndranáth Thákur* [*Tagore*]: *Osobnost a dílo*).

Even before the two Indologists, however, the first Czech visiting professor in Santiniketan was the painter, Jaroslav Hněvkovský (1884-1956). Hněvkovský travelled twice to India in the beginning of the 20th century. He undertook the first journey to South India with his dear friend, Otakar Nejedlý, also a painter, in 1911. After meeting Rabindranath Tagore in London, where Hněvkovský had a successful exhibition, he accepted Tagore's invitation and spent several months in Santiniketan in the spring of 1922.

Hněvkovský left not only paintings from that journey, including those of Rabindranath himself, but also an extensive and vivid diary which he later published. He recalls his reservations for the Bengal school, his dialogues with Nandalal Bose and Asit Kumar Haldar, the memories of the Tagore family, other foreigners at the campus, and, last but not least, his muse, a young student Rani or Ranu Mookerjee, who later became the first Chairman of the Academy of Fine Arts in Calcutta. In that capacity, she visited Czechoslovakia in 1970 on the occasion of Hněvkový's posthumous exhibition.

Among other Czech visitors to Santiniketan was a Czech-American astronomer and humanist, Karel Hujer (1902-1988), who went to India in 1935 and stayed for two weeks with Mahatma Gandhi at his ashram in Wardha and three weeks in Tagore's Abode of Peace.

The popularity of Tagore's works in Czechoslovakia did not diminish with time. The nineteen twenties and thirties saw more new translations and, although Tagore was blacklisted during the Nazi occupation, new editions came out just after the war. In the three years before his centenary, Dušan Zbavitel, Tagore's most prominent translator into Czech, prepared for print *The Works of Rabindranath Tagore* in three volumes, mostly out of his own translations. In addition, he wrote a Czech monograph, *Rabindranath Tagore: The Evolution of a Poet* (*Rabíndranáth Thákur: vývoj básníka*). Zbavitel's translation of *Gardener* from 1966 had a print-run of thirteen thousand copies and that of *Gitanjali* from 1978 of seven thousand copies. Although hardly any piece of literature in Czech reaches such volumes under free-market conditions, Tagore's Czech translations in a book format total 36 original editions and many are still available in new prints and widely read.

As Europe and Asia nowadays meet each other in a globalised world, Tagore's humanism and 'his faith in the common destiny of the East and West and all people on the Earth' regain importance.

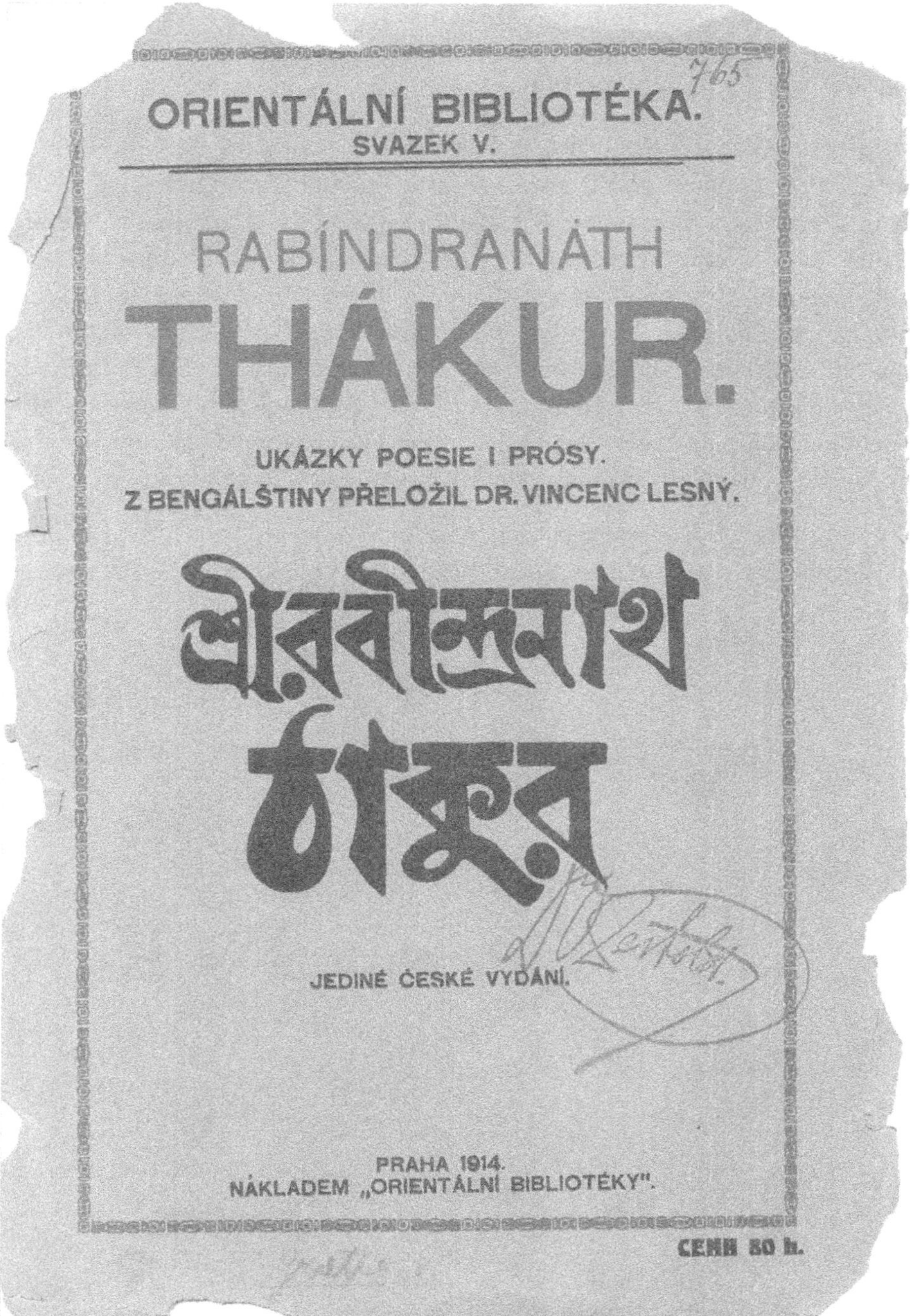

Figure 1a – A facsimile of the first European translation of Tagore directly from the Bengali original. "Examples of Poetry and Prose" translated into Czech by Vincenc Lesný in 1914.

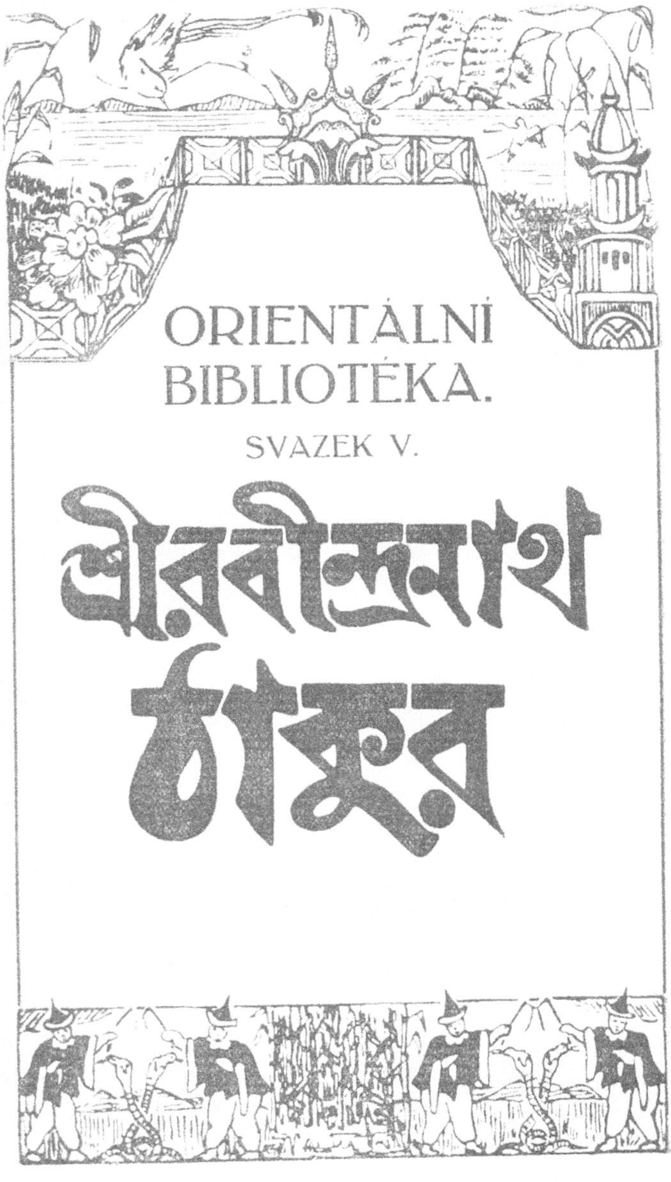

Figure 1 b

Figure 2 – A 1947 edition of Gitanjali.

Figure 3 – Winternitz, Tagore, and Lesný in Prague

When my mind was released
from the black cavern of oblivion
and woke up into an intolerable surprise
it found itself at the crater of a volcanic hellfire
that spouted forth a stiffling fume of insult to Man;
it witnessed the longdrawn suicidal agony of the time spirit
passing through a mockery of a monstrous deformity worse than death.
On its one side a defiant savagery and the growl of homicidal drunkenness,
on the other timid powers tied to the load of their carefully guarded hoardings,
meekly settling down to a silent safety of acquiescence
after miscalculated bursts of impatience.
At the old nations' council chambers plans and protests
are pressed flat between the tightshut prudent lips.
In the meanwhile across the sky rush with their blazing blasphemy
the soulless swarms of machinemade vultures-
carrying their missiles of ravenous passion for human entrails,
Give me power, O awful Judge, sitting on the throne of Eternity,
give me a voice of thunder, that I may hurl imprecation
upon this cannibal whose gruesome hunger
spares neither women nor children,
that my words of repugnance may ever rock
upon the heartthrobs of a history humiliated by itself,
till this age choked and chained
finds the bed of its final rest in its ashes.

Santiniketan,
December 25, 1937.

Figure 4 – A facsimile of the poem written by Tagore in response to the goodwill message by Karel Čapek and Vincenc Lesný and broadcast by Czechoslovak Radio on the Christmas Eve of 1937. (Rabindranath Tagore Correspondence Archive, Rabindra Bhavan, Vishvabharati University, Santiniketan, File 74, No. 119.)

Tagore's Visit to Hungary

Imre Bangha[1]

Tagore's visit to Hungary is an important episode of his European tour in 1926. The fact of the visit is widely known in Hungary since the bust of the poet and the tree planted by him in Balatonfüred are certain to attract the attention of the many visitors to one of the most popular tourist places in the country. However, many details of the sojourn are not easily available to modern readers.

The visit has been studied for decades. Investigation of Rabindranath's stay in the country goes back as early as 1943, when a person under the pseudonym, *Balatoni Krónikás*, 'Chronicler of Lake Balaton' published an article about the poet's days in Balatonfüred.[2] In 1960 an article by Kshitish Roy shed some more light on Tagore's contact with Hungary.[3] Another person to investigate the visit was the Hungarian wife of the poet Kanti Ghose, Eta Boglár (Ghose), who was librarian at the Visva-Bharati. She not only translated passages from the Bengali letters of Mrs. (Rani) Nirmal Kumari Mahalanobis,[4] companion of the poet on his tour in 1926, but also started to collect materials in Hungary in 1960.[5] Further research was done by Gyula Wojtilla who in a chapter of his book *Rabindranath Tagore in Hungary*[6] described Rabindranath's personal contacts with Hungary including his visit.

After Wojtilla's volume, as a result of the publication of the digests of some Hungarian newspapers, some materials came to light that show further details about Tagore's days in Budapest and Balatonfüred. The sources for this investigation are contemporary reports and writings published in major daily newspapers and magazines as well as letters written to or by Tagore and his companions. The two groups of sources help us to look at Tagore from almost an intimate distance and to see how the Hungarian public interpreted the visit of the 'oriental sage'.

The 1926 Tour

At the beginning of the 1926 European tour, Rabindranath mentioned that he would be happy to come to Hungary but he had not had any formal invitation.[7] Animated by this announcement, the Hungarian Feminist Society sent him a letter of invitation on 6 July.[8]

The new mayor of Budapest, Jenő Sipőcz, sent a formal invitation to Rabindranath. The practical organisers of Rabindranath's programme were Imre Kun, director of the *'Koncert' Hangversenyvállalat*, the largest enterprise for organising concerts in Hungary, and Viktor Huszka, a representative of the Tourist Office of the capital. It was Imre Kun who handed the formal invitation to Tagore in Berlin.[9]

Tagore in Vienna

On 15 October Tagore travelled to Vienna, where he broke down, cancelled his lecture and postponed his journey to Hungary. Later a Hungarian organiser visited Rabindranath in Vienna and he was successful in persuading him to come to Hungary.[10]

Tagore's programme in Hungary had already been fixed on 14 October. It would be a joint effort by different literary societies such as the illustrious conservative Kisfaludy Társaság and the Hungarian PEN Club. The famous Petőfi Társaság would provide Hungarian music and dance and the Society of Hungarian Playwrights would organise a dinner in the Fészek Klub, one of the most outstanding venues for literary events in Hungary.[11] These four societies were the best organised bodies of Hungarian literature at that time and represented different political and literary views. Their joint effort reflects the wide popularity Tagore enjoyed in 1926. It was, however, not until 23 October that Tagore's programme was finalised. The lecture would be shifted to Wednesday the 27th and on Thursday the capital would organise a large dinner with musical performances in honour of the poet. Tagore expressly asked to hear Hungarian music and see Hungarian dances for which he would have an opportunity that evening.[12] It was announced that the cultural minister might also come to the programme.[13]

In his interviews, Tagore never mentioned that he was going to Hungary to deliver a lecture but rather that he wanted to get to know the people and the culture. He did not speak about the theme of his lecture or about Indian philosophy although it may have been expected of him. However, he again and again expressed his aim that he wanted his university to profit from his tour. As can be seen from the interviews, this did not only consist of material resources.

An Overwhelming Welcome

On the afternoon of Tuesday 26 October, Tagore took a train from Vienna to Budapest together with his secretary, P.C. Mahalanobis, Mrs. Rani Mahalanobis and a German baroness. The Baroness was said to have joined Tagore's company in Berlin and did not let her name but her monogram, L P, be published in the papers.[14] Later a journalist found out that she was a 38 year old sculptor called Lisa Pott and belonged to a rich Austrian family.[15] She had had a long correspondence with the poet, had met him in London and joined his company in Vienna to accompany him to Santiniketan and take the chair of sculpture. It was also stated that she spoke five languages.[16]

The organisers of the programme joined the poet already on the train, together with Ferenc Zajti.[17] Journalists also attached themselves to them and described how Tagore's company occupied three first class compartments placed at their disposal by the Austrian government. It was on the train that Tagore gave his first interviews in Hungary. [18]

At the chief railway station of Budapest, the Keleti Pályaudvar, a crowd of enthusiastic people waited for the train that was about half an hour late. On appearing, the poet was given an immense ovation. First Alderman Géza Lobmayer welcomed him on behalf of the capital, and then a young girl tried to salute him in French on behalf of the Hungarian Feminists.

Tagore's appearance was not only that of a venerable sage. Some journalists compared the long-bearded poet arriving in a late autumn day to Santa Claus.[19] At the end of the welcoming ceremonies, Zsolt Harsányi, a conservative writer of historical novels, gave a speech on behalf of the PEN Club about the tribulation and toil of the Hungarian people, which was greeted by great applause.

Although the Indian poet was received with an ovation at the station, his reception was somewhat less enthusiastic than it would have been five years earlier. A journalist noted,

> An animated crowd of representatives of the capital, the youth of Hungary and of the press was present — but official Hungary, in the form of its governing bodies or its literary or academic societies, was missing. The Nobel Prize winning great poet, founder, maintainer and professor of the university in Santiniketan would perhaps have merited after all a welcome from either the Hungarian academic bodies or the literary societies so proud of their long history.[20]

From the railway station Tagore went to the Hotel Gellért. Rabindranath was tired, but still he received the representatives of the press in the hotel. When he appeared in the door, some drew a sketch of him while he was uttering a few words: 'I feel I haven't come among alien people. I am aware of the fact that your cradle swayed in the same place as mine: in Asia. I would be happy to make the eastern contacts of the Hungarians stronger by my presence.'[21]

Next day, he felt so tired that he decided to remain in his room. Though his Hungarian 'friends' the theologian János György, the Orientalist Ferenc Zajti, and Viktor Huszka from the Tourist Office urged him to take a morning sightseeing tour of Budapest, Rabindranath declined the invitation.[22] The animated company succeeded only in taking Tagore to see the Fishermen's Bastion and the nearby Matthias Church in the Castle of Buda.[23]

Tagore's lecture: Civilisation and Progress

Tagore's lecture, *Civilisation and Progress*, had a similar title to the one, *Culture and Progress*, that had been delivered in Hamburg on 10 September.[24] It might have been identical with the one delivered in Prague two weeks earlier under the title *Civilisation and Progress*.[25] The programme was held in the great hall of the Music Academy. The hall has space for more than eight hundred people but it was overcrowded and many were forced to wait outside hoping at the very least to have a glimpse of the 'Wise Oriental'.[26] Some representatives of the political elite were present at the lecture: Archduke József Ferenc and his wife; Emil Nagy, a former minister of justice[27] and Pál Petry a state secretary together with many other officials of the Cultural Ministry.[28] Apart from them there were some other elements of the Hungarian intelligentsia including the poet Dezső Kosztolányi and László Ravasz, a Calvinist bishop who wrote an article about Tagore[29], as well as teachers from the Musical Academy[30]. However, 'the majority of the audience… was recruited from among women… female students, learned women and ladies eager for eccentric pleasure.'[31]

Zajti in his introduction to the lecture spoke about the development of Bengali poetry and about Tagore whom he introduced as a poet and apostle.[32] When the Indian poet appeared on the stage he was welcomed with an applause that did not let him speak for minutes.[33] In his lecture Tagore explained: 'In the age of all-powerful machine when numbers are all that matter, men crab and kill each other… The only way for redemption and purification is if man makes peace with his own soul and having known his soul he follows its commands.'[34] 'Fire was a great achievement for man… It was a blessing… However, fire remained material and did not speak to the soul… Having faith in human solidarity, in love, resulted in man's finding the fire in his soul.'[35]

This lecture in English was translated into Hungarian by György Kemény.[36] The translation may not have been really accurate. Only the reporters of the dailies *Budapesti Hírlap* and *Pesti Napló*, who probably knew English, gave details of the train of thought. After his speech, Rabindranath recited one of his poems, 'The caged bird and the wild bird', followed by the recitation of its Hungarian version given by its translator, Vilmos Zoltán. At the end, Rabindranath sang their national song, which the audience heard standing, and tried to sing its refrain together with the poet

The audience venerated the distant prophet in Tagore, who could offer a solution to their problems. His presence was far more important than what he said.

Banquet in Hungarian Style

Next day at 11 in the morning, Rabindranath went to the studio of the photographer Sterlinszky[37], and then for a sightseeing tour in a car accompanied by Zajti and others. He visited the Opera, the Heroes' Square and the Parliament. Some journalists noted how much he knew about Hungarians.[38] Allegedly he spoke to Zajti about the Huns and the Hungarians' coming to the Carpathian Basin.[39]

In almost all of his interviews, the poet mentioned that he had come among relatives and seemed to speak more and more explicitly about the Huns as relatives of the Hungarians. Similarly, Hungarians did not fail to bring this theme up when speaking to Tagore.

Another frequent theme to be brought up was the truncation of Hungary : 'Two thirds of their country had been cut off and distributed to other countries. In the whole of Europe they and Austria are the poorest… All the people I have so far met are very sensitive about this. Everyone says the same, 'how unjustly they have reduced our country'.'[40]

At a certain point of the journey Rabindranath was presented with a metal seal with the following inscription on a central panel in the style of lettering on a headstone: 'In this metal seal is contained one particle of blood and tear-soaked soil of each of the dismembered provinces of the thousand-year old Hungary. Accept this memento Rabindranath Tagore. This seal is a token of all our sufferings and tribulations, our loyalty and devotion and of all our beautiful hopes.'[41]

That evening a reception was organised by the City Council in honour of Tagore. The evening programme was attended mainly by elements of the conservative literary and political elite such as Pál Petry, the State Secretary of Culture, poet Miklós Józan, who later became the Unitarian bishop of Cluj (Kolozsvár), Elek Petrovics, the Director of the Museum of Fine Arts and Gyula Pekár, President of the Petőfi Társaság. At first Ilona Hollós, a lady 'in Hindu dress' recited some poems of Tagore in Hungarian translation; then two opera singers, Izabella Nagy and Imre Palló sang Hungarian folk songs.[42] Imre Palló (1891-1978), one of the best singers of the time, who later became Director of the Hungarian National Opera House, sang a song in folk style, *Elment a rózsám Amerikába* 'My rose went to America.' Another folk song by Izabella Nagy made such an impression on the poet that she had to repeat it at his request.[43] The famous Gypsy band of Béla Radics also performed some Hungarian songs in folk style.[44] After the songs came the dance-performances with Hungarian solo and duo dances, *de Savaras* dance, and 'Hindu dance'. According to a report, the latter dances had a strange impression on the poet: 'He shook his head a great deal when naked ladies swept up dancing on the stage and he did express his disapproval… After that a Hindu dance was announced... a naked young lady… the poet looked at her in astonishment for a moment and then whispered something to his secretary. The secretary also started to laugh.'[45] The programme ended with traditional recorder music and a piano recital. By the request of the audience Tagore also recited one of his Bengali poems.[46]

The cultural programme was followed by a banquet, where Rabindranath was given traditional Hungarian food with stuffed turkey as the main dish. Jenő Sipőcz, the Mayor of Budapest and Gyula Pekár, the president of the Petőfi Társaság proposed toasts.

Health Under Siege

A visit to the East Asia Museum and to the Museum of Fine Arts was planned for the next day, 29 October.[47] However, Rabindranath felt so exhausted that he abandoned this programme. Since Tagore's address in a hotel was published in the papers on the first day, he had to receive crowds of visitors in the Gellért Hotel.

On Friday, Saturday and Sunday Tagore remained in his room with the exception of a visit to the Fészek Klub at 6 p.m. on Saturday where the PEN Club and the Society of Hungarian Playwrights organised a reception in his honour and he also gave a speech.[48] Here he and his company were received by Menyhért Lengyel representing the Society of Hungarian Playwrights, Ferenc Székelyhidi on behalf of the Fészek Klub and Antal Radó as a member of the PEN Club.[49] He could comprehend everything that he read as a child, but reading had left its influence on him. [50]

Rabindranath met Hungarian intellectuals not only at official receptions, but there were also spontaneous initiatives. It was one of his days in Hotel Gellért that the poet received a letter from the town of Pécs from a young scholar, Károly (Charles) Fábry, who later settled in India and became a famous art critic, backing Amrita Sher-Gil.

> At present (to the best of my knowledge) I am the only Hungarian Indologist.... I wish from the depth of my heart to see you and to speak with you. I am a poor university student (this year I will be doctor of Philosophy), living from what I earn with French, German and English lessons; but I would not spare expenses at all to go to Budapest if you would promise me to receive me and to speak with me.[51]

The letter was not sent in vain. The poet met Fábry in Budapest: 'My great Guru, The day on which I had the opportunity to speak with you was my most beautiful day since I live.'[52] In this second letter of his, the young scholar asked Tagore for a testimonial to one of the greatest scholars of Sanskrit, Professor Winternitz, because he wanted to get a scholarship abroad.

These days also saw the creation of a new poem, *Diner belāy bāśi tomār...*[53] dated 30 October in Budapest.

> In the daytime your flute resounded with various notes
>
> The touch of the song entered my life,
>
> the notions of myself and you stayed away.
>
> I inquired of everyone on the way: 'Who played this flute?'
>
> They allured me with different names
>
> and I wandered to different doors.
>
> Now the sky has shrunk and the tired daytime closes its eyes.
>
> If you make me roam in different roads,
>
> then I will be lost searching in vain.
>
> Having abandoned the outside, stretch your seat within;
>
> Come and play your flute in the inner apartments of my life.[54]

In a Sanatorium at Lake Balaton

'According to his original programme, the poet would already have been the guest of Masaryk in Prague but on Friday morning, he had a strange suffocating sensation. He felt his heart pounding'[55] wrote the papers. Viktor Huszka from the Tourist Office called the world-famous cardiologist, Sándor Korányi (1866-1944) to examine Tagore.[56] On Saturday at midday and on Sunday Professor Korányi called on the ailing poet.[57] 'A long and thorough medical examination diagnosed extreme exhaustion.'[58]

On the advice of Professor Korányi, Tagore decided to spend a few days in a sanatorium in Balatonfüred near Lake Balaton, one of the largest lakes in Europe. A huge crowd gathered at the railway station Keleti Pályaudvar to say goodbye to the poet.[59] This silent farewell might have been one of the most beautiful public events of Rabindranath's stay in Hungary. The doctors suggested the poet spend some ten days in the sanatorium. Having learned by the experience in Budapest, they did not allow anyone to disturb him, '[a]lthough Tagore's wish was that his door should be open to anyone'[60] In the resort of Balatonfüred, the poet at last found the long desired rest.

A report on 3 November announced that the condition of the poet had improved. From that day he was allowed to walk around outside on the shore of the lake. Tagore was recovering and was able to leave the sanatorium more frequently. On 5 of November, he laid a wreath on the Jókai-statue and planted a tree.[61] The poet's concern for trees is illustrated in some letters of the 1926 tour where he wrote to Mira Debi in Santiniketan about the trees that should be planted there.[62] In another letter from Vienna he said, 'I am all athirst to see the creeper in front of my porch burst into flowers — with the fullness of the sap of life running through the fibres of its being.'[63]

The annual festival of *Vriksharopana*, tree planting, was established in Santiniketan in 1928 and Kshitish Roy, the author of the article *Tagore in Hungary*, wrote that, 'This beautiful festival was initiated two years earlier in Hungary.'[64] Although people in Hungary understood it as an Indian tradition, they were not aware of the significance of tree planting in India as an example of an entirely unselfish act, and until today Tagore's *Vriksharopana* is interpreted simply as a gesture of friendship. Moreover, a journalist wrote that in the Indian tradition the one whose plant took root would see the tree's bursting into leaves.[65] According to the reporter on the daily *Pesti Napló* the poet said when planting the tree, 'Let this planting symbolise the fact that I, the Hindu poet, will gain new force in Hungary.'[66] In memory of this act Tagore composed a quatrain in Bengali and translated it into English,

He taru, e dharātale rahiba na yabe,. . .

When I am no longer on this earth, my tree,

Let the ever renewed leaves of thy spring,

Murmur to the wayfarer,

The poet did love while he lived.[67]

At this time, local people started to call on Tagore and brought small gifts to him.

It is wonderful how much the poet is loved here. His books are so popular even among the poorest that it cannot be compared to any other country. In other places the poet also received lots of affection but the affection of the Hungarians is somewhat unique.[68]

In the sanatorium the poet also had time for work. He told to a reporter of the magazine *Balatoni Szemle* that during his European sojourn he had never been able to work so much as in Balatonfüred.[69] The song *Pāntha pākhir rikta kulāy...*[70] was composed in Balatonfüred on 9 November.[71]

In the empty nest of the bird of passage on a hidden branch

It is watching with ears opened,

in the space between the leaves.

The sound of the wings' returning to the nest

has become finally silent.

and the mantra of the evening star

has arisen at the day's farewell time.

The Moon gave a thrill to the ocean's waves,

and the light's notes entered in the holes of the forest shade

The sleepless emptiness resounds in its chest during the watches,

As the bamboo-branches murmur in the night wind.[72]

It is in Balatonfüred that Rabindranath finished on 7 November his collection of 250 brief Bengali poems entitled *Lekhan* (Scribblings).[73] The poet wrote the poems by hand on special metal plates. These plates were sent to Berlin where the printing plates were prepared and sent to India for printing. It was published in book form in 1927 in Santiniketan.[74]

A large part of the English poems of the *Lekhan* together with a few other poems in English (the original of which were published as late as in 1946, in the posthumous *Sphulinga*) came out in the form of an illustrated English book entitled *Fireflies* in 1928 in New York.[75] The English versions in the *Fireflies* are sometimes different from those in *Lekhan*.[76]

One day Rabindranath went to the neighbouring villages to learn about Hungarian national costume.[77] An outing to the Benedictine abbey of the neighbouring peninsula, Tihany, was also planned for Tagore. On Saturday the 6th, Ottó Berkes, the director of the Premostratensian school and monastery in Keszthely, another town on Lake Balaton, sent a car for the poet to come for lunch to the monastery.[78] The poet was busy writing, so at first he declined and sent only the Mahalanobis couple, Lisa Pott and Zajti, who came to visit him in the Sanatorium. In the afternoon, however Dr. Schmidt took him to Keszthely and he joined the company for tea.[79] Here students of the grammar school sang Hungarian popular songs to him and gave a demonstration of Csárdás, the most famous Hungarian dance.[80]

> At tea a teacher gave a toast praising the poet and mentioning the tribulations of the Hungarian nation. Tagore was moved and said that on seeing the students he remembered his own school.[81] At another occasion the Roman Catholic bishop Nándor Rott also invited him to the nearby town of Veszprém.[82] These invitations and the fact that a sanatorium belonging to the Benedictine order hosted him show that even the Roman Catholic Church, was not united against the poet. Tagore's breakdown and retreat to the resort of Balatonfüred once more directed the attention of the international press towards the aged poet and various foreign newspapers reported it.[83]

The most detailed description of Rabindranath's activities in Balatonfüred came in *The Observer*.

> Dr. Tagore has planted a lime tree near the statue of Sándor Kisfaludy (a famous Hungarian poet who lived in the early part of the nineteenth century) 'in remembrance of the renewal of life-force' which came to him on Hungarian soil, and is reported to have written several poems extolling the beauties of Balaton, which he calls 'the lake of smiling quietude.' He has also laid a wreath on the memorial erected last year to Hungary's most famous novelist, Maurice Jókai, fifty of whose novels (translated into English) are to be included in the library of the university founded by the poet in Bengal.[84]

The Last Days in Hungary

In many of the countries visited Rabindranath met the heads of state. This became part of his programme in Hungary, too. The physical condition of the poet having improved quickly, he returned to Budapest in a car with Dr. Schmidt[85] on Wednesday, the 10th. They arrived back at the Hotel Gellért early in the afternoon and met the Regent the very same day. In the private audience,

> Tagore warmly thanked the Regent for the unforgettable welcome he was given in the land of Hungary... — I will seek the opportunity to return among the Hungarians as soon as possible. The Regent told him that he also had unforgettable memories of the poet's homeland because in 1893 he spent a long time in India. War-scarred Hungary — said the Regent — hopes in patience and forbearance that one day better times will dawn on her. — I hope the same from my heart — answered Rabindranath Tagore.[86]

It was in Budapest that a letter from Robert Bridges, the British Poet Laureate, reached Rabindranath inquiring whether he would accept an honorary doctorate of the University of Oxford. Edward Thompson, a Fellow of Oriel College wrote that because of Tagore's return of his knighthood it was believed that he would refuse any honour from a British source, therefore an inquiry was necessary from the side of the university before the final decision could be made. However, the wording of the letter was not very fortunate with sentences like 'I think your election is very uncertain.'

> Tagore not unnaturally took this as a suggestion to come to England on the chance of the university offering him a degree, which might make him look foolish if this did not happen and he replied with some abruptness that it was not convenient for him to come. Oxford's apparent coldness seemed all the greater by contrast with the warmth and enthusiasm which surrounded him in Budapest.[87]

On the morning of the 11th Rabindranath had another sightseeing tour in Budapest and on the 12th he took the train for Zagreb. In Zagreb Tagore was to give a lecture. 'From there he would go to Belgrade, than travel to Constantinople or Constanța from where he is to take the ship to return to India.'[88] About the rest of the tour Prasanta Chandra Mahalanobis wrote in a letter,

> Doctors advised him to avoid the colder climate of the north and travel by the eastern route to Egypt. So we decided to go through the Balkans and I insisted on refusing all engagements. It was no use. Men were coming from the different legations every day... so we went from one place to another

the poet lecturing every day, sometimes morning and evening in the same day. The weather was superb, the driest November in fifty years, and the poet naturally in the highest spirits all the time. We went to Zagreb (Croatia), Belgrade, Sofia and Bucharest.[89]

During the whole European journey Rabindranath was celebrated to such extent that a chronicle of Tagore's life gives the title *Like a Conquering Hero* to the chapter describing this tour.[90]

There seems to be some difficulty in specifying Tagore's date of departure, since the daily *Az Est* wrote that he would take a train to Zagreb on Thursday[91] 11 November. However a letter written from Budapest on the 11th by P. C. Mahalanobis to Rathindranath Tagore[92] gives details of their journey. The poet and his company would take the 22.40 train on the 12th, arrive in Zagreb next morning at 7.20, present a lecture there in the evening and leave for Belgrade at 12.20 on the 14th. They would arrive in the Serbian capital at 21.35 the same day and then would present a lecture on the 15th. Rathindranath, the poet's son would join the company in Belgrade.

On the day of his departure Rabindranath wrote a short poem to Hungary and translated it into English:

hāngerī, ātithya tumi dile pānthajane

hāngerī, birām tumi dile srānujane;

prabāsīre āpanāra kariyācha yabe

bicchede kakhano tār bicched nā habe.

Hungary, thou hast offered welcome to him

who travelled from a distant home

and restful shelter when he was weary and weak.

Since thou hast accepted him who was a stranger

the separation of distance

will never separate him from thee.[93]

Before his departure the poet said in an interview, 'I haven't found such warmth and such affability anywhere. I promise that I will be very grateful. From now on many Indians will be keen on coming to Hungary.'[94] It seems that he kept his promise because a few years later Rathindranath, the son of the poet,[95] the then Maharaja of Holkar and several other foreigners came to visit Balatonfüred on the advice of Tagore. He even sent a guest who spent thirty thousand *Pengős* in the Danube Palace in the course of 9-10 days (at a time when two hundred *Pengős* were a good monthly salary) and hired a whole hotel at Lake Balaton.[96]

Tagore's wish to have some Hungarians in Santiniketan was fulfilled. Shortly after his visit he invited Julius Germanus to teach at the fledgling department of Islamic studies in Santiniketan.[97] Two painters, Elizabeth Sass-Brunner and Elizabeth Brunner also spent some time in Tagore's university. In addition to them Ervin Baktay and Károly Fábry also visited Santiniketan.

On his 1930 European tour the poet expressed his intention to come to Hungary.[98] However he broke down at the beginning of the journey and had to cancel the visit.[99] Still he never forgot about his stay in the country and until the year of his death, he kept on sending greeting cards to the head physician of the sanatorium.[100] He also gave a warm welcome to the Hungarians visiting his house in Calcutta[101] or his university at Santiniketan.[102]

The Forming of a Legend

As we have seen, the press in Hungary[103] paid considerable attention to Tagore's visit to the country and afterwards the poet himself spoke with emotion about it.[104] However, the press coverage was less than adequate since several mistakes occur about the unusual visitor. For example, the fact that Rabindranath wrote on metal plates was noted with astonishment by a Hungarian journalist already in Vienna who thought that the poet was writing on silver plates.[105]

This journalist was not alone in making mistakes about Tagore. In several articles, Rani Mahalanobis was taken to be the daughter of the poet.[106] Another one took her to be his wife.[107] Similarly, several reporters announced that she would come to the reception in honour of Tagore in Hungarian national costume.[108] According to a journalist, Ervin Baktay, who at that time was in India, was a visiting professor of Tagore's University in Calcutta.[109] Another one wrote that on seeing Budapest, Rabindranath promised that he would suggest that traffic between India and Britain should be done via Hungary.[110] These blunders were caused by sloppiness or misunderstanding.

It was however not only in Hungary and not only in the twenties that people made mistakes about Tagore's visit. Under a photo of Rabindranath and Zajti in front of Hotel Gellért the following caption was published in India, 'Prof. V. Lesny receives the poet in Prague with Prof. and Mrs. P. C. Mahalanobis: 1926.'[111] The idea that Rabindranath visited Hungary several times returned again and again.[112] Commemorating the centenary of the poet's birth, a journalist wrote: 'In 1937 he visited Balatonfüred again and afterwards he spent one and a half days in Budapest. Then he came to visit the Puszta in Bugac in the Great Hungarian Plain. He came to our town (Kecskemét) on 16 July in 1937.'[113] After his journey to Sri Lanka in 1934, Tagore never left India. The incentive to this writing may have been a visit to the Csárda of Bugac by a rich Indian.

Another focus of the attention of many journalists was what the poet was writing in Hungary. The reporter of the *Esti Kurir* claims to know on the day following his arrival that, 'Rabindranath Tagore today wrote a poem about Budapest, the Danube and a small boat carrying vegetables.'[114] A journalist of the daily *Az Est* wrote that Rani Mahalanobis showed him Tagore's notebook containing a twelve-line poem dated the same day.[115] Another report wrote that the poet spent his first morning in Balatonfüred 'Writing poems about the beauties of Lake Balaton.'[116] Yet another piece of news tells us that in Balatonfüred Tagore was writing a drama:

> His most recent and perhaps his greatest drama is being written here… It is supposed to shed light on the blighted economic situation of post-war Europe… He points out (in a frame of an exciting plot) the desire for enormous wealth, treasures, new territories and conquests that rages in the Europeans… The accumulation of goods will never end because it is without rest and reason.[117]

Motives for Travel

The poet Kosztolányi was curious about the reason for Rabindranath's travels and offered a solution that had nothing to do with politics but with understanding the psychology of the poet. 'What could have made him, a sick man at the age of sixty-five, come here from the banks of the Ganges and decide to play the tiring role of a poet-prince in distant countries... In all probability he does this because he likes it. There should still be a lot of vitality in his well-made body.'[118] Nevertheless sometimes it is difficult to understand Rabindranath's decisions during his visit. If he felt weak, could he not have refused the tiring programmes?

This tractability of Rabindranath did not affect his discrimination since he did not invite Zajti, who was eager for an Indian journey to Santiniketan, but the more talented Julius Germanus. Probably he was amenable on his journeys because he did not want to pay so much heed to what was happening to him. This enabled him to find values regardless of organised programmes and it enabled him to recognise Hungarians as a distinct colour 'in the grey monotony of western civilisation'. He had a disposition to find advantages even in the inconveniences of a journey.

It was also Rabindranath's deep desire to experience expressions of humanity in different people all around the world. According to P. C. Mahalanobis Rabindranath used to say, while abroad: 'When people in foreign countries come to me, give me things out of love or do me some service, then I realize in the deepest way that I am a human being, and my human life has a meaning.'[119]

Among the inconveniences of celebrity and the innumerable cases of misunderstanding there were also some spontaneous moving events during the journey. In many instances meeting Tagore brought out the best in those who saw him and the dividing line between East and West was dissolved on the ground of common humanity as in the case of the Gipsy group leader Béla Radics's visit to the sick poet:

> That day when the poet fell ill and Baron Korányi gave an instruction in the papers that nobody should visit the poet, I was sitting alone in the antechamber of the poet's bedroom. Then the Gipsy King, Béla Radics entered and said, 'Excuse me, I did not come to disturb the poet but for his nursing.' When I did not understand what he meant, he said somewhat embarrassed but smiling gently towards me, 'During the war I was travelling with the soldiers in different hospitals. Where there were some suffering moribunds, they called me to play and give them some strength. . . . When I read in the newspaper today that our Tagore was tired and ill and Baron Korányi had cancelled all his engagements and ordered him to lie in bed and forbidden anyone to visit him, then I brought my violin and came here. I am not going

to the poet's room. I will play my violin to him from this neighbouring room for a few minutes and then take my leave. I am sure that this violin will make him feel better and he will sleep well at night. Saying this, he took his violin from its case and began to play. I told the poet who was playing and why. When he was asked to come to his room, the Gipsy King did not move at all saying that the doctor had forbidden it. He played near the door for about an hour and then as he came so he left without any notice. Before leaving, he said, '. . . . A respected guest such as Tagore fell ill in our country and all the doctors joined hands in healing him. If a common person like me could give some help in his nursing then what better fortune can I have. I am not able to give anything else. I heard that he loves music. I have offered him my best today.[120]

In summing up the results of the visit, Tagore himself and Rani Mahalanobis in their writings are full of praise for the country. After the visit Tagore cherished good memories of Hungary. He also found favourable circumstances to finish a remarkable book of poems and to write some new songs as well. The stay in Hungary also prepared the ground for Hungarian artists and scholars to visit Santiniketan.

It is not only the small town of Balatonfüred that keeps alive the memory of Rabindranath Tagore's visit to Hungary, but when Indian artists and politicians visit Hungary they also make a trip to the place associated with Rabindranath to pay their tribute his memory. In this way the fortnight Tagore spent in Hungary became one of the most important points of reference in the history of Indo-Hungarian contacts.

Notes

(Abbreviations in the notes: BpH.=Budapesti Hírlap; MH.= Magyar Hírlap; Mo.= Magyarország; NU.=Nemzeti Újság; PH.=Pesti Hírlap; PN.=Pesti Napló. When no year is mentioned after the names of journals and newspapers, then the year of publication should be taken as 1926.)

1 This article is a slightly revised version of the chapter 'The Visit of a Poet-Prophet in Hungary' in my book, *Hungry Tiger* (New Delhi: Sahitya Akademi, 2007), pp. 120-184.

2 'Balatoni Krónikás' 1943 p. 323–7

3 Roy 1959–60 pp. 28–34.

4 The diary and letters have later been published as a book (Mahalanobis 1969). Parts of this translated into Hungarian by Eta Ghose are published in Lipták 1961 pp. 35–47. See also Kshitish Roy: 'Tagore in Hungary.' *Indian Literature* III/1 (1959–60) pp. 28–34.

5 Lipták 1961 p. 37.

6 The Bengali translation of this volume is: Roy 1984. This work served as a basis for Pal 1991

7 See next note and *Pester Lloyd* 9 July.

8 This letter, today in the Rabindra Bhavan Archives, refers to an article published on the previous Sunday, which I was not able to trace.

9 Kun 1926 p. 1.

10 Kun and the daily *Magyar Hírlap* 19 Oct. mentioned the secretary of the *'Koncert' Hangversenyvállalat* (Kun 1926 p. 2.), though Kun did not remember his name. It is strange that the director did not know the name of his secretary. Probably Kun did not remember well and it was not the secretary but Huszka. See also *Mo.* 19 Oct.

11 *Mo*. 15 Oct.

12 *BpH.* 24 Oct.

13 *Mo.* 20 Oct.

14 *PN.* 27 Oct. See also *NU.* 27 Oct.

15 *PN.* 31 Oct. See also *PN.* 7 Nov.

16 *PN.* 31 Oct. See also *PN.* 7 Nov.

17 Kun 1926 p. 3.

18 *PN.* 27 Oct and *MH.* 27 Oct.

19 *BpH.* 27 Oct. and *Magyarság* 27 Oct. Actually his long beard attracted special attention. Later a nursery rhyme was published in a caricature with a perch in Lake Balaton terrified by Tagore's beard. See *Friss Újság* 7 Nov.

20 *Magyarság* 27 Oct.

21 *Magyarság* 27 Oct.

22 *Esti Kurir* 28 Oct.

23 *Az Est* 28 Oct.; *Mo.* 28 Oct. and *PH.* 29 Oct.

24 Mukhopadhyay and Roy 1961 p. 485.

25 Home 1941 p. 92.

26 Kun 1926 p. 6.

27 *MH.* 28 Oct.

28 *BpH.* 28 Oct.

29 *PH.* 29 Oct.

30 *MH.* 28 Oct.

31 *Mo.* 29 Oct.

32 *PH.* 28 Oct.

33 *MH.* 28 Oct. and *PN.* 28 Oct.

34 *BpH.* 28 Oct.

35 *PN.* 28 Oct.

36 Kun 1926 p. 6.

37 *Az Est* 29 Oct. Later Rabindranath sent an Indian gift to this studio. (*Hétfői Napló* 2 Nov.)

38 *Mo.* 29 Oct. See also the following note.

39 *Esti Kurir* 29 Oct.

40 Letter of Nirmal Kumari Mahalanobis to Prabodh Chandra Mahalanobis from Budapest dated on 5–8 November. See Mahalanobis 1969 p. 226.

41 Roy 1959–60 pp. 29–30. According to Kshitish Roy the present is preserved in the 'Rabindra Sadana (Tagore Museum)' in Santiniketan.

42 *Magyarság* 29 Oct.

43 *PN.* 29 Oct.

44 *BpH.* 29 Oct.

45 *Mo.* 30 Oct. This exaggeration may have been influenced by the contemporary misapprehension in Hungary that in India women dance naked. Perhaps the uncovered belly or back was considered indecent.

46 *PN.* 29 Oct.

47 Ibid.

48 Letter of Nirmal Kumari Mahalanobis to Prabodh Chandra Mahalanobis from Budapest dated on the 31 Oct. See Mahalanobis 1969 p. 218.

49 *PH.* 31 Oct.

50 *Új idők.* 28th Dec 1941.

51 This letter of Charles Fábry dated 13 Oct. 1926 is in the Fábry-file of the Rabindra Bhavan Archives in

Santiniketan.

52 This letter of Charles Fábry dated 2 Nov. 1926 is in the Fábry-file of the Rabindra Bhavan Archives in Santiniketan.

53 The song is Nr. 603 in the *Pūjā* section of *Gītabitān*. (Tagore 1973/74).

54 Translation mine

55 *Az Est* 3 Nov.

56 *PH.* 31 Oct.

57 *Esti Kurir* 3 Nov.

58 Ibid.

59 *Esti Kurir* 3 Nov.

60 *Esti Kurir* 3 Nov.

61 *PN.* 7 Nov. and Letter of Nirmal Kumari Mahalanobis to Prabodh Chandra Mahalanobis from Budapest dated 5–7 Nov. See Mahalanobis 1969 p. 226. Rani Mahalanobis's letter refers to the tree-planting as 'today' under the heading of the 5th. However the inscription at the tree erroneously gives the date as the 8th.

62 Letters written on 15 May from Bombay and on 19 May on board to Eden. See Saha 1984 p. 248.

63 Roy 1959–60 p. 31.

64 Ibid pp. 33–34.

65 'Balatoni krónikás' 1943 p. 324.

66 *PN.* 7 Nov.

67 Both versions are quoted under the date of 7 November in the letter of Nirmal Kumari Mahalanobis to Prabodh Chandra Mahalanobis from Budapest dated 5–7 Nov. See Mahalanobis 1969 p. 221–231. A photograph of these poems in handwriting is published in the same book (no page number). Here also 7 November is given as the date of composition.

68 Letter of Nirmal Kumari Mahalanobis to Prabodh Chandra Mahalanobis from Budapest dated 5–7 Nov. See Mahalanobis 1969 p. 225.

69 'Balatoni krónikás' 1943 pp. 323–327.

70 This song is Nr. 196 in the *Prem* section of *Gītabitān* (Tagore 1996 p. 349).

71 Mahalanobis 1969. The photo of the manuscript of these poems is included in the book without page number.

72 Translation mine

73 Wojtilla 1991 (Tagore balatonfüredi írása) pp. 189–194.

74 Ibid. pp. 192–3.

75 Tagore 1928.

76 Das 1994 pp. 616–7.

77 Lipták 1961 p. 44.

78 *Keszthelyi Hírek* 14 Nov.

79 Letter of Nirmal Kumari Mahalanobis to Prabodh Chandra Mahalanobis from Balatonfüred dated the 5–7 Nov. See Mahalanobis 1969 p. 229–230.

80 Lipták 1961 (Rabindranath Tagore) pp. 233–235.

81 *Keszthelyi Hírek.* 14 Nov. 1926.

82 *Mo.* 3 Nov.

83 E.g. 'Recuperating in Hungary' *Dundee Courier and Advertiser.* 10 Nov.; 'Poet's Breakdown' *Daily Record and Mail.* (Glasgow) 10 Nov.; 'Better' *Daily Express* 10 Nov.; 'Dr. Tagore in Hungary' *The Observer.* 14 Nov. p. 12; 'Tagore in Hungary' *The Inquirer: Journal of free religious fellowship and organ of Unitarian Christianity* (London) 20 Nov. (Cuttings of these articles are found in the '1926 — the Italian tour' file of cuttings at the Rabindra Bhavan Archives.)

84 *The Observer.* 14 Nov. This report seems to be erroneous since these books are not in the library of the Visva-Bharati University.

85 Mo. 11 Nov. See also Mahalanobis 1969 p. 231.

86 *Az Est* 12 Nov. and *Pester Lloyd* 11 Nov.

87 Thompson 1979 p. 286–287.

88 *Az Est* 12 Nov.

89 Letter to Edward Thompson dated 12 December 1926. Quoted in Thompson 1979. p. 276.

90 Home 1941 p. 92.

91 *Az Est* 12 Nov.

92 Letter in the P. C. Mahalanobis file of the Rabindra Bhavan Archives.

93 Mahalanobis 1969. The photo of the manuscript of both versions of the poem is included in the book without page number.

94 *Az Est* 12 Nov.

95 *Literatura* Aug. 1928.

96 Lipták 1961 (Rabindranath Tagore Balatonfüreden) p. 44.

97 *Az Est* 17 Nov. 1928

98 *Mo.* 3 Apr. 1930.

99 *Az Est* 17 May 1930.

100 Lipták 1961 (Rabindranath Tagore Balatonfüreden) p. 46.

101 One such instance is described in Geszty 1937 pp. 41–44. The author met Rabindranath in Calcutta on 5 March, 1933.

102 Baktay 1932 pp. 113–114.

103 Press in Hungary includes the German newspaper of Pest, the *Pester Lloyd*, which reported about the poet every day between 27 and 30 Oct.

104 *PN.* 31 Aug. 1930.

105 *NU.* 27 Oct.

106 *Esti Kurir* 29 Oct.

107 *Az Est* 29 Oct.

108 See previous note as well as *Magyarország* 29 Oct. and *Az Est* 29 Oct.

109 *Magyar Hírlap* 19 Oct.

110 *Magyarország.* 29 Oct.

111 Home 1941 p. 93.

112 'He also visited Hungary several times.' See *Művelt Nép* 13 May 1956. 'He came to Hungary in 1925 and returned the next year.' See *Nagyvilág* May 1961 p. 726.

113 *Petőfi Népe,* 6 May 1961.

114 *Esti Kurir* 28 Oct.

115 *Az Est* 31 Oct. p. 7. This poem must have been the *Diner belāy baśi tomār...* written on 30 October.

116 *Esti Kurir* 3 Nov.

117 *Mo.* 3 Nov. The description resembles the plot of the play *Raktakarabi* (Red Oleanders) written in 1923 but never translated into Hungarian.

118 Kosztolányi 1926 pp. 782–4.

119 Quoted in Mahalanobis 1985 p. 160 (Bengali section). On p. 47 (English Section) an inadequate English translation of this passage is found.

120 Letter of Nirmal Kumari Mahalanobis to Prabodh Chandra Mahalanobis from Budapest dated 7 Nov. See Mahalanobis 1969 p. 227–228.

References

Das, Sisir Kumar (ed.): *The English Writings of Rabindranath Tagore (Volume One: Poems)* (New Delhi: Sahitya Akademi, 1994).

Dutta, Krishna & Robinson, Andrew: *Selected letters of Rabindranath Tagore.* (Cambridge: Cambridge University Press, 1997).

Dyson, Ketaki Kusari & Adhikary, Sushobhan: *Ronger Rabindranath* (Rabindranath in Colours). (Calcutta: Ananda Publishers, 1997).

Geszty, Júlia: *'Rejtelmes Kelet – Egy magyar leány utazása Indiában, Sziámban, Jáva szigetén, Kinában, Japánban, Koreában, Mancsukóban.'* (Mysterious East – Travels of a Hungarian Girl in India, Siam, the Island of Java, China, Japan, Korea and Manchuria) Singer és Wolfner, Budapest, [1937].

Home, Amal: 'Rabindranath Tagore – A Chronicle of Eighty years: 1861–1941 (Eighty illustrations).' In. Home, Amal (ed.): *The Calcutta Municipal Gazette — Tagore Memorial Special Supplement.* 13 Sept. 1941 (reprinted in 1986) pp. 59–110.

Kosztolányi, Dezső: 'Érzések és gondolatok egy estélyen' (Impressions and thoughts in a soirée). *Nyugat* 16 Nov. 1926 pp. 782-4.

Kun, Imre: *The Nobel-Prize Winning Poet Rabindranath Tagore: Guest Appearance in Hungary.* A report in English in the Archive of the Hungarian Academy of Sciences MS 5137/25. [1926?]

Lipták, Gábor: (Rabindranath Tagore). *A Könyvtáros* 1961/5 pp. 233–235.

Lipták, Gábor: 'Rabindranath Tagore Balatonfüreden' (Rabindranath Tagore at Balatonfüred). In. Lipták & Somogyi 1961 p. 35–47.

Lipták, Gábor & Somogyi, Marianna: *Rabindranath Tagore.* Hazafias Népfront, Veszprém, 1961.

Mahalanobis, Nirmal Kumari: *'Kabir Sange Yurope'* (In Europe with the Poet). (Calcutta: Mitra o Ghosh, 1969).

Mahalanobis Prashantachandra: *Prasanga Rabindranāth.* Mahalanobis Trust, Calcutta, 1985

Pal, Shankar Narayan: 'Prabāse Rabīndracarcā: Parba Hāngerī' (The study of Rabindranath abroad: Chapter on

Hungary). *Korak (Sāhitya Patrikā)* Śāradīy 1398 (1991) pp. 44–54.

Pekár, Gyula: 'Rabindranath Tagore.' *Új Idők.* 14 Nov. pp. 564–5.

Peteanu, Annamaria: 'Hāngeriyān jātir janma ki Bhārate?' (Do Hungarians originate from India?) *Deś* 27 Dec. 1980 pp. 27–29.

Petrovic, Sveto: Tagore in Yugoslavia. *Indian Literature* XIII/2 (1970) pp. 5–29.

Radice, William (transl.): *Tagore: Particles, Jottings, Sparks: The Collected Brief Poems.* (Angel Books, 2001).

Roland, Romain: *Inde journal (1915–43). (*Paris-Lausanne-Bâle, Editions Vineta, 1951).

Roy, Kshitish: 'Tagore in Hungary.' *Indian Literature* III/1 (1959–60) pp. 28–34.

Roy, Vishvajit (transl.): *Hāngerīte Rabīndranāth.* (Calcutta: Navayug Publishers, 1984).

Saha, Panchanan: *Rabindranath and Germany.* (Calcutta,:Indo-GDR. Friendship Society, 1986).

Szász, Zoltán: 'Tagur.' *Pesti Napló* 30 Oct. 1926 p. 7.

'The Tagore-Elmhirst correspondence 1921–1941.' In. Krishna Dutta and Andrew Robinson (eds.): *Purabi: A miscellany in memory of Rabindranath Tagore 1941–1991.* (London: The Tagore Centre UK, 1991) pp. 72–126.

Thompson, Edward: *Rabindranath Tagore: Poet and Dramatist.* (Calcutta: Riddhi-India, 1979).

Wojtilla, Gyula: *Rabindranath Tagore in Hungary.* (New Delhi: Hungarian Information and Cultural Centre, 1983).

Wojtilla, Gyula (transl.): 'Rabindranath Tagore: Az irodalom dicsősége' (the Glory of Literature). *Veszprémi Napló* 3 Aug. 1991 p. 6.

Wojtilla, Gyula: 'Tagore balatonfüredi írása' (Tagore's writing from Balatonfüred). In. Várhegyi, Miklós és Kőszegi, Lajos (eds.): *Tagore: Szentjánosbogarak.* Veszprém Megyei Múzeumi Igazgatóság, Veszprém, 1991 pp. 189–194.

Zsolt, Béla: 'Tagore bukása' (Tagore's failure). *Magyar Hírlap* 14 Nov. 1926 p. 9.

Tagore's Bulgarian Birthday

Alexander Shurbanov

Rabindranath Tagore visited Bulgaria as part of an extensive European tour in late 1926. He spent two days, 17th and 18th November, in Sofia, where he met writers, artists, intellectuals and translators of his work, gave two talks about contemporary culture and had a reading of his poetry in a crowded theatre hall. Then he passed through Varna on the Black Sea coast and spent a night in the Danubian city of Ruse, where he had further meetings before leaving the country.

In that troublesome time for Bulgaria – in the wake of a coup d'etat and a great popular uprising in 1923, followed by a dastardly terrorist act at the heart of Sofia and a no-less cruel subsequent police repression against a wide range of left-wing activists and intellectuals –Tagore's visit turned into a great event in the spiritual history of the nation. His name and his work were not unknown to Bulgarians and his appearance in person was seen as almost a miracle promising a brighter future for the nation. It has remained a living memory to this day. This is how a contemporary writer recorded his impressions and sentiments directly after the event:

> Tagore is coming to Sofia! Isn't that truly incredible? The great spiritual leader of three hundred million Indians, the sage from the fabulous land of the jungles and the holy Ganges, the man who revealed in his poems the contemplative, perceptive Indian soul – he is now coming to Sofia. We, a group of messengers from the Arts Club and the press, went to meet him at the border. Here we are at the Pirot railway station.[1] We step into the saloon-car and suddenly we stand face to face with the sage-poet. His dark eyes met ours and seemed to unveil our souls. Then he lowered his eyes and smiled with the kind of smile that gives out love for humanity. The sage had won our hearts – they sensed his gentleness and kindness and surrendered to him...
>
> The poet may be tired, but his modesty will never say no—he agrees to be photographed, signs books and pictures handed to him. He is moved by the welcome of the rejoicing multitudes of Sofia. This outburst of affection for a poet is for him a sure sign that our people can hope for a bright future... Tagore has left our country. But his presence will remain for long impressed on our souls. Sofia has lived through two days of spiritual jubilation: such a stir in all circles we haven't experienced for years... Wisdom and beauty have been our guests and they have made us remember that we are human beings."[2]

The author of these emotional lines, Dr. Asen Zlatarov, was a central figure of the Bulgarian intellectual and artistic life at the time. Professor of Chemistry and Physics at Sofia University, as well as a poet and novelist, he was above all a fiery and brilliant orator, a cultural educator and a popular tribune who fought bravely against the political evils of the age: dictatorship, violence, racism, fascism and war.

A book published the next year and prefaced by Zlatarov was still full of the excitement of the event.[3] Its author, Vasil Stavrev, who had been another witness and participant in the improvised Tagore mela in Sofia, attempts a sketch of the poet from life:

> In order to make the features of this extraordinary man stand out more clearly, I shall take the liberty of saying a few words also about his so very interesting outer appearance.
>
> A tall, sturdy old man who, in spite of the great exhaustion of his constant travels, never stoops his shoulders and, like some Colossus, always towers above those around him. Long white hair crowns his broad and tall forehead like a halo. The features of his face are composed and absolutely regular—the features of a truly Aryan face. His dark coppery complexion, more yellowish than reddish, is characteristically offset by the surrounding whiteness of the hair and the long beard of a Biblical prophet, which completes his amazingly handsome portrait. Yet, the deep intelligent eyes piercing the bottom of one's soul, are the most striking part of this face. How penetrating they are, those dark, large eyes!.. And his mild, kindly smile too, as pure as that of a child! Isn't it the best expression of the childlike purity of his soul? And, lastly, his voice! If it is true that the voice can sometimes be the best interpreter of the inner spiritual manifestations, it is Tagore's voice that is the surest proof of that. The poet has such a sweet and melodious voice that even the commonest prose acquires with him the tone of a divine chant...
>
> What wonder then that, endowed as he is with all these rare qualities, dressed in his long attire, brownish-yellow in everyday life and ashy-gray on solemn occasions, he should carry, wherever he goes, fascination and joy? What wonder that, long after seeing him, you should continue to live with the beautiful memory you have kept of him?

Stavrev goes on to express the gratitude of the Bulgarian public for the warmth with which Tagore responded to their enthusiasm and the new confidence which this rapport inspired in the people at a dark moment in their history:

> In his peregrinations, he did not forget about our small country too. The readers know—whether from the papers or from personal participation in the festivities marking his arrivals and departures in Sofia and Ruse—what jubilant and cordial welcome he received here. According to the assurances of his companions and his own admission, he had seldom had such a warm reception as the one given him in Bulgaria.
>
> As a matter of fact, Tagore's own farewell words at the end of his last public lecture in Sofia bear out this assertion with characteristic eloquence. He is recorded to have said: 'Here, in your country, I am celebrating my birthday once again, for it is here that my ideals have found their full expression. I

have discovered my dwelling-place in your hearts. Now, that I am on my way, let me bid you my hearty farewell. I feel sad—it is painful—but I am taking away dearest memories of you. I am leaving you with the best of feelings for the Bulgarians, for the Bulgarian people.'[4]

The best feelings were on both sides. Tagore's visit was a revelation for his hosts—a myth had suddenly materialised in front of them. The reaction of the press was unprecedented. Over sixty news-items, stories and articles related to the occasion came out in the six weeks between the poet's arrival and the end of the year. Some periodicals dedicated special issues to this event. The culmination came with two books about Tagore written by Bulgarian admirers of his work.

The first of these, *Rabindranath Tagore's Wisdom*,[5] published in the same year and reprinted in the next, is really an attempt to outline the writer's philosophy and represents an enthusiastic reaction to his treatise *Sadhana*, that had recently appeared in Bulgarian translation,[6] as well as some other works such as *Nationalism*,[7] with which the readers were already familiar. The author of the book, Vicho Ivanov, a respected writer, critic and art historian, says that Tagore's doctrine, inherited from the ancient Indian philosophy, is 'a teaching of simple unity with the universe, achieved not through the external forms of cult and ritual, but through a love of life and the world.' The conclusion Ivanov draws from his survey is characteristically topical:

> In an age when people persecute and kill each other ruthlessly, when the stifling of the loftiest human aspirations has turned into a horribly destructive system, Rabindranath Tagore's appeal for peace and brotherhood in the world is a genuine revelation of man's still un-extinguished conscience. It is for this conscience that the great Bengali bard summons our love, for it is the way, the truth and the life.

Vicho Ivanov's book has an introduction written by the eminent poet, prose-writer, critic and translator Lyudmil Stoyanov, whose early symbolism had gradually given way to a more realistic mode of writing. This essay, entitled 'Rabindranath Tagore: A Teacher of Life,' makes the point that the aim of the poet's teaching is not a mere passive contemplation of the kind we have learned to associate with Eastern thought but an active participation in life, something that brings Tagore's philosophy close to the best traditions of the West. For 'he rates freedom higher than goodness and preaches not a denial of the world and Nirvana, but the affirmation of the world in joy.'

The second book, *Rabindranath Tagore: Life and Work*[8], written by Vasil Stavrev with an introduction by Dr. Asen Zlatarov and published in 1927, is the one I have already quoted from in connection with Tagore's visit and the immediate impression it produced on the Bulgarian intellectuals of that time. It is, in fact, a rather ambitious attempt to outline the entire development of a man of genius from early childhood into the mature years of creative work and then to the profound wisdom of old age. While the biographical materials are drawn from the poet's book *My Reminiscences* in its German translation, and the analysis of his major writings leans rather heavily at times on Heinrich M. Benfey's book on Tagore as well as some other sources in German, there are, in this thoughtful survey, some original conclusions as well, and it must be given the credit of being the most exhaustive account of Tagore's fascinating life and work that has ever appeared in

Bulgarian. An important asset of the book are its images - portraits of Tagore's father and himself at the age of sixteen as drawn by his brother Jyotirindranath, six photographs of the poet including a very good one, signed by Tagore himself and showing him in a dark fur cap, Tagore loaded with flowers during his sojourn in Sofia, two more views of the original school buildings at Santiniketan and a facsimile of a page from *Gitanjali* in the author's handwriting in Bengali.

Stavrev argued convincingly that, while contemporary European poetry had lost its original connection with music and had turned into a bookish art for the elect, Tagore's lyrics were sung throughout Bengal and had thus transcended the limits of personal creation, and turned into something like folk poetry. A similar contrast was drawn between the highly specialized, exclusive character of some aspects of Western philosophy and Tagore's teachings, which could justly be called 'a philosophy of life,' for they were closely connected with man's everyday existence and could play an important role in his moral perfection. Stavrev further paraphrases Benfey's interesting observation about the contrast between the prevalent 'idealism' in Tagore's poetry and the 'realism' of his prose. But he concludes, quite reasonably, that the idealistic vision of the poet is alive at the core of his novels as well despite their circumstantial life-likeness, and that was what endeared his prose to the Europeans. For Tagore was 'in the true sense of the word, a universal apostle of humanism in a dehumanized mercantile age.' The author closes his inspired study with the following outburst of genuine admiration:

> In general, Tagore is a rare phenomenon of our time, an uncommon personality, one of those great seers who are born once in a century to illumine the way of mankind with the light of the spirit and spiritual perfection, to raise it above the material poverty of the earthly and lead it towards the ideal purity of the heavenly.
>
> Having crossed the frontiers of his great and boundless land, he belongs already to the whole of mankind and, like an ancient prophet holding the sceptre of truth high above his grey head, he is calling us towards a loftier life, towards a more elevated kind of humanity, of which we, contemporary men, are in such need.

The impressive wave of Bulgarian writings about Tagore in 1926-1927 is but one aspect of the remarkable increase of interest in him triggered by his visit. We cannot also help noticing the equally striking upsurge in the publication of Tagore's literary and non-literary works in translation. More than fifty items appeared in the press.

The books were also numerous. *Sadhana* and *Nationalism*, upon which Vicho Ivanov commented in his 1926 book, had appeared in separate volumes shortly before. The translation of *Sadhana* belonged, in fact, to Vasil Stavrev, and it was apparently this book that inspired him to attempt his own comprehensive account of Tagore's life and work in the next year. The introduction to this work is written by Nikolai Rainov, the translator and commentator of *The Home and the World*. Rainov was a writer of poetry, prose and drama, literary critic and historian of art, philosopher and painter, a man indeed of almost as many talents as Tagore himself and keenly interested in the thought and culture of the East. He did much to recommend and introduce Tagore to Bulgarian readers. *Kabir's Poems*[9] and a small selection from *Whispers of the Soul*[10] were also released in 1926. And the next year saw *My School & The Woman*[11], *My Teaching,*[12] the first Bulgarian version of *Gora* from the German,[13] a collection of nine short-stories, named after one of them, *The Night of Coming True*[14] and prefaced by Dr. Asen Zlatarov again, new translations of *The Home and the World*[15] and *Gitanjali*[16] from the Russian as well as new editions of the earlier translations of both *Gitanjali* and *The Gardener*. It was also in 1927 that Ivan Altimirski turned for the first time to Tagore's English versions of his own poetry in order to render *The Grape-Gathering*[17] in Bulgarian from a text that had come from the poet's hand. Another translator, Maria Tsanova Stefanova followed suit with *The Crescent Moon/ Sishu*[18] an edition which has the added value of presenting eight illustrations by Nandalal Bose, Ashit Kumar Haldar and Surendranath Ganguli as well as a foreword of Stefanova telling us, in almost religious terms, about her meeting with the poet in Sofia.

As I have hinted earlier, Rabindranath Tagore had already been well domesticated in Bulgaria when he visited the country. For almost a decade and a half, his works had been recreated by translators and his exceptional personality commented upon with reverence. Tagore's fame first reached Bulgaria in 1913, when the press announced that he was the new winner of the Nobel Prize for literature. On that occasion, the authoritative journal *Suvremenna misul* published a biographical note containing some information about his writings as well as a brief appraisal of their character:

> Tagore is an optimistic poet, a poet of love and the joy of life. The basic mood of his songs is a cheerful pantheism. His verse is remarkable for its inherent simplicity and sincerity. It is permeated by a deep religious feeling, yet the object of this feeling is not outside nature but in nature itself, in its life, in the labour of man.[19]

A number of Tagore's poems from *Gitanjali* and *The Gardener* in Bulgarian translation appeared in the next few issues of the same journal. Other magazines, such as *Vecherna Poshta* and *Az Znam Vsichko,* had already published a couple of *Gitanjali* poems almost simultaneously with the announcement.[20] The next few years saw dozens of publications of this kind, and among the translators during that early period we come across the name of one of our first-rate poets, Nikolai Liliev, whose selection of three *Gitanjali* pieces (64, 78 and 94) can be found in the avant-garde *Vezni* magazine of 1919.[21] The same year is highlighted by the first separate book of Tagore's poetry in Bulgarian. This is the complete text of *The Gardener,*[22] translated from an Italian version, which, according to the translator's introductory note, had the sanction of the author. This pioneering edition is the work of C. D. Maksimov-Mircho, a leftist journalist who also wrote some short-stories and poems. In the already-quoted note, he looks down upon the preceding Tagore translations in the periodicals as 'several small fragments done for the most part from poor Russian translations.' But whatever the intermediates, the end product is not very different in quality.

The first Bulgarian edition of *Gitanjali* was brought out in the next year, 1920.[23] It was translated by another poet and journalist, Metodi Vecherov, again from the Russian. This translator also has an introductory note, which says among other things:

> 'Rabindranath Tagore writes music with his words. The variety and freedom, the originality of his experiences and his spontaneity are close to the people: they do not appear strange to anyone and are in no need of defence. And a whole nation, a whole civilization, so alien to us, comes alive in these visions together with the poet.'

The next important step towards the Bulgarianization of Tagore was taken in 1922 with the translation of his novel *Ghare Baire/ The Home and the World,*[24] done from the French version of the book by Nikolai Rainov. By the time the novel came out, the author's name was widely known in Bulgaria—some forty publications of his poems and short-stories including excerpts from *The Home and the World* had made their way into various periodicals. That early phase also saw a dozen articles about Tagore's astonishing range of activities and achievements. But Nikolai Rainov's rather lengthy introduction to his translation is of special value, not only because of the intriguing personality of its author, but also because of its thoughtfulness and depth of appreciation. Besides the detailed analysis of the novel from a psychological, social and philosophical point of view, the translator offers cogent comments on Tagore's other major works in the different genres, and makes some observations of the following kind:

> Can the portrait of such a many-sided genius be completed in a few pages? The harmoniously-developed personality is unknown to the West and that makes Tagore a miracle in our century. There is in this novel, as well as in Tagore's poems, the bright effervescence of a joy...which one experiences in the presence of cosmic life—that eternal, uninterrupted, inexhaustible essence striving towards some great aims, which are not always transparent to human intellect. This intimation of the universal life permeating man, the earth, every common weed and creature, is the keynote of Tagore's entire work...It is of this unity of the human soul with the world and the universal soul—and of this supreme love—that Rabindranath Tagore sings. And he sings as no one before him has ever done.

Rainov was not alone in his enthusiasm. In the next three years a number of Bulgarian journals showed an increasing interest in Tagore's work. Over thirty new publications appeared, a considerable part of which were non-fictional: pamphlets, articles, essays and speeches dealing directly with the philosophical, ethical, social and political matters of the contemporary world. This focusing on the writer's non-literary work is quite understandable if we remember that the period 1923-1925 was, as noted above, the time of great civil and political disturbances in Bulgaria and fraught with authoritarian tendencies. The democratic press was naturally looking to international authorities of Tagore's stature for support and encouragement. And, of course, translation is often called upon to do the work of suppressed journalistic protest in an age of growing terror and censorship.

Another proof of the rise of interest in Tagore can be found in the great flow of articles about him—their number in last three years was larger than that of the preceding nine. The slant of the new publications was more frequently ideological than aesthetic.

Yet, in spite of this general trend, attention to Tagore's literary work continued to grow steadily. The 1923-1925 period produced four new books: the dramas *Chitra* (1923)[25] and *Sacrifice/Visarjan*

(1924),[26] the anthology *Poems of Childhood* (1925)[27] and the novel *Chaturanga* (1925).[28] The last work requires special discussion, because it was translated by Konstantin Konstaninov, a writer renowned for his excellent style and a very successful translator of French and Russian fiction. This translation is itself based on the French version of Madeleine Rolland and includes Romain Rolland's introduction as well as an appended note about the development and growth of Tagore's novels. Here, it is said that the predominant influence until 1890 was that of 'the great Bengali writer Bankim Chandra Chatterji,' but this was overcome in the novels about the life of the peasantry and the middle class of Bengal written during the last decade of the century, after which came the mature social novels *Naukadubi/The Wreck*, *Chokher Bali/Eyesore*, *Gora*, *Ghare Baire/The Home and the World* and *Chaturanga*. The note closes with the statement that 'To this must be added a considerable number of comical sketches, still un-translated from the Bengali, which show Tagore as one of the major Indian humourists.'

The upsurge in the reception of Tagore's writings during his visit to Bulgaria and immediately thereafter did not die out in the following years. After the *Night of Coming True* collection of 1926, another bunch of short-stories came out in 1929 in a book called *Light and Shadows,*[29] and three years later the cycle of poems *Golden Boat/Sonar Tari*[30] followed. In the meantime, 1930 saw the full version of *Whispers of the Soul*[31] and the text of another drama, *Saniyasi*, or *The Ascetic.*[32] The press sustained a steady interest in Tagore during the thirties and the early forties, but no more books of his were published until the end of the Second World War if we do not count the reprint of the 1927 *Gitanjali*, which came out in 1942, probably as a homage to the poet immediately after his death.[33]

There have, of course, been no further surges of Tagore publications on the scale of the late twenties, but a new peak in his popularity was reached three decades later, when Bulgaria started preparing for the celebration of the poet's centenary in 1961. A new novel, *The Wreck/ Naukadubi,*[34] was added to those already published in Bulgarian, in 1958. It was translated from the English by one of our most experienced translators, Sider Florin. The volume also included the memoir, *My Boyhood Days,* and the first selected bibliography of Tagore's writings available in Bulgarian. A new collection of a dozen short-stories under the already familiar title, *Light and Shadows,*[35] appeared in 1960. It was translated from a recent Russian edition by one of Bulgaria's finest poets and translators, Atanas Dalchev. The expert introduction of E. Borovik's was also borrowed from the source-book. The year of the centenary, 1961, saw a new, masterly translation of *Gora,*[36] done by Sider Florin from the English and accompanied with a special address to the Bulgarian readers by the then Indian Minister of Culture, Mr Humayun Kabir. And Tagore's *Russian Letters/Rashiar Chithi,*[37] which had created quite a stir in our press in the early thirties when they were written, were now translated anew from the authoritative Soviet edition of Tagore's collected works and published for the first time under a separate cover.

To sum up, on the eve of Rabindranath Tagore's 125th anniversary in 1985, Bulgaria had twenty-seven books of his most representative works, some of them published in more than one edition. The translation of his poetry covered two independent and complete versions of *Gitanjali*,

The Gardener, *The Fruit-Gathering*, *The Crescent Moon*, *Poems of Childhood*, *Kabir's Poems*, *Whispers of the Soul* and *Golden Boat*. Of the novels, *The Home and the World* and *Gora* had two different translations each; *Chaturanga* and *The Wreck* were also available as well as the memoir, *Boyhood Days*. The writer's dramatic legacy was represented by *Chitra*, *Sacrifice* and *Sanyasi*. Some two dozen short-stories appeared in three separate collections. There were, lastly, five non-literary books including the following works: *Sadhana*, *Nationalism*, *My School*, *The Woman*, *My Teaching* and *Russian Letters*.

Two Bulgarian books about Tagore's life, his thought and art, were an eloquent proof of the nation's serious interest in his inspired message to the human race. Over two hundred publications of the poet's works in the different fictional and non-fictional genres had found their way into various periodicals, and there was a similar volume of journalistic material dedicated to this great man, informing the readers about his amazingly varied activities and achievements as a writer, artist, musician, actor, philosopher, politician, educationalist, etc. Some of the most distinguished Bulgarian writers and translators had helped to bring Tagore close to the people by conveying to them the many enticing facets of his life and work. Some of them had shown a lasting attachment to the Prophet from the East, returning to him again and again in the course of their careers, drawing attention to the promises of his humanistic vision. One in particular, Vicho Ivanov, is the author not only of the book about Tagore's philosophy, but of twelve more publications dedicated to him and strewn over a period of nearly four decades. Nikolai Rainov also showed an abiding attachment to the writer.

And it was in 1985 that Bulgaria's largest and most renowned publishing house for literature in translation, Narodna Kultura, prepared a representative three-volume edition of Rabindranath Tagore's selected works.[38] These included two of his most popular novels: *Gora* (in Sider Florin's translation) and *The Home and the World* (in a new, third, Bulgarian version, belonging to the chief editor of the entire selection, Vladimir Ganev). The edition also contains a large body of over two dozen of Tagore's best short-stories, translated by Ganev again, as well as over five hundred poems, including the complete texts of *Gitanjali*, *The Gardener* and *Stray Birds* plus a fair sample of *The Fruit-Gathering*, *The Crescent Moon*, *The Fugitive* as well as other books and cycles, all translated anew by the writer of this paper. Three of Tagore's most successful plays – *Post Office/ Dakghar*, *Red Oleanders/ Raktakarabi*, and *The Chariot of Time/ Kaler Yatra*, *Rather Rashi* – are presented here in Bulgarian for the first time. The poet's *Reminiscences/ Jivansmriti*, his *Russian Letters/ Rashiar Chithi*, and some of his non-fiction writings are also included. Most of the materials are translated from English and some from Russian.

Vladimir Ganev is the author of an inspired introduction entitled 'Rabindranath Tagore: a Great Indian and World Writer.' But what makes the edition particularly valuable for us is its opening address to the Bulgarian readers by India's late Prime Minister, Indira Gandhi, written shortly before her tragic death.

There has certainly been no more ambitious and comprehensive attempt to offer a single body of the choicest Tagore writings in the different genres to the innumerable lovers of his genius in Bulgaria. As a jubilee edition, it marked the climax of a long process of accumulation that started more than seven decades ago, in the remarkable year 1913, when Europe became suddenly aware of a great presence in the East.

This process, of course, has not come to an end yet. Five years after the 125th anniversary boom, in 1990, two of the highlights of the jubilee edition, *Gitanjali* and *Gora,* came out once again in a handsome volume prefaced with an essay about Tagore's importance in the modern history of ideas and aesthetics, called 'The Other Name of India' and written by the translator of *Gitanjali*, who is also the author of this survey.[39] And more recently, in 2009, an impressive volume of Tagore's poetry was brought out by Zahary Stoyanov Publishers,[40] including the representative selection of the 1985 jubilee edition in combination with some exquisite engravings of Nikolai Rainov's on Eastern motifs.

In the meantime, Tagore's life and work has been represented and commented upon in a number of publications including an Indian travelogue titled *The Land of Living Eternity* (1990)[41] and the volume of a 1997 conference proceedings, *India in Bulgarian Scholarship.*[42] In 2008, Dr. Stefania Dimitrova's book *Rabindranath Tagore: The Mythical Sentinel*[43] appeared as a necessary follow-up after eighty years of Ivanov's and Stavrev's earlier attempts at an in-depth interpretation of the poet's life and work, endeavouring to take stock also of the by then long and eventful history of his reception in Bulgaria. I also received information about a book published almost simultaneously with Dr. Dimitrova's one, in India, and covering much the same topic. The title is: *Rabindranath Tagore and the Bulgarian Connection: Facts and Documents*, and it is the work of two Bulgarian expatriates with a long-standing interest in Tagore.[44]

Needless to say, 2011 gave a new boost to Bulgarian Tagoreana, with celebrations of the great Indian's 150th anniversary across the country but with two particular focal points marking his 1926 itinerary, Sofia and Varna. In both cities, solemn public events were organized along with speeches, research papers, poetry recitals and concerts. An exhibition of Tagore's paintings, sent by the Indian government, travelled to many cultural centres and was greatly appreciated everywhere. The Bulgarian Friends of India Club initiated a project of commemorating Tagore and his connection with this country by naming an alley after him in a central Sofia park and putting up a plaque on the public building where he gave his first talk during his unforgettable visit.

Perhaps the most telling testimony to Rabindranath Tagore's growing popularity in Bulgaria can be found on the internet where one can see how many people keep discovering Tagore's immortal legacy with genuine enthusiasm and admiration. They are impatient to quote and share their exciting encounters with his work. Tagore's poetry, above all, continues to inflame new readers every day, especially among the young. They seem to find in it that which is most necessary for survival and advance in today's troubled world – a faith in the resources of the human heart and mind, and a striving towards the dream of universal unity that stems from the depths of the age-old, ever-vibrant Indian civilization and permeates everything bequeathed to us by this extraordinary thinker and artist.

Notes

1 Pirot was the name of a city in Yugoslavia (Serbia) at the border with Bulgaria.

2 Dr. Asen Zlatarov's preface to Vasil Stavrev's book *Rabindranath Tagore: Life and Work* (1921) (see n. 3 below).

3 Vasil Stavrev. *Rabindranath Tagore: Zhivot i tvorchestvo* (Rabindranath Tagore: Life and Work), Sofia, 1927.

4 D. Boshnakov. *Rabindranath Tagore v Bulgaria* (Rabindranath Tagore in Bulgaria), Sofia, 1961.

5 *Mudrostta na Rabindranath Tagore*, Sofia, 1926.

6 *Sadhana: Putyat kum suvurshenstvoto*, Sofia, 1926.

7 *Natsionalizum*, Sofia, 1926.

8 See n. 3 above.

9 *Poemi na Kabir*, Sofia, 1926.

10 *Iz shepoti na dushata*, Lom, 1926.

11 *Moeto uchilishte. Zhenata*, Sofia, 1927.

12 *Moeto Uchenie*, Sofia, 1927.

13 *Buntovnikut Gora* (Gora the Rebel), Sofia, 1927.

14 *Nosht na Sbudnovenieto*, Sofia, 1927.

15 *Dom i Sviat*, Sofia, 1927.

16 *Gitanjali: Zhertveni pesnopeniya* (Gitanjali: Song Offerings), Sofia, 1927.

17 Grozdober, Sofia, 1927.

18 *Lunen Surp*, Sofia, 1927.

19 *Suvremenna Misul*, III, No. 9-10, 1913.

20 *Vecherna Poshta*, XIV, No. 31, 1913; *Az znam Vsichko*, No. 3, 1913.

21 *Vezni*, I, No. 5, 1919.

22 *Gradinaryat*, Sofia, 1919.

23 *Gitanjali*, Sofia, 1920.

24 *Dom i Sviat*, Sofia, 1920.

25 *Chitra*, Stara Zagora, 1923.

26 *Zhertvata*, Aitos, 1924.

27 *Poemi na detstvoto*, Sofia, 1925.

28 *Na chetiri glaca*, Sofia, 1925.

29 *Svetlina i senki*, Sofia, 1929.

30 *Zlatnata ladiya*, Sofia, 1932.

31 *Shepoti na dushata*, Sofia, 1930.

32 *Sanyazi/Asket*, Sofia, 1930.

33 *Gitanjali*, Sofia, 1942.

34 *Krushenie, Moeto detstvo* (Wreck, My Boyhood Days), Sofia, 1958.

35 *Svetlina i senki*, Sofia, 1960.

36 *Gora*, Sofia, 1961.

37 *Pisma za Rusiya* (Letters about Russia), Sofia, 1961.

38 *Rabindranath Tagore: Izbrani tvorbi v tri toma* (Rabindranath Tagore: Selected Works in 3 vols), Sofia, 1985.

39 *Gitanjali. Gora*, Sofia, 1990.

40 *Rabindranath Tagore, Dumi s rituma na vechnost* (Rabindranath Tagore, Words with the Rhythm of Eternity), Sofia, 2009.

41 Alexander Shurbanov, *Zemyata na zhivata vechnost*, Sofia, 1990.

42 *India v bulgarskata nauka*, Sofia, 1997

43 Stefania Dimitrova, *Rabindranath Tagore: The Mythical Sentinel*, Sofia, 2008.

44 Anna Nikolaev and Nikolay Nikolaev, *Rabindranath Tagore and the Bulgarian Connection: Facts and Documents,* Visva-Bharati Publishing House, Kolkata, 2009.

Rabindranath Tagore and Angelos Sikelianos – Two Universal Thinkers: A Comparative Study

Andreas L. Katonis

I have preferred to use the words 'thinkers': though Tagore and Sikelianos may be familiar to us more as national poets and writers, both are something more. They are comparable, and their work and their universality connect them. The Greek Indologist and historian of philosophy Vassiliades remarked that they shared similar inspirations. Both of them founded, e.g., spiritual movements to promote global understanding (2000: 174). The two lives (Tagore: 1861-1941, Sikelianos: 1884-1951) ran, indeed, almost in parallel. Tagore, to be sure, is better known worldwide; he came from a big country and he travelled a lot, whereas Sikelianos' Greece was and remains small, and if the country is familiar, then its reputation is generally reserved for Ancient Greece. This was, in a sense, also the concern of the Greek thinker helped by his first, American wife: the revival of certain aspects of Ancient Greece, not for his actual homeland only but for the whole world. This was his 'Delphic Idea', still not forgotten: efforts are being made in our days again, to give his dream a modern shape. Now admired in his country and in Europe, Sikelianos is among the most renowned modern Greek poets, integrated by some critics, together with Kavafis (Cavafy) and Seferis, into a 'Modern Greek Triptych' (Ivanovici 1979).

As we know from different sources, Tagore visited Greece in 1926, during a series of travels in European countries.[1] Some evidence on his arrival and visit could be collected from the contemporary press: he arrived in Piraeus in late November, coming from Rumania on a steamer, attended by his wife, grand-daughter, P.K. Mahalanobis, Professor of the University of Calcutta, the professor's wife, and others. He was given a warm welcome by Kastriotis, the Director, and by the members of the Piraeus Literary Society, and later by the Committee of the Society of Greek Writers, led by Mrs. Eleni Negroponti. He came for a two-day visit, and the next day he left for Alexandria, Egypt. During his short stay, he visited the Acropolis and its Museum, the Stadium (one of the ancient monuments rebuilt), where a patriotic event took place, the workroom of the artist Von Peske where he was given a portrait by the latter, and had different talks with Greek fellow writers and officials. One is impressed to read that Tagore knew not only about ancient Delphi but also about the plans for the cultural revival of the place and would have liked to visit Greece, again, in the next spring to attend the first Delphic Festivals (cf. the issues of the newspaper *Vradini*, Thursday 25, and Friday 26, November 1926).

This paper is dedicated to a comparative study between Tagore's and Sikelianos' thinking and literary work rather than to investigating the details of Tagore's visit to Greece and of his initiatives, for which an extended domestic and international archival research is needed. Such a research would be most welcome, however, both for the case of the Greek-Indian relations and the Delphic Idea which never ceased entirely to be an inspiring source, and which, in our time, is the subject of new efforts to revive and further develop the Sikelianian ideals.[2]

Tagore's *Song Offerings* or *Gitanjali* had already been translated by the time he came to Greece, and this cycle was republished, for the last time in 1978.[3] The collection contains 103 poems, so that it may be assumed that it relies on the English corpus that consisted largely of the translations made by Tagore himself. The Greek title, like the English one, may not render satisfactorily the strong devotional connotation of the Sanskrit *añjalí*, something like a 'prayer offering', but, to be sure, the religious tenor is well apprehensible in all texts. Trikoglidis, who approaches his work with great respect, goes as far as to place Tagore's figure between Jesus of Nazareth and Plato (Tagore 1978: 5).[4] He adds that the Indian poet, with his calmness and his serene beauty, 'paved for the Europeans the way to the orient', and 'graced the entrance to the Oriental temples with flowers; the exhausted western spirit found a new source of inspiration' (*ib.* p. 6).

There are several points where a comparison between the Indian and Greek poets is relevant. Tagore may have been more successful in reaching his objectives but the *universalism*[5] of the two authors in thought and program is similar. Both of them travelled a lot. Sikelianos travelled mostly to the West whereas Tagore visited several other countries as well. Both of them originated from a respected family: a wealthy one, self-evident for a Brahmin family like in the case of Tagore, whereas Sikelianos was born into a good and educated milieu (his father was a language teacher having his own school) with an aristocratic attitude and traditions (Anton 1988b: 253), on the western Greek island of Leukas. Both of them had the possibility to read and learn at home, and, strikingly, both of them began to read law: prospective barrister. Tagore left University College London to explore Shakespeare and more; Sikelianos enrolled at the Faculty of Law of Athens University in 1901. By 1902, it was clear that he took a much more intense interest in the cultural and literary life of the Greek capital, and he never transferred to the Bar. Tagore won the Nobel Prize for Literature as early as 1913, and was the first non-European to receive the award. The Greek poet was not as lucky, but he was nominated for this prize three times, in 1945, 1947 and 1949 (Frangou-Kikilia 1993/2001: 47). Tagore, in the wake of the family tradition, founded an ashram, a spiritual hermitage, which later became the Visva-Bharati University in Santiniketan (Vassiliades 2000: 191[17], Sen 2005: 114ff.). 'Santiniketan' which means 'Abode of Peace' (cf. Sanskrit *shanti* or *śāntin* 'peace, rest, calmness, tranquillity, bliss') expresses Tagore's intentions well. Sikelianos did much to develop the so-called *Delphic Idea*, part of which would have been an international 'Delphic University' (Sikelianos 1930b). If he had been successful the university would have been, in continuation of the Delphic Oracle that was in a sense a centre in classical times, an educational establishment, open to the whole world (*ib.*, e.g. p. 7), with a focus on art, literature, drama, dance and similar practices. One may recall that from the 6th c. B.C. onward, a famous oracle of international reputation had developed in Delphi, on the site of very ancient cult centres, , and it was literally believed that Delphi was the centre ('the navel') of the Earth. The *omphalos*, symbolizing the centre, an ancient religious stone artifact, is now on exhibition in Delphi Archaeological Museum. Like the Olympic Games, the Pythian (i.e. Delphic) Games were important and had a panhellenic character. These were, however, not only athletic games but also various cultural events where drama and music had a paramount role.

Sikelianos was helped by his wife, Eva, with her contributions, and in an altruistic fashion, with her whole fortune. Eva's contribution is set out in a large autobiography (Palmer-Sikelianos 2010). Tagore was happier: the Visva-Bharati University in Santiniketan provides education on Indian culture both to Indians and foreigners. The Delphic University would have been also part of a universal *Delphic Union* (on which cf. Sikelianos 1932). There is no direct information on any relationship between Sikelianos and Tagore. But, to be sure, Tagore, when he visited Greece, was aware of the plans for the Sikelianian Delphic Games, and he wished to be able to visit Delphi the next spring in order to attend the Games.[6]

They may not have known each other personally but certainly, the Indian tradition and culture is present in the Greek thinker's work. E.g., he writes on cultural and moral 'revolution' and he acknowledges Gandhi's achievements (1932: 3-4). He also invites East and West to meet at Delphi (*ib*. p. 16), and does not forget about Eastern and Western music either, for the cultivation of which he had ordered a special organ for his wife from Germany capable to analyze and play also Eastern tunes (*ib*. p. 17). One remembers that Tagore was a composer, too. He is the composer of 'Jana Gana Mana', the National Anthem of India. The composition, first sung in 1911, was officially adopted by the Constituent Assembly as the Indian National Anthem on January 24, 1950. In his treatise on the *Delphic Union*, in quest for origins, Sikelianos alludes to 'the clean foundations of primeval societies', and Vedic India, together with 'Orphic Greece', is among them (1932: XI). Surprisingly, he cites Dhan Gopal Mukerji who wrote that 'Europe cannot attract India any more, only if Greece is with Europe; Greece, indeed, speaks to the heart of the Indians' (*ib*. p. 2). Last but not least, *love* is essential in the poetry of Tagore. Sikelianos may not have written on love and affection so often if compared, but in his *Delphic Appeal* this is, beside universality, very emphatically present: 'Mankind needs love', he repeats several times (e.g. 1930a: 4-5). Similarly, an important thought is expressed in his work on Digenis Akritas, a medieval Greek epic hero, when he declares that thought is unable to come up because love had come before (cited by Ivanovici 1979: 67).[7] Whereas Tagore's religiousness and spirituality cannot be questioned, his Supreme Lord was neither a Hindu god nor the Christian one but a universal spiritual being; similarly, as Keely and Sherrard remark, 'Sikelianos felt no embarrassment in invoking "my Christ and my Dionysos" in a single breath' (Sikelianos 1979: XVI). Both of them were prolific writers, and in any case, Sikelianos was much more prolific than Kavafis (Cavafy) and Seferis, his two comparable fellow-travellers (ib. XIX). Last but not least, the 'Delphic Idea' can be compared with the 'Idea of India', at least in Tagore's interpretation, who thought that the idea 'militates against the intense consciousness of the separateness of one's own people from others' (Sen 2005: 72). Sen adds: 'Celebration of Indian civilization can go hand in hand with an affirmation of India's active role in the global world'. On Tagore, he writes further that 'Tagore put the rationale well, in a letter to C.F. Andrews: 'Whatever we understand and enjoy in human products instantly becomes ours, wherever they might have their origin' '(*ib.* p. 86). From the Greek side, we add to this that the newspaper *Vradini* greets Tagore on his visit as the 'noble apostle for the brotherhood between peoples' (25 November, 1926, p. 1.).

Having a look at Tagore's Nobel Prize winning work, the *Gitanjali*,[8] the Western reader is impressed by the similarity between his poems and works of Christian inspiration. This might be a 'departure' from the prescribed course of Indian thinking. We know, however, that the moral worth of Hinduism and the Christian values are close. Sikelianos also 'departed' from accepted values, but this was not a real departure: he remained in his native culture and native land. He only expanded various dimensions and stressed things differently. Hinduism and Christianity, two spiritual commitments, have universality in common, and similarly, the two creative geniuses, Tagore and Sikelianos, were committed to *universality*. Besides the universality of the two spiritual commitments, Hinduism and Christianity, there is also the universality, strongly felt, both in Tagore and Sikelianos. One may consider Tagore's short poem 'Who is This?' meaning the Lord in whose company the poet is 'ashamed to come to [His] door' - this finds a parallel in Christian liturgy. This may be compared to a New Testament passage: 'When Jesus entered Capernaum, a Roman officer met him and begged him for help: "Sir, my servant is sick in bed at home, unable to move and suffering terribly." "I will go and make him well," Jesus said. "Oh, no, Sir," answered the officer. "I do not deserve to have you come into my house. Just give the order, and my servant will get well" (Matthew 8: 5-8).' In the Catholic liturgy, during the Holy Communion, the passage beginning with 'Just give the order' is adapted as a parable where the word 'soul' substitutes the original 'servant'.

Here is the famous 'Prayer' by Sikelianos, where *love* is also present, comparable not only with Tagore's 'Give me Strength' using the word 'prayer' but first of all with his 'Face to Face'. Sikelianos writes:

Naked the soul prays to You. Stripped of joy,
of suffering and pleasure,
naked the soul prays to You, Creator, with its
uncreated voice alone,

that voice which, before entering my flesh, in Your breast –
as a cicada hidden in the olive tree –
beat in my heart as Your will, crying 'Victory,
victory in all things,

and it was not my voice, it was Yours, Lord; with that
alone I pray to You: release in me

the secret purpose I tasted deeply outside time,
so that I may love, may love

beyond human images and all created things, beyond the single
pulse that throbs inside me,
one now for the living and the dead: grant me,
grant me deliverance,

to feel again the uncreated Eros
filling my breast,
and to be to all, to things near and far away,
as the wind's sound and breath.'[9]

Tagore's 'Face to face' from the *Gitanjali*:

Day after day, O lord of my life,
shall I stand before thee face to face.
with folded hands, O lord of all worlds,
shall I stand before thee face to face.
Under thy great sky in solitude and silence,
with humble heart shall I stand before thee face to face.
In this laborious world of thine, tumultuous with toil
and with struggle, among hurrying crowds
shall I stand before thee face to face.
And when my work shall be done in this world,
O King of kings, alone and speechless
shall I stand before thee face to face.[10]

The 'Signet of Eternity' by Tagore expands the horizon both in time and space: the poet waits for 'the happy moment [when he is going] to see', whereas in the 'Brink of Eternity', at which he has arrived to see 'the allness of the universe' one feels an affinity with Christian hymns. Let's stop here: this is not the place to analyze the amazing passage by Sikelianos on 'The death of Digenis'. Besides 'love' that 'overtakes thought', there is an addition. Thought says 'you can't enter', but the poet is defiant: 'I'll enter; and I enter free, and I will step out free' (Ivanovici 1979: 67).[11]

Tagore did not visit Greece again and could not attend the Delphic Games. Nor did Gandhi. Yet, they were and they are here, in Greece, in many ways. Leaving Greece, the Indian writer said goodbye with these words: 'With my good wishes for Greece and my sympathy for her people'.[12] During 2011 and 2012, Tagore was warmly celebrated and evaluated in a series of commemorative programs, with a Festschrift (Vassiliades 2012) included.[13]

Sikelianos, who did so much for Delphi, the Delphic Games, for bringing peoples together culturally and in their physical reality, was not lucky enough either in being able to continue the games or in establishing a university.

Like Tagore, he 'did not fit', either. Yet Tagore gained a unique professional and social standing. For Sikelianos, there was indifference instead. This may happen to great spirits. In such cases, it is Posterity that must discover them, give them a just evaluation, and carry their work further. That is our duty, indeed.[14]

Notes

1 A first source for this date is to be found in the Preface by Tsewang Topden, Ambassador of India in 2012 (see Vassiliades 2012: 4).

2 It would be exciting to search in Greek sources for details of the occasion when Tagore was decorated with the Order of the Redeemer by the Greek Government. This event took place, according to Kripalani's Tagore biography, on 25 November, 1926.

3 This is an edition by the Athenian publisher Iridanos, under the title *Lyric Offerings*, with the subtitle *Gitanjali*, containing the 1921 Foreword by the translator K. Trikoglidis, and a new note by Iridanos. The same publishing house, in the same year, also issued *Indian Short Stories* by Tagore, containing e.g. the story on the Ganges riverside, that of Kouzoum. The second edition of the *Lyric Offerings* translated by Rena Kartheou (Athens, 1954) was included as no. 51 in the series *A Hundred Immortal Works* . The volume contains also *The Gardener* (published by Tagore in 1913), translated by Kostas Kartheos, and *Stray Birds* (published by Tagore in 1916), translated by Kostas Kartheos and Rena Kartheou). This book gives, in Greek translation, W. B. Yeats' Introduction to *Gitanjali* (pp. 5-14). A number of other works of Tagore have also been translated into Greek.

4 Indeed, in the contemporary press, like in the two numbers of the newspaper *Vradini*, this impression recurs: the first report, written by the editor-in-chief, writes on his 'beautiful shape' and compares him

with Jesus of Nazareth (Vradini, 26 November 1926, p. 7), whereas in the issue of the next day, the poet's appearance is found 'patriarchic', and a portrait made by the artist K. Klonis is published there, too (p. 1).

5 Sen (2005: 95) relates that 'Tagore's many sided writings' did not fit 'into the narrow box Yeats wanted to place - and keep him'. Whereas he admired the *Gitanjali* and supported its English edition, Yeats had come to denounce Tagore by 1935.

6 We are informed on this by the 25.-26.11.1926 issues of *Vradini* (pp. 1 and 7 of the first, and p. 1 of the second). What is even more impressive, is that Tagore had the program for the next year's Delphic Games in his hand and showed it to the editor of the newspaper. The second adds that Mahatma Gandhi, too, knew about the Games, and was interested in attending them. The editor also remarks that he had a correspondence with Eva Palmer-Sikelianos. This gives the impression that a sort of indirect contact must have existed, if not between Sikelianos and Tagore, at least between Tagore and Sikelianos' wife. Further archival research may add more evidence to this exciting subject.

7 On Sikelianos' religiousness, cf. Palmer-Sikelianos 2000: 143.

8 There is one more Greek tribute to the Indian poet: a recent Festschrift edited by D. Vassiliades, with the Preface by Tsewang Topden, the Ambassador of India in Athens, containing contributions by such names as Supriyo Tagore, also active in Santiniketan, Amartya Sen, Professor at Harvard University, and others, with a CD added on the history of *Gitanjali* by Malaya Bhattacharya (Vassiliades 2012).

9 For the Greek original, see Sikelianos 1979: 40, where also the English translation, cited here.

10 For the Greek rendering of this poem, see Tagore 1978: 56-57 (no. 76). For the English translation, here, as well as elsewhere, I used online sources that could be considered reliable.

11 As far as I know, there is no English translation of this play. I cite the whole passage in the French rendering:

'Je me suis endormi dans un profond vertige,

Comme l'abeille une nuit entière dans la rose,

Et je viens de rouvrir les yeux pour veiller

Sur mes derniers moments, là où la pensée n'arrive pas,_

Mais où l' amour parvient,|

Là où la pensée écrit: «tu ne peux entrer», à quoi je réponds

«J'entrerai», car j'entre et sors comme je veux…' (Sikelianos 1960: 117-118). For the Greek original, cf. Sikelianos 1948: 76).

12 *Vradini*, 26 November, 1926, p. 1.

13 D. Vassiliades recently released a book on Mahatma Gandhi, published by ELINEPA, the Greek-Indian Society, in Athens.

14 On the contemporary lack of understanding about Sikelianos' ideas, cf. Samoladas 1982: 25. Still, in a sense, posterity decided in favour of the Greek poet. Now, an International Delphic Council exists, the founder of which is the German Christian Kirsch. He could win the support of the late Minister of

Education Melina Mercouri, whose death in the early 90's was a serious setback. Yet international cultural events called the Delphic Games regularly take place in recent years. But Delphi is not the only scene of the modern games: see some remarks on the ancient and modern importance of Delphi by M. Laser (1995) and by the archaeologist F. Lang (1995), in German, with parallel Greek texts.

15 This first edition was followed by a second one in 1950 in the 2nd volume of works collected under the title *Thymele* with an added part to the title: *Christ Unbound or The Death of Digenes*, and so was the French translation: *Christ délivré. La mort de Digénis*. In its final form, the work was placed in the 3rd volume of *Thymele*, which, again, forms part of the *Apanda*, the Collected Works of Sikelianos, © Anna Sikelianou 1951, 1955, 1971.

References

Anton, J.P. Eva Palmer-Sikelianos (in Greek). *World Biographical Lexicon* 9A (Athens: Ekdotiki Athinon, 1988[a]), pp. 255-256.

Anton, J.P. Angelos Sikelianos (in Greek). *World Biographical Lexicon* 9A (Athens: Ekdotiki Athinon, 1988[b]), pp. 253-255.

Frangou-Kikilia, Ritsa. Sikelianos, Angelos (in Greek). *Encyclopedia Papyrous-Larousse-Britannica* 54: 1993/2001, pp. 46-49.

Ivanovici, V. *A Modern Greek Triptych*: Kavafis (*Cavafy*), *Seferis, Sikelianos* (in Greek), (Athens: Exandas, 1979).

Lang, Franziska. Die Pythischen Spiele von Delphi. *Chronika* 2, Nr. 4: 1995, 1-2.

Laser, M. Wiederbelebung: Delphische Spiele der Neuzeit. *Chronika* 2, Nr. 4: 1995, 1995,3-5.

Palmer-Sikelianos, Eva. 2000 *The Sacred Panic* (*Autobiography*) (in Greek), (Athens [Koropi Attikis]: 2000).

Militos (Translated and complemented with an Introduction from *Upward Panic*: *The Autobiography of Eva Palmer-Sikelianos*, Routledge, 1993, by J.P. Anton)

Samoladas, Z. *Sikelianos, the Delphic Idea, and its Future* (in Greek). Thessaloniki: Graphic Arts ASE, 1982.

Sen, Amartya. 2005 *The Argumentative Indian. Writings on Indian History, Culture and Identity*. (London etc.: Penguin, 2005)

Sikelianos, A. *The Delphic Appeal* (in Greek). Delphi: ©Author, 1930(a).

Sikelianos, A. *The Delphic University* (*A Preliminary Draft*) (in Greek). Delphi: ©Author, 1930(b).

Sikelianos, A. *The Delphic Union* (*An Intimation*) (in Greek). Athens: ©Author, 1932.

Sikelianos, A. *The Death of Digenes* (in Greek). Woodcuts by Spyros Vassiliou, (Athens: Ikaros, 1948).

Sikelianos, A. *Poèmes akritiques. La mort de Digénis. Tragédie*. Adaptation française par Octave Merlier. Bois de Spyro Vassiliou. Athènes: IFA, 1960.

Sikelianos, A. *Selected Poems*. Translated and Introduced by Edmund Keeley and Philip Sherrard. (Bilingual Edition). (Princeton, New Jersey: Princeton University Press, 1979).

Tagore, R. *Lyric Offerings. Gitangali* (in Greek). Translated by Kostas Trikoglidis. (Athens: Iridanos, 1978).

Tagore, R. *Gitanjali: Selected Poems. 'Song Offerings'.* Translations made by the author from the original Bengali. 2010 (Online source: http://www.schoolofwisdom.com/history/teachers/rabindranath-tagore/gitanjali)

Vassiliades, D.Th. *The Greeks in India. A Survey in Philosophical Understanding.* (Delhi: Munshiram Manoharlal, 2000).

Vassiliades, D.Th. *Rabindranath Tagore* (A Tribute to His Memory) (in Greek). Ed. by D.V. Indo Hellenic-Society for Culture & Development, (Athens: 2012).

Tagore's Historical Visit to Greece

Demetrios Th. Vassiliades

Arrival and Reception

The Romanian steamship, Imperator Tragiano, with Rabindranath Tagore and his company on board, arrived in Piraeus at 7 in the morning on 25 November 1926 and sailed the same day at 4:30 pm for Alexandria, Egypt, the poet's next stop in his long and tiring voyage back to India.

Greek journalists seem to have been well-informed of the lectures and the tour of the eminent Indian poet in the Balkans and covered them extensively from Belgrade, Sofia and Bucharest.[1] Upon completion of the series of lectures, Tagore started his voyage back to India, passing by Istanbul, Athens and Alexandria.

A Greek newspaper of Istanbul, 'To Fos', in its first-page article entitled 'Rabindranath Tagore in Our City,' reported that the steamship approached the berth of Galatas, but the poet stayed on the deck till sunset enjoying the view and did not go ashore. He stated to the journalist of the newspaper who approached him that he was making the voyage to Europe to get acquainted with its countries and their peoples, their pain and woes. He spoke only English and translation by an English-speaking traveler made the conversation difficult. Hence the discussion was limited to learning about his impressions of Bucharest, which were admirable.[2]

The Greek Drama Writers Society undertook the initiative of inviting and receiving Tagore in Athens; a day earlier they sent the following telegram to him: 'The Greek Drama Writers Society seeing you as a representative of the world intelligentsia welcomes you to the Greek waters and would consider it a privilege if you could attend the lunch we arrange in your honor.'[3]

The steamship of the royal Romanian line with Tagore (accompanied by his son Rathindranath, his daughter-in-law, Pratima Debi, and his granddaughter, Nandini, as well as his secretary and joint secretary of Visva-Bharati, Professor Prasanta Chandra Mahalanobis, together with his wife Rani Mahalanobis) aboard, sailed from Istanbul for Piraeus where she arrived one day late due to a heavy fog in the Bosporus.

A boat with young people[4] was first to approach the ship. They were the members of the Piraeus Literary Society with the representatives of the Piraeus Journalists Union who obtained the permit from Harbour Master, Mr. Papatheodorou to go onboard and greet Tagore. A few journalists who had been waiting since dawn went with them.

The deck officers showed the visitors to the dining-hall of the first class where they met Tagore who had already been up since dawn and was enjoying his morning hot drink. 'Serene with radiant eyes, he turns his head and looks at us. A Christian figure - the way he is dressed in a tunic – he looks like Russian philosopher Tolstoy.'[5] His personal secretary, Mr. Mahalanobis undertook a few necessary introductions. Tagore greeted them warmly shaking hands with the first Greek intellectuals who went to receive him. In her turn, Ms. Rena Avgeri respectfully presented him with a few laurel and olive boughs and greeted him as follows: 'The members of the Piraeus Literary Society, united by the flaming love to art and human ideals offers through me, a simple daughter of Greece, a seeker of the good, these laurel and olive boughs. These boughs are offered as a recollection of the sacred trees of Athena and Apollo, great Greek Gods of Wisdom and Art, to a wonderful wise man, a sublime poet.'[6] The next one to greet him was General Secretary of the Society Mr. Kastriotis, who did it in French.

Tagore, obviously deeply moved emotionally, held the boughs and thanking them for their reception, he expressed his enthusiasm about stepping on the sacred soil of Greece and visiting Athens, the city of light and culture.[7] Then he asked if there was anyone who could speak English. Then the editor-in-chief of the afternoon newspaper *Vradini* spoke, who did not lose the chance to ask the first questions. From the article he published a few hours later entitled 'Rabindranath Tagore in Athens – First Interview in Greece' we learn that the poet talked emotionally about the enthusiastic reception he enjoyed in Belgrade, Sofia and Romania and about the purpose of his journey, which was to proclaim to the peoples the religion of love and brotherhood. He also replied to the question put by the editor- in-chief regarding the revolutionary movement: 'Every nation has its individualism and mission, but I cannot imagine mankind divided in groups. For me humanity constitutes a single unit.' He avoided answering questions on politics saying: 'I am not interested in political issues.'

During the conversation, he kept expressing his joy at being there and he wished that he had more time during that trip to stay and get to know Greece, the country whose culture and history he admired so much. 'I am sorry I cannot stay longer in Athens. I would like to get to know Greece better. In particular, I would like to meet Greek students and talk to them on the issues of interest for the students of the whole world. I would like to pass them regards from their Indian colleagues. Unfortunately, my health condition does not allow me to stay in Greece. I have been travelling for almost six months in Europe and that trip has tired me a lot. Now I follow my doctor's instruction and go back home to rest.'[8]

Besides, he wanted to learn about the Delphic Festivals of the following spring and the editor-in-chief of *Vradini* answered that 'the great Gandhi has already shown his interest in the Festivals. He has already written three times to Mrs. Sikelianos, and is very enthusiastic about that.' Tagore asked again about the Festivals and the time of the celebration, he was holding with special care a white roll with the programme of the Delphic Festivals. Thereafter, the discussion turned to the Italian people about whom Tagore talked with great enthusiasm saying: 'I know that they are revolutionaries by nature.'

When asked about the purpose of his trip, he responded: 'My mission is to preach the love of the people. During my trip I have seen many peoples, I studied their lives and through my lectures I worked for global friendship.'[9] To the same question put by another journalist, Tagore replied: 'I have travelled to get to know Europe and European culture and I have the best impressions of my trip. The European peoples have many great spiritual virtues. I do find a great spiritual kinship with the peoples of the East which lets me hope that the creation of a new culture with the closest contact between the two peoples is feasible.'[10]

More philosophical were the questions made by partner of *Eleftheros Typos* newspaper, Mr. K. Bastias, who the following day published the interview together with a profound reflection, caused by the apostolic and ascetic image of Tagore, on both extroversion and materialism that dominated modern Greeks while becoming strangers to the spiritual life.

'Your question is too general,' answered Tagore to the question put by Mr. Bastias about modern culture and continued: 'People who cannot see the general lines, often talk about as many cultures as there are the states since the onset of the world. But what they call culture is a combination of the external phenomena which are presented in different forms in different countries. I never viewed this phenomenon from this perspective. For me, there exist two conflicting currents over the period of many centuries. One current tries to lure mankind towards an external superficial side of the world, towards a materialistic perception of life, and the other tries to enlighten the divine inside all of us. There are people who prefer the first one, but I condemn it and follow the path of the second current. When I recommend abstinence, when I talk about purity of the body, my opponents should not get surprised. Since I believe that deep in the soul there is the temple of God, how can I advocate the ousting of the ark of the spirit and becoming prey to human passions?'

- Teacher, I presume that at this point you are entirely in agreement with your compatriot Mahatma Gandhi.

- Mahatma Gandhi is a compatriot and a friend of mine and there is no differences between us as far as cosmogonic views go. We disagree on politics, which constitutes for him an exclusive course, while for me this is something strange, something which often has no place in my thoughts. It is a little detail of life, and I care only about the basic lines upon which a man should proceed.

- You are said to have no liking to western culture and that science has never won your affection.

- We should not be categorical. What I always say is that Europe is not on the right path. It despises the soul and considers only the body. Like anything else, its science is external and, therefore, unable to save us. Take a look at medicine. It believes that only the mechanical interference and treatment of the body will solve the major problem of the patient. Yet the patient could offer much to himself, and perhaps sometimes even more than science. Is it not possible that we are concentrating on the body and forgetting that there is a soul which has the strongest effect on the body? And then the West wants to mechanize everything. It wants to erect altars to unrestrained selfishness. It wants to ostracize love and kindness. This is not civilization, but misunderstanding and ignorance of life.

- And what is your idea of ancient Greece?

- I see the land that nurtured great people, or rather great souls. I have studied ancient Greek culture and I dare divide it into two categories, or two periods. There was a period during which, in Greece, the issue of the soul and the cosmogonic issue were considered as the more tangible issues of life. Yet, there was another period during which the soul was forgotten and the so-called materialistic version of life prevailed. That period might have produced remarkable monuments but it did not produce notable figures. There are intellectuals who admire both Aeschylus and Lucian. What an illusion! The one saw God and the other was unable to see him, nor himself, and he laughed and mocked the world. The first one was the creation and the other one – a decay.

- And Sophocles?

- He was not at the divine level of Aeschylus. As for Euripides, he attributed too much importance to logic, and that was unnecessary. The world's greatest problems have always been placed before the great people for the clarity of a solution. The way they express their concerns or provide a way-out of spiritual anxieties is a secondary issue. Never is a great soul worried much about which form to give to its content.[11]

At that point, the conversation was interrupted as some voices were heard on the ship's stairs. The inflow had begun. Journalists, reporters, photographers and illustrators came all together into the hall and went directly to the poet. The pencils of the illustrators and the cameras began to work quickly. Tagore, confused, turned his eyes upwards to the sky as he sees that there is little hope that the people will feel sorry for him, 'and at that moment his beautiful figure looked like the image of Christ when he exclaims to his Father – Eli, Eli, Lama Sabachthani'.[12] Tagore rose quietly and went to his cabin. After a little while, he emerged dressed in a black mantle and a black velvet cap and agrees to pictures being taken together with the journalists.

Around 9 am, a delegation of the Greek Drama Writers Society arrived onboard, with its President, Mr. Nikolaos Laskaris, and the members Messrs. D. Bogris, Petros Haris and Mrs. Eleni Negroponti (Alkis Thrilos), who welcomed him on behalf of the Society and invited him to visit Athens and have lunch together with them in the historic hotel Grande Bretagne in the centre of the city. The poet thanked them and expressed his wish to visit the antiquities.[13]

Visit to Acropolis

Rabindranath Tagore boarded the boat of the port authorities which brought him to the pier where he was received by crowds of people. There he got into a car together with his family, along with the secretary with his wife and the members of the Society and went to visit the Acropolis.

As Petros Haris informs us in his article, published the following day,[14] in the car, Tagore answered various questions. He spoke about the festive reception he enjoyed in the Balkan countries and on the fact that his poetic work was well known, and that it gave him great pleasure, since he aspired to be liked and have an influence as a lyric poet. He added as well that he planned to write many poems as soon as he returned to India. Lyric poetry seemed to interest him more than anything else as it is there that his deep satisfaction lay and his future attempt to express himself via art.

Then the conversation went on to the poet's ideology, his social and political views which greatly impacted the peoples of the East. 'I am working,' he said answering a particular question, 'for the renaissance of the folk art: music, philology, pottery and handicrafts. I am trying to make the people enjoy their passions and emotions via their own feelings. That is the basis of our long-term efforts from which I expect much.'

To the question regarding Gandhi, he replied: 'There is a great distance between me and Gandhi. Yet we work with the same spirit, with the difference that Gandhi seeks revolution of the eastern peoples, a political revolution, while I am only a poet. I confine myself to the area of art and thought and don't deal with politics.'

Immediately afterwards the poet's view of western civilization was discussed and his position on the ideals of the eastern peoples. He said: 'Upto a certain point the Indian people must accept machines and western civilization, to the point that it does not harm. The extraordinary capitalization and mechanization of everything is a disaster.'

To the question if he was a communist, the poet responded negatively saying: 'I believe that the owners of wealth should use it in a good manner - for the development of the people. If they don't use it that way, one day they will have to deal with the pressure of public opinion. People do good, make sacrifices during the war, not because they are pressed and forced by state power, but owing to public opinion. You get dressed not because you will be taken to prison if you go out naked, but because it is imposed on you by public opinion. And that is my purpose – the creation of a strong public opinion which would strengthen the good and prevent the bad.'

When the car passed by the Zappeion, the writers asked the poet if he would like to attend the ceremony of oath-taking by the recruits at the Kallimarmaro Stadium. The answer was in the negative and, therefore, they proceeded to the Acropolis where they arrived shortly afterwards. The car stopped at the stairs of the Propylaea (entrance), but the sixty-five-year-old Tagore, exhausted by the trip, said that he was not able to climb up to the sacred rock and see the Parthenon. His secretary remained with him and they admired the Acropolis from the footsteps while his son and two Indian ladies, accompanied by the writers, went to the top of the hill.

Then the German artist, Vinko Von Peschke, took the opportunity to invite him to his studio in Anafiotika, where the poet stayed for half an hour 'to see modern Greek art,' as he said. He stayed there alone, ate jam and accepted, with great emotion, a few pieces of Greek embroidery given to him by Von Peschke as a gift. A little while later, his companions returned from the Acropolis. The poet enquired about the kinds of pots manufactured in Greece and, thereafter, they all left together for the hotel where the poet and his family rested till lunch time.

Lunch in Athens

At 1:30 pm, the Greek Drama Writers Society gave a lunch in honour of the distinguished guest in the Grande Bretagne hotel which, besides the poet, was attended by Mr. Mahalanobis and his spouse Tagore's daughter-in-law, President of the Society, Mr. N. Laskaris along with Ms. Laskari, the Rector of the Athens University, Professor Simos Menardos, professors Orlandos and Andreadis, as well as learned scholars of his works, Mrs. Alkis Thrilos and Messrs. Th. Sinodinos, Daralexis, K. Bastias, Babis Anninos, Potamianos, Delikaterinis, Bogris, Pant. Horn and Petros Haris.[15] The Acting Minister of Foreign Affairs, Mr.Argiropoulos, who was supposed to preside, did not attend owing to being delayed at the German Embassy, nor did the Mayor of Athens, Mr. Patsis. The President of the Greek Poets, Kostis Palamas, could not attend owing to prior commitments.[16]

The lunch was hosted in a most hospitable manner, and when fruits were served the participants started to deliver their addresses. Mr. Laskaris was the first to speak: 'The Greek Drama Writers Society,' he said – 'is proud to host the eminent contemporary poet. No need to talk much of that matter. We would rather like to hear the voice that inspires all civilized people today.'

Tagore responded as follows: 'There is a saying in my country – "After the feast behave like a king". I would like to be able to enjoy once again the pleasure of today's feast. Unfortunately, I don't have enough time for that and I will have to be brief. I thank you for the honour you did to me by inviting me to this festival. My country-men will be happy when they learn about the hospitality I enjoyed in this country, where great culture once flourished. There used to be a close contact between Greece and India and the fruits of that contact were of great significance. I hope that the ties between our two countries have not been extinguished completely and that in the future they will re-gain their old strength. Greece and India both face the same problem. Both of us have a great past and the works of our ancestors are always in front of us. Our ancestors happen to be our biggest competitors. Both young Greeks and Indians always have to struggle to live at the level of our glorious past. Yet, it is not enough to ponder on the past alone. We have to live in the modern reality. We have to create works which will make us immortal like our ancestors. Both peoples suffer from similar political problems but they will get resolved as time goes by. Each of us has to contribute to the culture of our countries - Greece and India, and give constant value to it.'[17]

Then professors Menardos and Andreadis spoke. The first one, drinking a toast to the guest, said: 'Please let me tell you how happy we are to see you here in the classical land of Greece. At this moment we are not only representatives of certain philological unions but of thousands of Greeks who would be very happy to hear you or just see you. We imagine that Sophocles had a similar Olympian image like you. In any case, your voice is no less sweet that his. I wish humanity can hear it for a long time.'[18]

Finally, Mr. Menardos showed Tagore the translation of his poems into Greek. The poet responded that it was very interesting that his poems had been translated and expressed his gratitude for the work done.

After lunch and a few minutes before departure, the Chief of Protocol of the Ministry of Foreign Affairs, Mr. Lelis, arrived at the hotel. After explaining the reason for the absence of the Minister of Foreign Affairs at the lunch, he gave a long leather box to the Rector, Professor Menardos, who in turn, gave it without any further formality to Rabindranath Tagore and announced that H.E. the President of the Hellenic Republic, Mr. Pavlos Koudouriotis, had decorated Tagore with the order of the Grand Commander of the Redeemer, one of the highest orders of the country.[19] Tagore thanked them for this great honour. After a little while, accompanied by the writers and journalists, Tagore left for the port of Piraeus, where he boarded the steamship which was getting ready to sail.

Taking advantage of a few minutes of the car ride to the ship and before its departure, the editor of *Empros* newspaper had an opportunity to ask him a few more questions.

> I have read a lot about the ancient and modern Greece, Tagore answered, - and I am happy that I have managed to visit it even for a few moments. For sure, one cannot form an opinion about a people so quickly. But that's why my secretary will stay here to collect all information regarding your spiritual life.

- Have you visited other Balkan peoples?
- Yes, I was well received in Serbia, Bulgaria and Romania. They are enthusiastic peoples.
- How did you like Europe?
- Europe is a place still upset by the war. It is a place which reached the heyday of culture, and then declined. A proof of its decline is its distance from the ideals of freedom.
- The League of Nations does not seem the salutary institution to you?
- Not yet. Distrust and profit dominate there. It may do something later. And yet, how easy it would be when all nations could unite and make a union with a lasting peace. Some wish and say love one's neighbor; love of man for man is everything.
- Teacher, is there a discussion about a Pan-Asian Society of Nations?
- Yes, I have read something about it. There is such a movement in India as well, with Gandhi at the head. However, I don't approve of it. A union of peoples should not have political criteria but humanitarian ones.

We said goodbye to him and wished that he come again to Greece.

'Who knows,' he said, - 'I was told about some Festivals that will be held in spring in Delphi. I have the programme of the Festival. If God wishes, I will come then.'[20]

The last moments that the poet spent in Greece were recorded by the editor-in-chief of the *Vradini* newspaper, who had taken the first interview of Tagore: 'There, on board the ship, Tagore was relaxed and talked to his escort about the beauty of the Attica he saw; the next moment, he turned to me and said spontaneously without being asked: 'I will never forget this view. Here, where the sky is always clear, one can really see God. What a divine day! I greatly wish and hope that I manage to come again to Greece to get to know it at leisure. I might manage to go to Delphi in spring.'[21]

The following day, *Vradini* published on top of the first page the handwritten greeting of the poet - 'With my good wishes for Greece and my sympathy for the Greek people. Rabindranath Tagore' together with a picture of Tagore and Mrs. Mahanalobis as well as two sketches of his face in the way the artist, Von Peschke, and the illustrator Kl. Klonis saw him. Tagore had seen the first sketch earlier onboard the ship and he had found it very beautiful and put his signature on it in Bengali. The article ended with a personal description of the last moments of farewell as the Editor-in-chief of *Vradini* describes it, full of emotion:

> The steamship lifted anchor and was ready to sail. I was the last one who stayed there. I greeted the great poet and his company, boarded the boat and left. My eyes were still looking at the beautiful figure bathing in the light of the ready-to-set sun. A moment later, the poet raised his right hand slowly and greeted me in the manner traditional to his Motherland. It was something between a greeting and a blessing, so beautiful, so impressive. At the same moment, his son with his wife greeted me as well in the same way, and little Nandini was waving her scarf vigorously pushing her hand through the railings. I arrived at the port and they delivered to me the copies of *Vradini* and a few gifts that were to be given to the poet. Alhough the ship had started sailing, I did not despair. With two good rowers we went chasing the Romanian ship. It was sheer adventure. Under the danger of being washed away by the whirlpool made by the propeller, we approached the ship. The ladder had been lifted, and so a rope was thrown to us. We clutched it and I managed to give the package to a sailor standing at 'the entrance door.' Afterwards we left the rope and it felt as if a huge hand threw us far away, as if the ship itself gave us a strong kick. After a little while, I saw Mrs. Tagore, the wife of Tagore's son, opening the gifts and the great poet holding *Vradini* in his hands. A few moments later the ship turned. The beautiful picture vanished. It immersed in the world of reminiscences.

Tagore and the Greek Community of Alexandria

In Alexandria of Egypt, Tagore arrived in the morning of 27 November and was warmly received by the large Greek community and officially as well by the Greek Consul, Mr. Marinos Sigouros,[22] member of the honorary reception committee.

At 11 in the morning, the five-member committee of the city's Greek Literary Society 'Nea Zoe' consisting of Messrs. Pavlos Petridis, Petros Magnis, K.N. Konstadinidis, Pilos Zagras and G. Pilavakis visited the eminent poet in the villa of Mr. Suares[23] where he stayed. There they met the Greek Consul, famous writer Mr. Nomikos, and translator of *The Gardener,* Mrs. Efi Apostolidou-Lagopoulou with her spouse. After the introductions, Mr. Pilavakis addressed the Indian poet and presented him with an authentic parchment tied with a ribbon in Greek colors. On the parchment the salutation of the 'Nea Zoe' Society to the great intellectual on the occasion of his arrival in the glorious city of the Ptolemaic Dynasty was written in Greek along with its translation into English.

Tagore, who remained standing, while he was listening with great attention and pleasure to the address of Mr. Pilavakis, responded: 'I am greatly impressed seeing that, with such brilliance, you continue the beautiful tradition of the city of Ptolemy. Alexandria has always been an intermediate station between the civilizations of the East and the darkness of the West. It is the city where literature, thought and art were promoted to an unimaginable degree to be spread in the whole world'.

Then the Poet sat down and to the question - why had he not stayed in Athens longer - answered: 'I had a great desire to visit Athens for a longer period of time. Unfortunately, I could stay only for a few hours. The ship got delayed in the previous ports and arrived in Athens only on the day she had to sail. During that short period the English Embassy[24] offered me lunch where a few intellectuals were present. But even that was in a rush and had to be limited to standard addresses. Depending on time, I hope to stay in Greece longer during the next trip.'

Regarding his impressions of the trip, he said: 'The trip to the Balkans and the East made an exceptional impression on me. I found the peoples, who are known for their aggressive behaviour and who are said to fight with their neighbors all the time, have cultures that are especially unique, with simple thoughts, deep feelings and an almost oriental sensitivity. And if they err, they do it due to an excessive enthusiasm and not like modern Europe with full awareness. These countries gave me so much rest after Western Europe. I almost believe that I have found my home-country again. And that is why, now that I have got to know them, I will go there again, first of all to Greece and Egypt. The speed of transport has created the category of those who circle the world - the globetrotters, who generally inspire the idea that if one stays in a country for a few hours, he can get to know it. What a mistake! The mentality of a country is deeply hidden and it is not revealed but to the one who seeks and who will love the country.'

Mr. Pilavakis explained in a few words the purpose and the work of 'Nea Zoe' and the Poet added: 'The way modern life has become the fight to dominate, a fight which destroys every high ideal, you, the young people, who have so much love for art and for beauty should not abandon that devotion. Continue your work and be persistent, since only together with mankind, the passion will be saved, something much higher than the pursuit of material aims.'

The members of the Literary Society and other visitors were also greeted by Tagore's son, his daughter-in-law, a pretty brunette who was wearing an Indian gold-embroidered, silk dress, and his cute grandchild.[25]

Later, the Consul of Greece, Mr. Sigouros presented Tagore with the unpublished Greek translation of *Gitanjali* by Mr. Apostolos Leontis. Besides, Mrs. Lagopoulou gave the poet the well-known beautiful translation of *The Gardener* in a golden cover.[26]

Impressions of Tagore's Trip to Athens

The journalists and writers congratulated the Greek Drama Writers Society for the initiative to invite and host the eminent poet, but at the same time criticized the weakness of the officials to respond to the occasion and duly honour him with a more official and wider reception. His delayed arrival and the current political instability contributed to it, since the national elections of 7 November 1926 did not give the necessary majority to any party, and intense consultations led to the oath-taking by the Ecumenical Government under the premiership of Alexandros Zaimis on 4 December, a week after Tagore's departure from Athens. Besides, Tagore's refusal to visit the stadium during the oath-taking by the young soldiers did not facilitate his contact with the political leadership that was present there.

Although Tagore declared that he was not interested in politics, his international position on the world brotherhood of the peoples enjoyed considerable acceptance, especially by all those who were skeptical about the prevailing trend for the creation of the Pan-Asian Union of Muslims and Hindus against the colonial policy of the Christian West.

The whole press focused on Tagore's references to ancient Greek culture and the interest he showed in the attempt to revive the Delphic Festivals, and the many times he expressed his wish to participate in them the following spring. There were a few exaggerations as well. The visit of the Indian Nobel Prize winner to Greece and his phil-hellenic message gave a good reason to some journalists to stress the national spirit, depressed after the Asia Minor catastrophe. For example, the Greek edition of the New York newspaper, *National Herald*, on the day of Tagore's arrival in Athens, carried in bold capital letters its main title - 'Tagore praises Greek culture,' and the subtitle - 'The Cultured are only those initiated in Greek wisdom.' According to what the newspaper's correspondent from London wrote in the telegram - Tagore stated that he considered it his sacred duty to make his pilgrimage to Greece, the country that gave light to all nations of the world. He said that he was a student of the great Greek teachers of mankind and that the Greek language was indeed divine, and that no one could be considered a cultural person if he was not baptized in the springs of wisdom of the Greeks. The article ended with the caustic note of the editor about those Greeks of America who neglected teaching their children the Greek language.

The biblical appearance of the philosopher-poet wearing a mantle, as well as the gold-embroidered silk garments of the ladies accompanying him, attracted the interest of all friends of Eastern mysticism. The *Athinai* newspaper described the poet in the following words: 'Rather tall, slim with the silver crown and the patriarchal beard, with the appearance that reminds one of an ascetic who lives far from the world, Rabindranath Tagore looks like a living incarnation of someone from his philosophical poems who preaches the Gospel of peace and love. The divine purity of first apostles is reflected in his eyes, and his gaze emanates peace and tranquility.'[27]

At the same time, a few short tributes to his philosophy and work were published. The *Dimokratia* newspaper in its article 'Who is Tagore?' writes besides other things: '...Wrote poems, short-stories, dramas permeated with deep love to man, loyalty and admiration to his Motherland – India and an absolute faith in the Divine. It is due to that faith that his work is an inexhaustible source of peace and rest for man's soul... Tagore is not only a writer of great works. He is the God of his Motherland, a man who, together with Gandhi, commands the souls of the peoples of the

East today. A greatest figure in the history of man, to that great poet and intellectual, Greece feels enormous admiration.'[28] In a similar vein, the *Elliniki* newspaper characterized Tagore as a poet who managed to unite melodically the grand cultural heritage of his country with the anxieties of the modern Europe. 'That link erupts in each of his poems and especially in his short-stories, in his theatrical works, in the extraordinary variety of his work. Neither a magician entirely, nor philosopher, he is a poet who declares that, in order to flourish, art needs the soil of buoyancy and non-profit'[29] The *Estia* newspaper pointed out the political value of his work, - 'He struggles with all his undisturbed internal balance for the merger of two great cultures, the old Eastern and European which is stranded by the mechanization of life, the worship of individuality and the cultivation of group and national egoism. He seeks and preaches for this merger for the salvation of European intelligentsia from suffocating egoism.'[30]

It is particularly important to stress here that in the state of Bengal and especially in its metropolis Kolkata, where Tagore was born and grew up, a strong Greek community had settled as of the beginning of the 18th century. Pioneering Greek Indologist, Demetrios Galanos (1760-1833), emerged from that community,[31] and that event could not be ignored by the Indian poet. Since 1947, when India became independent, many state and other high-level visits have taken place in both directions, but that short visit of the great poet of India to Athens was recorded on the golden pages of the long cultural history of Greek-Indian friendship which has continued for 2500 years now. The message of Tagore from Athens about the importance of presenting the ideas of the great civilizations of the West and the East as relevant values for our times continues to inspire and guide spiritual people till today.

Notes

1 See: *Eleftheros Typos*, 24-11-1926, p. 1 and *Proia* 25-11-1926, p. 1.

2 *To Fos*, 23-11-1926, p. 1. The article was re-published by the Piraeus newspaper *Simaia* in its issue of 25-11-1926, p. 4. From another interview we learn about Tagore's ideas of modern Turkey: 'The Turkish revolution will create a new era not only for Turkey, but for all peoples of the East as well. The religious fanaticism draws the East backwards, but Turkey engraves the way to follow.' *Tahidromos-Omonoia*, Alexandria, 26-11-1926, p. 3.

3 *Athinai*, 25-11-1926, p. 1.

4 Famous Piraeus litterateurs Messrs. N. Hatzaras, D. Avgeris, V. Kastriotis, Iconomidis, Katrimados, Asteriotis, Voulodimos and poetess Ms. Rena Augery as well as journalists, Messrs. Tzonakas, Levandas and Boukouvalas boarded the boat. *Apogevmatini*, 25-11-1926, p. 1.

5 *Ibid.*

6 *Ibid.*

7 'Tagore arrived today – enthusiastic anthem to the ancient culture' *Athinaiki*, 25-11-1925, p. 8.

8 *Eleftheron Vima*, 25 November 1926, p. 3. The editor of the *Vradini* newspaper, Ms. Karavia, informs us of Tagore's health condition. She spoke to the ship's doctor Dr. Zakharov, who told her the following: 'The poet is sleeping as the rough sea disturbed him. During the trip he ate little, talked little and seemed busy with the thoughts that bothered him. She talked further with the Greek doctor from Constanta, Mr. Sarafidi who travelled with us'.

9 *Eleftheron Vima*, 25 November 1926, p. 3.

10 *Esperini*, 25 November 1926, p. 5.

11 'Yesterday Athens hosted for a few hours the great Indian Teacher with the Religious Appearance – A conversation with him – The temple of God inside us'. *Eleftheros Typos*, 26-11-1926, p.1. The newspaper published on its first page the photograph of Tagore with Mr. K. Bastias and the sketch made by a colleague from the newspaper, Mr. Protopatsi, along with the Poet's signature.

12 *Vradini*, 25-11-1926, p. 7. A phrase from Matthew 27:46 which means 'My God, my God, why have you forsaken me?'

13 *Politeia*, 26-11-1926 p. 3.

14 'One day with Rabindranath Tagore', *Dimokratia*, 26-11-1926, p.1.

15 *Ibid.* p. 2 and *Eleftheros Typos*, 26-11-1926, p. 1.

16 *Empros*, 26-11-1926, p. 1. Later, there developed a mutual appreciation with the great Greek poet and composer of the Olympic Anthem, as can be seen from the letters they exchanged. See for example, the letter dated 6 April 1930 from Tagore addressed to Palamas after receiving his books in France:

> Dear Poet, I cannot thank you very much for your books of poems that you have sent me. Most of modern poetry has become like your own lagoon - a shroud weed-woven has wrapped it, the weeds that are wild technique and individual idiosyncrasies. We need world poets today, for the whole world has opened itself to our vision. Such poets must come from those countries which have leisure and light, whose soul is not smothered under the load of prosperity and mental horizon densely misted with political obsessions. And you are such a poet - and it seems natural for you to belong to your own country as well as all countries of the world. My cordial greetings to you. Rabindranath Tagore. P.S. Your books being redirected from India, I am late in acknowledging your kindness.' *Selected Letters of Rabindranath Tagore*, Cambridge University Press, Cambridge 1997, p. 376.

Another testimony to the love and reverence that Palamas bore for the poet was his sponsoring of the publication of *The Golden Book of Tagore* that was published on the occasion of his 70th birth anniversary in 1931. The other four sponsors were writer Romain Roland from France, Albert Einstein from Germany and Mahatma Gandhi and Jagadish Chandra Bose from India. 'Rabindranath Tagore - A Universal Voice,' *The Unesco Courier*, December 1961. In *The Golden Book of Tagore* Palama's handwritten letter in Greek was published along with an English translation which reads as follows:

> In December 1927 at Salonica, where I was invited to celebrate my literary jubilee, I said: 'A world-famous poet, Rabindranath Tagore, in one of his splendid utterances (transmitted to us by Romain Rolland in his book *Mahatma Gandhi*) tells us what a great thing is the poet's seclusion within his natural precincts, and his ever-active asceticism. What a great thing indeed is the 'poet's corner,' to use his own words. The poet's song is like the song of a bird. And the bird - what does it do when the rosy-fingered dawn puts life into it? 'Its awaking,' he tells us, 'is not disturbed by the necessity of finding its food. Its wings never fail to respond to the call of heaven. Songs of joy it sings to the new morning light.' A great lesson, from this great Wanderer from India, the spiritual ambassador whom Asia has accredited to Europe, as he has been happily described. Blessed be the poet who is satisfied in his own little corner. It

is the crib where Poetry made flesh is born.' Thus did I speak then.

And I add now: The poet, who embraces within his inner being a world of joy, no matter wherever he stands, diffuses over the whole universe the charm of his little corner, and his message is like a song. Rabindranath Tagore whether he sings a lyric or delivers a message, like another apostle of the venerable Mother to her far-away daughter, he is always the Gardener who attends to the flowers which grow in the garden of the Queen. His words breath forth the imperious mysticism of prophecy and the cool softness of an idyll, nay, he knows well how to loiter and pick up the best of flowers from the forests of Kalidasa and Valmiki and the flower-gardens of Shakespeare and Shelley.

The dedication to Rabindranath Tagore of a *Golden Book*, on the occasion of his Seventieth Anniversary, is but a very modest token of our gratitude to him. It would only remind us of the Temple of Spirit which we owe to his divine genius. I very humbly offer him my respects under the canopy of light poured out upon me by the sky of Attica; and my soul, in its flight to imprint a kiss of love and homage upon his venerable hand, reminds me of one of those stags which, as he himself depicts in his *Religion of the Forest*, 'runs about in the wilds to kiss the hands of hermits.' Athens, Kostes Palamas. 'The Poet's Corner,' *The Golden Book of Tagore*, Edited by Ramananda Chatterjee, The Golden Book of Tagore Committee, Calcutta 1931. pp. 13-15.

17 *Dimokratia*, 26-11-1926, p.2 and *Eleftheros Typos*, 26-11-1926, p.2.

18 *Elliniki*, 26-11-1926, p.1.

19 *Elevtheron Vima*, 25 November 1926, p.3.

20 *Empros*, 26-11-1926, p.1.

21 Tagore was not able to fulfill his wish to be present at the first Delphic Festivals organized by Greek poet Angelos Sikelianos and his American wife Eva Palmer-Sikelianos on 9 and 10 May 1927. Besides revival of the ancient performances, the Festivals were aimed at promotion, via the art and knowledge, of 'the Delphic Union', the world union for fraternity of peoples, and of 'the Delphic University,' whose purpose was to provide a wider education on the cultures and traditions of all peoples. Rabindranath Tagore seems to have shared a similar vision, founding the Visva Bharati School in Santiniketan. See also Andreas Katonis 'Tagore and Sikelianos – Two Poets, One Vision' in the *Rabindranath Tagore – Commemorative Volume to his 150th Birth Anniversary*, edited in Greek by Demetrios Th. Vassiliades, Indo-Hellenic Society for Culture and Development, Athens, 2012.

22 *Tahidromos – Omonia*, Alexandria, 25-11-1926, p.1 and 27-11-1926, p.6. During his visit, Tagore gave interviews to the main Egyptian newspapers and spoke positively about Zaghoul Pasha, whom he characterized as the leader of the Egyptian renaissance. *Ibid.* 29-11-1926, p.3.

23 Carlo Giuseppe Suarès (1892–1976)- French editor of the 'Messages d'Orient' magazine and organizer of the lecture of Tagore in Alexandria. The speech of Tagore, given on 28 November, was focused on Indian philosophy. Tagore characterized himself as a visionary poet differentiating himself from the mental perception of a philosopher, 'Don't expect that I will talk to you as a philosopher since I am not one. I am just a poet who is at the same level with you and not above you. I am a poet for whom the only gift is the spiritual vision and I see the truth without being taught like philosophers... My profession is to see things and convey my impressions in the best possible way... I find the Truth everywhere with spiritual vision.' *Ibid.*

24 The name of the hotel - Grande Bretagne seems to have confused Tagore who thought that he had lunch in the building of the British Embassy. The improper information in this respect was criticized by the *Filotekhnos* magazine, 1st Year, November – December 1926, p.141.

25 *Tahidromos - Omonia*, Alexandria, 30-11-1926, pp. 3-4.

26 *Ibid.* 29-11-1926, p. 4.

27 *Athinai*, 26-11-1926, p. 1.

28 *Dimokratia*, 25-11-1926, p.1.

29 'Our Great Stranger - Rabindranath Tagore - his Work and Thought,' *Elliniki*, 25-11-1926, p.1.

30 *Estia*, 25-11-1926, p.1.

31 See Vassiliades, Demetrios Th., *The Greeks in India - A Survey in Philosophical Understanding*, (New Delhi: Munshiram Manoharlal, 2000).

Tagore as an Envoy of India (and the Rest of the World) to Russia

Sergei Serebriany

India and Russia, in spite of the much advertised globalisation, are still very different worlds. The conference, organised by ICWA on the 9th and 10th of May, 2013, is telling evidence of this fact. In today's Russia it would be inconceivable and utterly unacceptable to organise a conference on the 9th of May (which is an official holiday since 1965). I wonder how many people in India would recall what kind of date the 9th of May is. But in Russia now it is of one of the most important historical dates, an occasion both for great joy and great grief. As a popular song says, it is 'a holiday with tears in one's eyes' ('Это праздник со слезами на глазах'). In the night from the 8th to the 9th of May in 1945 German generals signed in Berlin the complete capitulation. Thus ended what was probably the most awful war in Russia's history. It is till today a matter of controversy how many compatriots of mine perished during that war. Official figures kept changing (increasing) through various periods of our post-war history. By the end of the Soviet era it was usually said that the overall loss had amounted to about 26 million people, but now some historians claim that this figure is an understatement and that only among the actual combatants the loss has been more than 30 million. An acquaintance of mine, a colonel of the Soviet army, once told me: 'One more victory like this, and we would have ceased to exist at all'. The baneful effects of that war and that victory are felt till today in my country.

For India, WWII has been certainly a different experience.

But before WWII, as well as before WWI, India and Russia had as different histories and cultures. It is all the more amazing how Rabindranath Tagore, his personality and his works, succeeded in crossing political and cultural boundaries and in impressing Russia's psyche. In fact, Tagore, as far as I know, was the first modern Indian author to gain wide popularity among Russian readers.

The story of Tagore's reception in Russia, of his one time great popularity in my country, is part of at least two larger stories which so far has not been told in full.

The first story, not yet told properly, is that of Tagore's world fame after receiving the Nobel prize in 1913. To mark the hundredth anniversary of the award, an international project has been launched which may result in the publication of a book this very year.[1] The papers which constitute this volume describe the reception of Tagore, his works and his personality, in many languages and many countries of the world. The papers show that Tagore's fame almost immediately after 1913 spread all over the world like a forest fire (which later subsided from time to time and from place to place, but, nevertheless, is still burning). I wonder if any other 20th century literary figure could be compared to Tagore in this respect.

The other story in question is the post-1913 history of Russia. This history as a whole and certain events and processes in particular are, in today's Russia, subjects of heated controversies. Many arguments cannot be settled because, among other reasons, many historical documents are not available for scholarly investigation (either because they have been destroyed, or because these are kept secret). And so the yet untold story of Tagore's relations with Russia[2] is just a small part of the yet untold 20th century history of Russia in general.

Speaking about Tagore as an envoy of India to Russia, we may distinguish two aspects of the poet's 'envoying' ('invasion') into this vast territory. First, he came (in 1913 and after) by way of translations, came in the form of poems, short stories, essays, novels, etc. Second, he came once (in 1930) in person. His appearance in the form of literary texts was a great success (which later had its own chequered history). But his coming in person has been rather a failure. To understand the 'why's' of this seeming paradox, one has to take into consideration the history of Russia since 1913.

By 1913 Russia (the Russian Empire) was one of the 'great powers' which next year started the suicidal world war. As compared with other 'great powers', Russia was presumably rather less advanced and even rather archaic, as far as its political and social structures were concerned. But since the second half of the 19th century, after the 'great reforms' of Alexander II, Russia has been developing quite quickly, even if (as the later history showed) not quite harmoniously. World War I and the disaster of 1917 put an end to this process of development. And here I may paste a paragraph from an earlier paper of mine:

> 1917 was a decisive and tragic watershed in the history of our country. In February/March, the monarchy collapsed under the weight of its own vices and popular discontent. Hopes were high that democratic developments would follow the collapse of the former Russian empire. But the combined strain of the war and the revolution proved to be destructive. In October/November 1917, the Bolsheviks, a small party of bigoted extremists, staged a coup d'etat, a kind of counter-revolution, which brought the country back to its archaic authoritarian traditions. Of course, the Bolsheviks projected themselves as 'Marxists' and the harbingers of progress and social equity, and probably some of them sincerely believed in their ideology. But the results of their take-over, in the long run, have been disastrous for Russia and most other parts of the former empire. The Bolsheviks undertook their 'revolution' under the premise that it would be but a spark to ignite what they called the 'world revolution', as, according to the Marxist orthodoxy, Russia in 1917 was not at all ripe for a 'socialist revolution'. Vladimir Lenin seems to have believed till his death in 1924 that the 'world revolution' was round the corner. It was only in 1925 that the Bolsheviks officially acknowledged that this was not the case. Stalin put forward the slogan of 'Socialism in one country', which was absurd from the viewpoint of orthodox Marxists, but sounded very pragmatic. Stalin's formula, in a way, was similar to the 'National Socialism' of Hitler. By the end of

the 1920s, Stalin became almost the undisputed ruler of the country and launched his so-called Third revolution to build the 'Socialism in one country'. This excursus in history will help to understand the history of Tagore's reception in Russia and the Soviet Union after 1917.[3]

But now, for a while, I will come back to 1913. By that time there was in Russia quite a sizeable upper layer of educated people who were open for the reception of foreign, even quite exotic, cultural products. It was by this (and mostly for this) upper stratum of European educated people that the classical Russian literature of the 19th century (roughly, from Alexandre Pushkin [1799 – 1837] up to Anton Chekhov [1860 – 1904]) had been created. All through the 18th and 19th centuries educated Russians (both writers and readers) looked for ideas and inspiration mostly to Western Europe rather than to any other part of the globe. But by the end of the 19th century and more so at the beginning of the 20th century, attention of some Russians turned to non-European cultures, India, China, and Japan in particular. Among those Russians was, for instance, Leo Tolstoy (1828–1910) who by the end of his long life took considerable interest in Chinese and Indian thought. Tolstoy read about Ramakrishna and Vivekananda[4] and even corresponded with about twenty Indians (the last in time and the most famous of them was, of course, M.K. Gandhi). [5]

Nevertheless, before 1913 hardly anybody in Russia (including Leo Tolstoy) heard the name of Rabindranath Tagore. For that matter, hardly anybody in pre-revolutionary Russia knew Bengali. But the general cultural atmosphere was obviously ready for receiving the information about the new literary star from India. When Tagore got the Nobel Prize in 1913, almost immediately he became very popular in Russia. Translations of his works (from English) appeared one after another within a very short span of time.

It is often said and written that Tagore got the Nobel Prize for his English *Gitanjali*. This statement is not quite correct. The official formula of the Nobel Prize committee did not explicitly mention *Gitanjali*. The formula announced that Tagore had been given the prize 'because of his profoundly sensitive, fresh and beautiful verse, by which, with consummate skill, he has made his poetic thought, expressed in his own English words, a part of the literature of the West.'[6] In the Presentation speech, delivered (on the 10th of December, 1913) by the Chairman of the Nobel Committee of the Swedish Academy, the Swedish historian Harald Hjärne, five (!) books of Tagore in English translations were mentioned and discussed. These were three books of poems: *Gitanjali* (1912), *The Gardener* (1913) and *The Crescent Moon* (1913); a book of short stories, *Glimpses of Bengal Life* (1913); and a book of lectures, *Sādhanā: The Realisation of Life* (1913).[7]

All five of these books – and some other works as well – were translated into Russian (some of them more than once) between 1913 and 1915. It was *Gitanjali* and *The Gardener* that attracted the biggest number of translators. Between 1913 and 1919, not less than seven different Russian renderings of *Gitanjali* and not less than five different renderings of *The Gardener* were published.

Two *Collected Works* of Tagore in Russian translation were undertaken from Moscow before 1917. One of them was in six volumes and the other in nine. The first one started running a second edition in 1916, but the project was stopped, evidently due to the events of 1917. [8]

Indians are often inclined to think that Tagore's poetry impressed the Nobel Prize committee and other Westerners as an exotic voice from the unknown depths of Oriental wisdom. This is, again, not quite the case. Thus, the above mentioned Swedish historian Harald Hjärne, the Chairman of the Nobel Committee of the Swedish Academy, must have known quite well India's cultural history. He understood quite well the multicultural roots of Tagore's literary achievements. In his Presentation speech Hjärne praised Tagore's poetry as, among other things, a product of cultural synthesis. Hjärne said: 'Concerning our understanding of this poetry, by no means exotic but truly universally human in character, the future will probably add to what we know now. We do know, however, that the poet's motivation extends to the effort of reconciling two spheres of civilization widely separated, which above all is the characteristic mark of our present epoch and constitutes its most important task and problem'. [9]

In the Russia of 1914, for some readers Tagore's poetry might have been indeed an exotic voice from the mysterious East. But there were as well quite sophisticated readers and critics who, like Harald Hjärne, appreciated Tagore exactly as a universalist, as a poet who embodied a cultural synthesis. The writer Konstantin Paustovsky (1892–1968) wrote later, in his memoirs about those pre-revolutionary years: 'Rabindranath Tagore dominated our minds'. [10]

Even in V.I. Lenin's personal library in the Kremlin (where he settled in the spring of 1918) there were five books by Tagore, all in Russian translation: *Personality*, two different translations of *Nationalism*, *The Home and the World* (a translation published in Berlin), and an anthology published in Russia in 1923. [11] We do not know whether or not Lenin actually read those five books. If he did, we do not know what he thought. But further research may yield some information on this matter.

It is obvious that Tagore was perceived differently in different circles. Thus, for instance, in the writings of Alexander Blok (1880–1921), one of the most important Russian poets of the 20th century, we find not a single reference to Tagore. In his personal library there was no book by Tagore. The same applies to another important poet of the time, Osip Mandelshtam (1891–1938).[12] It seems that Tagore appealed more to common readers rather than to high-brow littérateurs.

During the 1920s, many cultural processes which began before 1917 continued one way or another. Thus, in the decade 1917–1927, the popularity of Tagore with the Russian reading public did not diminish, but probably even increased. Up to the late 1920s, many other works of the poet, in Russian translation, appeared both in the Soviet Union and in the emigré publishing houses in Western Europe. Some works were again translated more than once. For instance, the novel *Ghare Baire* came out in the early 1920s in at least three translations (all from English), one of them published from Berlin. In the reviews of the translations, Russian critics noted the similarity between the Indian events and personalities (of the Swadeshi movement) described in the novel and the events and personalities of Russian literature and Russian revolutions.

In the 1920s, the grip of the official ideology was not yet as tight as it became later. Some critics of the 'Communist' type denounced Tagore as a 'mystic' and an 'idealist', and a representative of 'exploitive classes', but this did not prevent publishers from publishing his works, as it did not prevent many readers from enjoying his poetry and prose. Thus, my grandfather's younger brother (born in 1900) told me that while wooing his future wife (it must have been in the 1920s), he recited poems from *Gitanjali* to her.

In the 1920s, the first straight translations of Tagore's works from Bengali into Russian were undertaken. The translator was Mikhail Tubiansky (1893–1937), a student of the well-known scholar, Fyodor Shcherbatsky (1866–1942). He tried his hand at making translations from *Gitanjali* in the original, but was not too successful because he was not a poet himself. (In fact, even now there are hardly such people in Russia who would be able to make poetic translations straight from Bengali into Russian.) But M. Tubiansky was quite successful as a translator of Tagore's prose works. He translated *Jivansmriti*, among others, and this translation has remained the standard one till today. M. Tubiansky, together with the philosopher, I. Kolubovsky, brilliantly translated Tagore's *Nationalism* (the translation published in 1922 and since then never republished). M. Tubiansky also wrote a number of prefaces to Tagore's other works, translated by other hands. Those prefaces have retained much of their value till today. But in 1927, M. Tubiansky went to Mongolia to study Buddhism there. He came back in the mid-1930s to a different country and was executed in 1937, at the height of Stalin's terror.

The publication of Tagore's works in Russian was at its height in 1927, but practically stopped by 1929, the year of the 'Great Breakthrough', when Stalin started his 'Third Revolution' in earnest. Till 1955, practically no works of Tagore were published in the Soviet Union.

During the 1920s, Tagore was more than once invited to visit the Soviet Union, but for one reason or another he repeatedly failed to come. When he finally came to Moscow in September 1930, it was too late. The acme of his popularity had passed. The regime was obviously very suspicious of the Indian guest, and the reception was rather restrained. No big boss met the poet, nor any eminent literary figure. The whole story of that visit has not been told yet and will probably be told only when the archives of the Soviet secret police are open to historians.[13] But it seems safe to say that the immediate impact of Tagore's visit upon the cultural life of the Soviet Union was negligible.

Soon after coming home Tagore published *Russiar Chithi* (*Letters from Russia*) in Bengali. This slim book, as far as I know, has not been given yet the kind of critical attention it deserves. An adequate interpretation of this book would require a cooperation of Russian and Indian (preferably Bengali) scholars. To assess the merits and demerits of the book, a good knowledge of 20th century history of Russia, on the one hand, and an intimate understanding of Tagore's psyche, on the other hand, are needed. Here I can only outline some approaches to the problem.

First of all, *Russiar Chithi* is a very ambiguous and ambivalent book. As Amrit Sen has put it, 'this text is a fractured one, alternating between admiration, critique and anxiety'.[14] Next, it is all too clear that this book is not so much about Russia, as about India. Tagore uses the idialised image of Soviet Russia (the image which had a very remote relation to the reality) to criticise the realities of Indian life, the British attitudes to India in the first place.[15] A tricky question is whether Tagore himself believed in the reality of the image of Russia that he presented in his 'Letters' or just used that image as a rhetorical device.[16]

For today's Russians, *Russiar Chithi* makes awkward reading. Most (if not all) praise of Soviet 'achievements' now sounds obviously wrong and sometimes grotesque. For instance, in Letter 3 Tagore exclaims: 'They are striving to destroy the power of the powerful and the wealth of the wealthy. Why should we be afraid of that?'[17] But now we know that Lenin and his comrades did indeed destroy the power of the former powerful in Russia only to establish their own power which in many ways proved to be more ruthless and in fact anti-human ('cannibalistic', in Tagore's own words – see further).

Tagore rather naively admired the 'achievements' of education in the Soviet Union. But most probably he did not know that after 1917 the whole system of education in the country, from the secondary one up to universities, had been badly damaged [18] and the educated elite had been already decimated during the 1920s (still more awful things happened in the 1930s and later).

An English translation of *Russiar Chithi* came out only in 1960,[19] with some cuts, the nature of which is still to be investigated. [20] The first Russian translation came out only in 1956, in the year of the 20th Congress of the CPSU. But the Russian translation was heavily abridged on ideological grounds. Thus, letter number 13 (the unlucky number!) was omitted, evidently because it contained the following passage:

> Nevertheless, I do not believe that they have been able to draw the proper line of demarcation between the individual and society. *In that respect they are not unlike the fascists*. For this reason they are loath to admit any limit to the suppression of the individual in the name of collectivity. They forget that by enfeebling the individual, the collective being cannot be strengthened. If the individual is in shackles, society cannot be free. They have here the dictatorship of the strong man. The rule of the many by one may produce good results for a time, but not forever. It is impossible to have a succession of competent leaders. [21]

The 'Conclusion' to the *Letters* has never been published in Russian either. I suppose the most 'criminal' passage was this:

> It is not improbable that in this age Bolshevism is the treatment, but medical treatment cannot be permanent; indeed the day on which the doctor's regime comes to an end must be hailed as the red-letter day for the patient.[22]

Needless to say, the 'Communist' rulers of the country could not possibly tolerate such suggestions, because they thought they would rule forever.

On September 25, Tagore gave an interview to a reporter from the governmental newspaper, *Izvestiya* (*News*). Among other things, he said:

> There must be disagreement where minds are allowed to be free. It would not only be an uninteresting but a sterile world of mechanical regularity if all of our opinions were forcibly made alike. If you have a mission which includes all humanity, you must, for the sake of that living humanity, acknowledge the existence of differences of opinion. Opinions are constantly changed and rechanged only through the free circulation of intellectual forces and moral persuasion. Violence begets violence and blind stupidity. Freedom of mind is needed for the reception of truth; terror hopelessly kills it. [23]

By 1930, the terror now associated with the name of Stalin had not yet deployed its killing capacity at its fullest, but such words as Tagore dared to utter to his Moscow hosts were already considered subversive heresy. The farewell interview was published in Russian, by *Izvestiya*, only in 1988, under Gorbachev. [24]

Now, one more interesting fact. On January 15, 1931, *The New York Times* published a letter signed by three Russian émigrés: Ivan Ostromislensky (1880–1939), the chemist, Sergei Rachmaninoff (1873–1943), the famous composer, and Ilya Tolstoy (1866–1933), the writer, a son of Leo Tolstoy. The letter was titled 'Tagore on Russia' and contained bitter remarks on an interview given by Tagore on his arrival to the USA. Here is that letter:

To the Editor of *The New York Times*:

> The Circle of Russian Culture, the aim of which is to foster intellectual intercourse among the Russian immigrants in New York, feels compelled to comment on a recent interview given by Rabindranath Tagore.
>
> He visited Russia, and many of his statements concerning that country have appeared in different periodicals, both in this country and elsewhere. Much to our surprise, he has given praise to the activities of the Bolsheviki, and seemed rather delighted with their achievements in the field of public education. Strangely, not a word did he utter on the horrors perpetrated by the Soviet Government, and the Ogpu [the Soviet secret police – S.D.S] in particular.
>
> Time and again statements similar to his have been given out to the press by persons who, officially or otherwise, have been kept on the payroll of the Communist oppressors of Russia. The value of such public utterances is well known to every thinking man or woman. Nor is it possible to answer every one of these misstatements individually.
>
> Tagore's case is different: he is considered among the great living men of our age. His voice is heard and listened to all over the world.
>
> By eulogising the dubious pedagogical achievements of the Soviets, and by carefully omitting every reference to the indescribable torture to which the Soviets have been subjecting the Russian people for a

period of over thirteen years, he has created a false impression that no outrages actually exist under the blessings of the Soviet régime.

In view of the misunderstanding which may thus arise, we wish to ask whether he is aware of the fact that all Russia is groaning under the terrible yoke of numerically negligible but well organized gang of Communists, who are forcibly, by means of Red Terror, imposing their misrule upon the Russian people?

Does he know that, according to statistical data disseminated by the Bolsheviki themselves, between 1923 and 1928, more than 3,000,000 persons, mostly workers and peasants, were held in prisons and concentration camps which are nothing but torture houses?

He cannot be ignorant of the fact that the Communist rulers of Russia, in order to squeeze the maximum quantity of food out of the peasants, and also with the intent of reducing them to a state of abject misery, are, and have been, penalizing dissenters by exiling them to the extreme north, where those who by a miracle are able to survive the severe climate are compelled by force to perform certain work which cannot be compared even with the abomination of the galley of olden times. These unfortunate sufferers are being daily and systematically subjected to indescribable privations, humiliations, suffering and torture.

At the very time of his visit in Russia, forty-six Russian professors and engineers were executed by the Ogpu without any pretence of trial, on the alleged ground that they dared to interfere with, or doubt the wisdom of, the notorious five-year plan.

At no time, and in no country, has there ever existed a government responsible for so many cruelties, wholesale murders and common law crimes in general as those perpetrated by the Bolsheviki.

Is it really possible that, with all his love for humanity, wisdom and philosophy, he could not find words of sympathy and pity for the Russian nation?

By his evasive attitude towards the Communist grave-diggers of Russia, by the quasi-cordial stand which he has taken towards them, he has lent strong and unjust support to a group of professional murderers. By concealing from the world the truth about Russia he has inflicted, perhaps unwittingly, great harm upon the whole population of Russia, and possibly the world at large.

Iwan I. Ostromislensky, Sergei Rachmaninoff, Count Ilya L. Tolstoy

New York, Jan. 12, 1931.[25]

This letter seems to have never been republished in Russia, though now its utterances would be accepted as true by most Russians.[26]

I do not know whether or not Tagore read this letter and, if he did read it, to what extent it influenced the poet's views on the Soviet Union. But some of his later utterances about the Soviet regime were even more critical than the 'Conclusion' to the *Letters from Russia.*

For instance, in a letter to Nitindranath Tagore dated 31 July 1931 Rabi-babu wrote:

> 'As famine spreads in Europe, fascism and Bolshevism are gathering roots. Both are symptoms of disease. Men with a healthy rationality can never believe that you can cause human benefit after stifling the freedom of thought ... I fear Bolshevism might spread in India ... whatever you do, do not mix or associate yourself with this band of cannibalistic predators ... some in our country, especially Bengalis, can imitate even if they do nothing else ... now they have started to imitate this disease. Avoid the contact of this false epidemic. I am sure there are these possessed people where you stay – do not associate with them'.[27]

To this we may add a quotation from Edward Thompson's biography of Tagore:

> Two years before he [Tagore – S.D.D.] died, he told an English visitor that he would welcome even a wave of atheism over India, for it would act like a forest fire that sweeps away dead stuff and rubbish. When I was last in India, in October 1939, after the outbreak of war, he said, and only half in jest, that he was sorry he would not see the Soviet Union conquer India, for it was the only Power which would drop a bomb impartially on those two symbolically contiguous buildings, the Moslem mosque and the temple of Jagannath at Puri![28]

During the conference in Delhi, at ICWA, in May, 2013, I had an interesting talk with Ms. Supriya Roy from Santiniketan, who had edited a number of Tagore's travelogues in English translation for Visva-Bharati. We agreed that it would make sense to bring out a new, corrected and enlarged edition, in English, of *Letters from Russia* which may include additional materials (like the letter and the passage quoted above), as well as reproductions of those paintings which Tagore drew during his stay in Moscow.

Soviet authorities could hardly know about critical pronouncements of Tagore in his private correspondence or in his private talks (though some Bengali knowing agents might have reported to Moscow about the critical remarks in *Russiar Chithi*). But since the 1930s and up to Khrushchev's Thaw, Tagore was evidently a *persona non grata* in the Soviet Union.

It was only in the mid-1950s, when Khrushchev 'rehabilitated' India, that Tagore got another chance in Russia. Since 1955, new translations started being published. The 1961 jubilee was celebrated officially. Two more *Collected Works* (eight volumes in 1955–1957 and twelve in 1961–1965) were published, rather in a hurry, especially the first series of eight volumes: the first volume was brought out to greet Jawaharlal Nehru on his first visit to the Soviet Union.

The volumes of those *Collected Works* contained several pre-1917 translations and some translations done in the 1920s (e.g., the translation of *Jivansmriti* by M. Tubiansky). A number of novels, dramas, short stories and essays were translated anew or for the first time straight from Bengali. By that time, there appeared a number of people who could translate prose straight from Bengali into Russian.

As for translating the poetry of Tagore, the procedure followed was one which was widely used for translating the poetry of non-Russian Soviet poets into Russian: a person who knew Bengali would prepare a word-by-word translation of a poem, then this translation would be passed on to a 'professional' poet who was supposed to transform this 'raw stuff' into a finished poetical 'product'. Among those employed in this work were some great figures of 20th century Russian poetry, e.g., Anna Akhmatova (1889–1966), Boris Pasternak (1890–1960), Maria Petrovykh (1908–1979), and David Samoilov (1920–1990), to name only the greatest.

Boris Pasternak contributed several translations to the eight volumes of the *Collected Works*. These translations appeared in volume 7, in 1957, before the scandal ensuing from the publication in Italy of Pasternak's novel, *Doctor Zhivago*. Pasternak had been commissioned for this work because he was deservedly considered one of the leading Russian poets of that time. But he entrusted some of the translations to Olga Ivinskaya (1912–1995), who is (rightly) considered the prototype of Lara in *Doctor Zhivago*. In fact, it was Pasternak who had taught Ivinskaya the art of translating poetry, and for her it became a source of income. In Ivinskaya's memoirs, there are two or three pages in which she tells the story of her apprenticeship and how she translated Tagore's poems under the tutorship of Pasternak. [29]

Anna Akhmatova was also commissioned to translate some poems of Tagore for the *Collected Works*. All in all, she translated more than twenty poems. Translations for her were certainly a source of income, because for many years her own poems were not published and she was constantly short of money. But it is on record that Pasternak and Akhmatova were not too much inspired by Tagore's poems in the word by word translations that they worked with. There is a private letter of Pasternak in which he explicitly stated that he failed to appreciate Tagore's poetry. [30]

As far as I can judge, no poem of Tagore in a Russian translation, published in the bulky volumes of the *Collected Works*, became a great Russian poem, or at least a very popular Russian poem. Russian readers still have to rely on the opinions of others who tell them that Tagore is a great poet. The situation is somewhat similar to that with Alexander Pushkin, who is a great poet for Russians, but not for foreigners.

Nevertheless, Tagore's second coming to Russian readers in the mid-1950s and later was, probably, as culturally important as his first coming before 1917. In the mid-1950s, after Stalin's death, the Soviet Union was a very closed country, ideologically oppressed and culturally impoverished, isolated from the rest of the world. Tagore's works came to Russia then as indeed a revealing voice from a larger world. Those who had ears could hear and learn from his poetry that there were in the world many other ideas and ideals beyond the limited confines of 'Marxist' orthodoxy. I may refer to my personal experience. In 1961, the issue of *The UNESCO Courier*, which was devoted to Tagore,[31] became for me, a teenager, a very strong and lasting impression.

It may be also mentioned that a poem of Tagore (from *Śesher Kavitā*) in a Russian translation by Adelina Adalis (1900–1969) was set to music and used in a Soviet film in the 1970s. [32] Tagore even became a butt of popular jokes among Russians, along with Lenin and Brezhnev. One joke ran like this: 'The name of which great poet is composed of the names of three wines?' — 'Rubin, Granat, Kagor (Cahor).' (These were indeed the names of popular and mostly cheap wines in the Soviet Union in the 1970s and early 1980s).

It should also be recalled that from the late 1950s, Tagore's works were translated into various languages of the Soviet Union, other than Russian, sometimes from English, but mostly from Russian translations. In the available bibliographical lists, we find translations of Tagore's works into about fifteen different languages of the former Soviet Union. But it is, of course, difficult to judge how successful those translations have been.

In 1981–1982, four volumes of the *Collected Works* of Rabindranath Tagore were published from Moscow. But the edition did not contain any new translations. In 1988, a very representative selection from *Gītavitān* saw the light of day. But after 1988, and till today, no new translations of Tagore's works have appeared in Russian. In fact, since the late 1980s, very few books by Tagore have appeared at all. And whatever was published was in the form of new editions of earlier books.

Symbolically, it was in 1991, several months before the collapse of the Soviet system, that a monument to Tagore was inaugurated in Moscow. This event coincided with the visit of the Indian Prime Minister, V.P. Singh, to Moscow, and probably had something to do with the 50th anniversary of the poet's death or the 130th anniversary of his birth.

As new books by and about Tagore are rare and far between, and as Tagore is no longer on any political agenda, younger generations in Russia have a very vague (if any) idea of a person and author by the name of Tagore. Some years ago, I took a class at my university, the Russian State University for the Humanities, one of the top institutions of higher education in the humanities in this country. My students were in the third year and specialised in Western European languages and literatures. Just by chance, I asked them what they knew about Rabindranath Tagore. To my surprise, I found that not one of the students had even heard the name of Tagore. So, out of print, out of mind.

Some time in the 2000s, Vladimir Putin, who was then, as again now, the President of the Russian Federation, had to pronounce the name of Tagore on a certain ceremonial occasion. His speech was broadcast over TV. Paper in hand, Putin struggled for some time with the name 'Rabindranath' and in the end, failed to articulate it properly. This was especially telling, as Putin usually articulates even difficult Russian words very well (unlike most of his predecessors, who were notorious for their funny pronunciations). That mishap demonstrated once again that the name of Rabindranath Tagore had practically disappeared from the mental horizon of most Russians.

In 2011, the 150th anniversary of Tagore went almost unnoticed in Russia. Hence, it is time to acknowledge that the poet and his works have to be introduced anew to most Russian readers.

The number of people who know Bengali in today's Russia is rather modest. So probably one of the first tasks is to educate a new generation of connoisseurs of and translators from the Bengali language. Then probably sometime somebody will translate anew Tagore's poems into Russian in such a way that Russians will know him indeed as a great poet. A somewhat easier task is to translate Tagore's prose into Russian. Now a colleague of mine, Vladislav Korovin, has done a new translation from Bengali of Tagore's *Ghare-Baire*. We hope to get it published soon with detailed commentaries. This novel may be read as very topical in today's Russia. The problems of religion vs politics and ethics vs politics, discussed in *Ghare-Baire,* are very important problems for us now.

We can learn a lot from the English writings of Tagore as well, for instance, from his *Nationalism*. Now in Russia some people again feel the temptation of 'great power' games which twice in the 20th century brought the country to a collapse. Various kinds of nationalism again may tear the country into pieces. In my opinion, a new edition of a Russian translation of Tagore's *Nationalism* should be included in a list of compulsory reading for all politicians in today's Russia. [33]

Clearly, the potential of Rabindranath Tagore as an envoy, to Russia as elsewhere, of Indian and universally human culture is not yet exhausted.

Notes

1 Uma Das Gupta, Imre Bangha and Martin Kämpchen eds., *Rabindranath Tagore: One Hundred Years of Global Reception 1913-2013* (work in progress).

2 I have tried to draw general outlines of this story in my three papers in English. The first one ('Tagore's Reception in Russia: Yesterday and Today') was originally presented at a Tagore conference in Dhaka in 2011 and now has been published in the book: Imtiaz Ahmed, Muchkund Dubey and Veena Sikri, eds., *Contemporarising Tagore and the World* (Dhaka: University Press Limited, 2013, pp.115-130). Another paper ('What can we learn from a reception history? Tagore in Russia') was presented at a Tagore conference in Budapest in 2012 and is going to be published in the book: Imre Bangha ed., *Rabindranath Tagore Beyond Bengali Literature: Proceedings of the International Conference held at Eötvös Loránd University, Budapest on 19-20 March 2012*, New Delhi: ICCR (forthcoming). Yet another, much longer and detailed paper, has been written for the international project mentioned in the previous footnote: Uma Das Gupta, Imre Bangha and Martin Kämpchen eds., *Rabindranath Tagore: One Hundred Years of Global Reception 1913-2013* (work in progress).

3 Serebriany S. 'Tagore's Reception in Russia: Yesterday and Today' in: Imtiaz Ahmed, Muchkund Dubey and Veena Sikri, eds., *Contemporarising Tagore and the World* (Dhaka: University Press Limited, 2013, p.115-130 Further in this paper I will use parts of that paper without specifically referring to it.

4 For more details see my paper 'Leo Tolstoy Reads Shri Ramakrishna' in Marietta Stepanyants, ed. *Russia Looks at India. A Spectrum of Philosophical Views* . New Delhi: Indian Council of Philosophical Research and D.K.Printworld, 2010, pp. 325 –351.

5 My Moscow colleague Tatyana Zagorodnikova has prepared most of those letters for publication

(both from Indians to Leo Tolstoy and from him to the Indians). I hope she will be able to publish this correspondence in the foreseeable future.

6 Cf. http://nobelprize.org/nobel_prizes/literature/laureates/1913/

7 Cf. *The Nobel Prize Award Ceremony Speech:* <http://nobelprize.org/nobel_prizes/literature/laureates/1913/press.html>

8 For more details see my papers mentioned in footnote 2.

9 See note 7.

10 Cf. Паустовский К. *Повесть о жизни* (The Story of [my] life). In: Паустовский К. *Собрание сочинений.* Т. 3. Москва: Гослитиздат, 1958 (Collected Works. Vol. 3. Moscow: Goslitizdat, 1958), p. 325.

11 Cf. *Библиотека В. И. Ленина в Кремле. Каталог.* М.: Издательство Всесоюзной книжной палаты, 1961 (*The Library of V.I.Lenin in the Kremlin. A Catalogue.* Moscow: All Union Book Chamber, 1961).

12 This information I have got from my colleagues, specialists in 20th century Russian literature.

13 A.P. Gnatyuk-Danil'chuk in his book, written at the beginning of Gorbachev's 'Perestroika', wrote (referring to archives of the Soviet Foreign Ministry, published in the 1960s) that 'Tagore was supposed to stay in the Soviet Union not for two weeks but for over a month, up to the end of October 1930. The programme included, besides Moscow, visits to Leningrad, Crimea and Caucasus. Tagore was to leave the Soviet Union via Siberia, and, as he planned, to go to Japan and then the USA' (A.P. Gnatyuk-Danil'chuk, *Tagore, India and [the] Soviet Union. A Dream Fulfilled.* Calcutta: Firma KLM Private Limited, 1986, p. 221). Further on the same page the author writes that 'Tagore decided to curtail his visit' for reasons of health. But from Moscow he travelled to Berlin and reached the USA not 'via Siberia' and Japan, but via Western Europe and the Atlantic. It may be suspected that the poet's visit to the Soviet Union was 'curtailed' for reasons other than health.

14 Here I quote a very interesting, but yet unpublished paper of Dr. Amrit Sen (Santiniketan) titled "Never so wholly stirred my mind': Rabindranath and *Russiar Chithi*'. Dr. Sen has kindly sent me an electronic version of his paper.

15 Cf. 'Every day I compare conditions here with India: what is and what might have been!' (Tagore, R. *Letters from Russia.* Tr. by Sasadhar Sinha. Calcutta: Visva-Bharati, 1960, p. 4)

16 Amrit Sen (in his paper mentioned in footnote 14) writes: 'I argue that Rabindranath's construct of Russia was a multifaceted utopia which he was continuously and consciously evaluating against the contemporary realities of India. At the same time he was deeply aware of this Russia presented to him as a construct, and he attempted to see through such projections'.

17 Here I quote the translation of Krishna Kripalani from the book *Tagore: A Life*. 2nd ed. Published by the author. 1971, p. 202. Cf. R. Tagore. *Letters from Russia.* Tr. by Sasadhar Sinha. (Calcutta: Visva-Bharati, 1960), p. 14.

18 Here is just one telling detail. My mother (born in 1915) went to school in the 1920s. She used to tell me with a sigh that during her school years there was no such subject as history. They were taught only 'the history of international labour movement'.

19 Rabindranath Tagore. *Letters from Russia.* Tr. by Sasadhar Sinha. Calcutta: Visva-Bharati, 1960.

20 All in all about four pages of the Bengali original have been omitted from the translation. It is not clear whether or not those cuts were authorised by the author himself.

21 Tagore, *Letters from Russia*, p. 92.

22 Ibidem, p. 122.

23 Ibid., p. 215-216.

24 Cf. K. Datta, K. and A. Robinson. *Rabindranath Tagore: The Myriad-Minded Man*. New York: St. Martin's Press, 1995, p. 297.

25 Cf. http://select.nytimes.com/gst/abstract.html?res=9E05E2DA173FE637A25756C1A9679C946094D6CF&n=Top%2fReference%2fTimes%20Topics%2fSubjects%2fC%2fCulture Cf. also S. Bertenson, Leyda J. *Sergei Rachmaninoff: A Lifetime in Music*. (New York: New York University Press, 1956), p. 272.

26 I thank Dr. Amrit Sen for drawing my attention to this letter in 2011. It is from Dr. Sen that I learnt about the very existence of this letter.

27 Rabindranath Tagore. *Chithipatra*, p. 179 (translated from the original Bengali by Amrit Sen). I quote from the paper of Amrit Sen "Never so wholly stirred my mind': Rabindranath and *Russiar Chithi*' mentioned earlier.

28 Edward Thompson. *Rabindranath Tagore. Poet and Dramatist*. With an introduction by Harish Trivedi. Delhi etc.: Oxford University Press, 1991, p. 281. When, in the early 1980s, at the library of of Moscow State University, I looked into a copy of the first edition of this book (1948), I found that this passage was covered with a censor's black ink.

29 Cf. an English translation of the book: Ivinskaya, O. *A Captive of Time. My Years with Pasternak*. Tr. by Max Hayward. London: Harvill Press, 1978.

30 A letter dated May 30, 1958 to the French poet René Char (1907–1988). The letter was written in French. For a Russian translation, see Pasternak, B. *Complete Works with Commentaries in 11 volumes. Vol. VI. Poetic translations*. (Moscow: Слово/Slovo, 2005), p. 625.

31 Now this issue of the journal is available online: http://unesdoc.unesco.org/images/0006/000643/064331eo.pdf

32 Cf. http://en.wikipedia.org/wiki/Alexey_Rybnikov. About the film cf. http://ru.wikipedia.org/wiki/%D0%92%D0%B0%D0%BC_%D0%B8_%D0%BD%D0%B5_%D1%81%D0%BD%D0%B8%D0%BB%D0%BE%D1%81%D1%8C_(%D1%84%D0%B8%D0%BB%D1%8C%D0%BC)
The song may be listened to on the Internet: http://dl.dropbox.com/u/79279702/ %D0%9E%D0 %B1%D1%89%D0%B5%D0%B5/%D0%90.%20%D0%A0%D1%8B%D0%B1%D0%BD%D0%B8%D0%BA%D0%BE%D0%B2%20-%20%D0%9F%D0%BE%D1%81%D0%BB%D0%B5%D0%B4%D0%BD%D1%8F%D1%8F%20%D0%9F%D0%BE%D1%8D%D0%BC%D0%B0%20(%D0%98%D0%B7%20%D0%BA_%D1%84%20'%D0%92%D0%B0%D0%BC%20%D0%B8%20%D0%BD%D0%B5%20%D1%81%D0%BD%D0%B8%D0%BB%D0%BE%D1%81%D1%8C').mp3

33 I would include in such an edition also the brilliant introduction of Ramachandra Guha to the recent Penguin Classics edition of *Nationalism*.

Rabindranath Tagore and the United States

Sugata Bose

It was a late autumn evening in 1930. New York and the world were in the throes of the Great Depression. Undeterred by the gloomy economic scenario, Henry Morgenthau had invited some five hundred guests at the Biltmore Hotel to meet Rabindranath Tagore, among them Franklin Roosevelt, then governor of New York, and Sinclair Lewis, the latest winner of the Nobel prize for literature and the first American to receive that honor. Tagore was staying in midtown at 1172 Park Avenue. 'The age belongs to the West,' Tagore said in the course of his speech, 'and humanity must be grateful to you for your science.' 'But,' he added, 'you have exploited those who are helpless and humiliated those who are unfortunate with this gift. A great portion of the world suffers from your civilization.' A week later while addressing a gathering of three thousand people at Carnegie Hall he expressed warm appreciation for the ideals of liberty and freedom of the West, but deplored its inability to live up to them in the East. He was particularly saddened by the failure of many in America, despite its great anti-colonial tradition, to recognize the justness of the cause of Indian independence.[1]

Tagore's lament must not lead us to believe that he staked his intellectual position on a claim to irreconcilable difference with Western modernity. He was certainly no believer in a clash of civilizations. He was a modern man with universalist aspirations of his own. Tagore always drew a distinction between the nobler quest for universalism or a global vision and the haughtier expression of political and cultural imperialism. He saw the world as an arena of multiple and competing universalisms, which he felt ought to be in conversation with one another.[2] Earlier in the year he had held a dialogue in Germany with Albert Einstein, with whom he would meet again in the United States. But it was art, not science that was his primary subject and mode of communication with Europe and America in 1930. 'Of late I have suddenly been seized with the mania of producing pictures,' he had written to his friend William Rothenstein on March 30, 1930.[3] A renowned Japanese artist and some European painters visiting Santiniketan had liked them and he had brought along some four hundred of these paintings on his final voyage to the West. With the enthusiastic help of the Argentinian literary figure, Victoria Ocampo, an exhibition of Tagore's paintings opened to critical acclaim at the Galerie Pigalle, Paris, on May 2; Gallerie Ferdinand Moller, Berlin, held a show in July. After being displayed in Birmingham, Munich, Dresden, Moscow and Boston, the paintings came to the Fifty-Sixth Street Galleries of New York in November, 1930.

'People often ask me about the meaning of my pictures,' Tagore wrote in the catalogue for the New York exhibition. 'I remain silent even as my pictures are. It is for them to *express* and not to *explain*.'[4] Even though the poet took to painting very late in his life, his interest in art can be traced back to the turn of the twentieth century. The spirit of Asian universalism in the domain of art was brought to India by two turn-of-the-century ideologues—Okakura Kakuzo and Sister Nivedita. Okakura had been deeply influenced in his early years by the Harvard scholar of Japanese art, Ernest Francisco Fenollosa, whose collection of Japanese and Chinese paintings he later cataloged for the Boston Museum of Fine Arts. Okakura's blend of Japanese nationalism and Asian universalism was appealing as a potential model for Indian intellectuals and artists in the era of the Swadeshi movement (1905-1908). Okakura first came to India in 1902 on the eve of the publication of his book, *The Ideals of the East*, for which Sister Nivedita, the Irish-born disciple of the Hindu sage Swami Vivekananda, wrote an introduction. Once Nivedita introduced Okakura to the Tagore clan, a formidable cultural bridge was established between East and South Asia. Japanese artists Taikan Yokoyama and Shunso Hishida soon followed Okakura's trail to Calcutta. By observing Taikan, Abanindranath Tagore learned the Japanese wash technique, of which his famous painting *Bharatmata* (*Mother India*, 1905) is a prominent example; the Japanese brush-and-ink style was more deeply imbibed by Abanindranath's brother, Gaganendranath Tagore. Rabindranath was a participant in the animated conversations about art that took place on the southern verandah of their Jorasanko mansion, even though his own contributions to the cultural productions of the Swadeshi era took the form of poetry, patriotic songs, short stories and novels. 'You will cut the bond decreed by Providence, you are so powerful, are you?' he had challenged Curzon, who had dared to partition Bengal.

Tagore, the poet, truly came to the attention of the Western world once he embarked on a voyage to Europe and America in 1912 with a manuscript of *Gitanjali* that won him the Nobel Prize for Literature in 1913. Tagore's devotional verse at that time struck a chord with a section of Europe's literati, including William Butler Yeats, disenchanted with the crass materialist excesses of Britain's Edwardian age and seized by a sense of impending apocalypse in the immediate pre-World War I era. A fifteen-year old Subhas Chandra Bose wrote about Tagore to his elder brother Sarat on September 17, 1912, lamenting 'how indifferent Bengal has been in showering laurels upon him and has suffered his genius, super-human though it is, to lie in the shade of neglect, whereas a foreign people, speaking an alien tongue and cherishing ideas and sentiments, diametrically opposed to ours in some cases, have lifted him up from this shade to sunshine and have extolled him as the greatest poet the world has produced.'[5] Tagore had composed most of his *Gitanjali* poems during a period of deep despondency brought on by an unrelenting series of deaths of loved ones in his private life and the disillusionment bred by the decline of Bengal's Swadeshi movement in the public domain. The rare, austere beauty of about half of the *Gitanjali* collection established Tagore's fame as a 'mystic' in engrossed conversation with the Supreme Being and laid the basis for a blinkered, one-dimensional view of his genius.

During the months that the first edition of *Gitanjali* was enchanting readers in Europe the poet was actually in the United States. 'We have landed in New York this morning,' Tagore wrote to Rothenstein on October 27, 1912, from Herald Square Hotel, 'and passed through the ordeals of the custom house. My turban attracted the notice of a newspaper interviewer and he attacked me with questions but I was almost as silent as my turban.'[6] From New York he went to Urbana-Champaign and later lectured in Chicago and at Harvard. The United States had made an unfavorable first impression on Tagore. 'I must say I do not like it,' he wrote to Rothenstein on October 31, 1912. 'America, like an unripe fruit, has not got its proper flavor yet. It has a sharp and acid taste.'[7] He began to change his opinion upon reaching Urbana-Champaign. Ezra Pound had sent what he described as 'the scoop of the year' – six poems by Tagore – to Harriet Monroe, the editor of the *Poetry* magazine in Chicago. These were duly published in December 1912 along with Pound's critical appreciation. Tagore came to Boston in the New Year, what was to be his year of triumph, to see his old friend Okakura, who was then at the Museum of Fine Arts. It can be surmised that it was Okakura who was instrumental in having Tagore invited to Harvard.[8] At the formal invitation of James Haughton Woods, a philosophy professor, Tagore delivered four lectures at Harvard in February 14 and 17 and April 7 and 9, 1913. T.S. Eliot, then a student at Harvard, was most likely a member of the audience. The lectures were later incorporated in Tagore's book titled *Sadhana: the Realization of Life*, published in November 1913.

By the time Tagore returned to England in April 1913, he was already a celebrity. On August 24, 1913, he composed a song on his new-found fame at his temporary home on Cheyne Walk in Chelsea:

> This jeweled chain is not for me –
>
> It is painful to wear, wounds me as I struggle to tear it off.
>
> It stifles my voice, silences my melodies –
>
> Distracts my mind from my work.
>
> That is why I am waiting,
>
> If only I could offer you this necklace, I would be saved.
>
> Accept me, bind me to you with a garland of flowers –
>
> I cannot show you my face, shamed by this string of gems.[9]

'The Nobel prize for literature,' the Harvard Crimson reported awkwardly in November 1913, 'has recently been awarded to the British Indian poet, Mr. Rabindranath Tagore. This is the first time that the award has been made to other than a member of the white race. Last spring Mr. Tagore gave a series of lectures in English at Emerson Hall, dealing with subjects of far-eastern philosophy.'[10]

While the First World War raged in Europe and the Middle East, Rabindranath Tagore set off on a global oceanic voyage from Calcutta on May 3, 1916, aboard the Japanese ship *Tosamaru*. From Burma the *Tosamaru* traveled further east towards Penang, Singapore and Hongkong. The poet felt a sense of joy observing the strength and skill of Chinese laborers working at the port in Hongkong and made an uncannily accurate prophecy about the future balance of power in the world. 'The nations which now own the world's resources,' Tagore argued, 'fear the rise of China, and wish to postpone the day of that rise.'[11] On May 29, 1916, the *Tosamaru* reached the Japanese port of Kobe. Tagore's three-month sojourn in Japan represented the fulfillment not just of a personal quest, but the search for an Asian universalism that had begun at the turn of the twentieth century. Tagore was as impressed by Japanese visual arts as he was unimpressed by Japan's tendency to imitate the worst elements of European nationalistic imperialism. It was only after rebuking Japan on that count that Tagore undertook the long Pacific crossing to North America on September 7, 1916. While he was in Japan, Tagore had received an invitation to visit Vancouver. He flatly rejected it. Well aware of the turning back from Vancouver harbor in 1913 of the *Komagatamaru* carrying Indian immigrants, Tagore refused to travel to Canada and Australia until these dominions withdrew their discriminatory immigration policies.

Ironically, the ship that carried Tagore across the Pacific was named *Canadamaru*. The poet disembarked in Seattle and then traveled from the west coast to the east, delivering powerful strictures against worshiping the new god called Nation. On November 21, 1916, he came down from Yale to give a lecture at Carnegie Hall in New York titled 'The Cult of Nationalism' which the New Haven Courier reported to have been 'a memorable day for the city... all New York proclaimed that the lecture was one of the most remarkable ones, from many standpoints, ever heard in New York.' 'The Nation [with a capital N],' Tagore had declared, 'with all its paraphernalia of power and prosperity, its flags and pious hymns, its blasphemous prayers in the Churches, and the literary mock thunders of its patriotic bragging, cannot hide the fact that the Nation is the greatest evil for the Nation, that all its precautions are against it, and any new birth of its fellow in the world is always followed in its mind by the dread of a new peril.' It must be remembered that Tagore was in America during a Presidential election season at the end of which Woodrow Wilson emerged victorious by a whisker with his slogan 'He kept us out of war.' Tagore's views on nationalism might have had a very different reception after April, 1917, once the USA had entered the Great War.[12]

Tagore was not averse to the idea of love for the homeland, a kind of old-style patriotism to be distinguished from the modern ideology of nationalism. In fact, during the Swadeshi movement in Bengal of 1905, he had composed his own pious hymns to the motherland. What he saw of the hubris of nationalism in the early twentieth century, especially European rivalries culminating in the carnage of the First World War, confirmed him in his position as a powerful critic of the western model of the territorial nation-state.

It was during another trip to Europe and America in 1920-1921 that Tagore invited the young Austrian art historian Stella Kramrisch to Santiniketan. At Kramrisch's initiative in 1922, a major exhibition of 175 paintings by Bauhaus masters, including a few by Paul Klee and Wassily Kandinsky, opened in Calcutta. The following year, the sociologist, Benoy Kumar Sarkar, helped organize an exhibition of 113 paintings of the Bengal School featuring Abanindranath, Gaganendranath, Nandalal and Sunayani Devi, among others, in Berlin's National Gallery.[13] The full flowering of the late phase of Tagore's creativity as poet and painter would begin in 1924.

That year, Tagore traveled once more by sea to Burma, China, and Japan. On this trip Tagore preached the virtues of close interaction among Asian cultures. Stung by the passage of the Immigration Act of 1924 (sometimes referred to as the Orientals Exclusion Act) in the United States, some of Tagore's admirers even established an Asiatic Association in Shanghai to foster solidarity among all Asians. In Japan, Tagore's entourage had the privilege of being hosted by Rabindranath's friend, Taikan, and to be introduced once more to masterpieces of Japanese art.

Yet it was Latin America that liberated the latent artist in Rabindranath Tagore. What began as elaborate doodles on his *Purabi* manuscript would metamorphose over the next few years into his drawings and paintings. What he observed on two further journeys – to Germany and Central Europe in 1926 and to Southeast Asia in 1927 – influenced the paintings he made in a burst of creative frenzy between 1928 and 1930. The 1926 trip afforded a direct encounter with German Expressionism, while the voyage to Java enabled an innovative departure in what came to be called modern primitive art.

In a letter to Rothenstein he drew a comparison between the reception of his paintings in 1930 and the enthusiasm over his poetry in 1912-13:

> In Germany my pictures have found a very warm welcome, which was far beyond my expectation. Five of them have got their permanent place in Berlin National Gallery, and several invitations have come from other centers for their exhibition. This has a strange analogy with the time, which followed the *Gitanjali* publication – it is sudden and boisterous like a hill stream after a shower and like the same casual flood may disappear with the same emphasis of suddenness.[14]

Having learned of Ludwig Justi's interest in acquiring a few items from his Moller Gallerie exhibition for the Berlin museum, Tagore had generously gifted five original watercolors. They remained in the collection of Berlin's National Gallery until 1937 when they were removed. Tagore was in the very good company of Kandinsky and Kokoschka in having his work labeled 'degenerate art' by the Nazi regime. Three of the paintings were returned to Tagore and two traded at a Berlin art dealership for pieces of European art.[15]

While Tagore toured Europe and America in 1930, the civil disobedience movement launched by Mahatma Gandhi raged in India. The Indian peasantry suffered because of the price collapse and credit crunch of the Great Depression. The economic woes of the jute cultivators of Bengal made Tagore reluctant about spending money on another picture exhibition being planned by Rothenstein in 1931.[16] The most famous poem in his book of poems *Parishesh* (1932) titled 'Query' ('Prashna') shows how far Tagore had travelled from the devotionalism of his *Gitanjali* days. A searing indignation came through in its final verse:

Stifled is my voice today,

My flute has lost its song,

And my world swallowed up in a nightmare

By the dark night's prison.

My God, O my God, I ask you in tears,

Those that are poisoning your air,

Those that are putting out your light,

Have you given them your love, have you forgiven them all?[17]

'I feel the present ferment in India's social and cultural life should open up new vistas before our artists,' Tagore wrote to Rothenstein on November 26, 1932, 'giving them abundant material wherewith to enrich their art. Nandalal and others have already felt the need of a new orientation in their technique and subject matter and their recent works show a vigorous departure from traditions and that spirit of creative adventure.'[18]

The intimate relationship between Tagore's songs and paintings has so far not been recognized. There seem to be two reasons for this. First, Tagore did not trust the ability of his own compatriots to appreciate his paintings. He went to the extent of saying in August 1930 that he would not take his paintings back to India as he could not brook insults from those who were ignorant. Second, Henri Bidou in a review of the exhibition at Galerie Pigalle led European critics to believe that in producing his poetry, Tagore worked as a painter and that as a painter he worked like a poet. In other words, Tagore had no conception beforehand what his paintings were going to be. Bidou's interpretation of Tagore's new work at 'the borderline of two arts' underplayed the form elements in Tagore's paintings and was oblivious to the creative process that produced Tagore's songs.

'Tagore's most powerful and original [visual] work relies on two units of form,' Stella Kramrisch noted perceptively in the catalogue for Tagore's New York exhibition, 'the drop and the angle. These linear organizing elements are filled in his paintings with color evoking a texture as if of a luminous body dappled with the darkness of its substantiality.'[19] Beyond these two premeditated form elements, yes, the moment of making paintings and composing songs was 'play time' for Rabindranath. 'Songs,' Tagore wrote in *The Diary of a Westbound Traveler*, 'are entirely the play of creation. As the rainbow appears, like the magic creation of rain and sun, a built-up archway of two whimsical temperaments in the sky, an exquisite moment proceeds on its triumphal journey through this very archway. In an instant the play is over, the moment passes by waving its colored garment; there is nothing more! Song-poems are nothing but this temperamental play of color. If the poet of the rainbow were accosted and asked, 'What is one to make out from all this?' the reply one would meet with would be, 'Nothing!' 'Then?' 'It just pleases me!'- the pleasure of the pure form, to every challenge, to creation, this is the last answer."[20]

Tagore came to the United States with his critiques of narrow nationalism but also bearing his gifts of song-poems and paintings. The American composer, John Alden Carpenter, set several of his *Gitanjali* song-poems in English into music. One of the songs he chose for his musical setting may serve as an especially poignant remembrance of the poet himself.

> On the day when death will knock at thy door, what wilt thou offer to him?
>
> Oh, I will set before my guest the full vessel of my life;
>
> I will never let him go with empty hands.
>
> All the sweet vintage of all my autumn days and summer nights, all the earnings and gleanings of my busy life, will I place before him at the close of my days when death will knock at my door.

Notes

1 See Prabhat Kumar Mukhopadhyay, *Rabindrajibani o Rabindrasahitya-prabeshak Volume 3* (Kolkata: Visvabharati Granthalay, 1952), pp. 288-291.

2 On Tagore's different universalism, see Sugata Bose, *A Hundred Horizons: the Indian Ocean in the Age of Global Empire* (Cambridge, MA: Harvard University Press, 2006), pp. 233-271; Sugata Bose, 'Different Universalisms, Colorful Cosmopolitanisms: the Global Imagination of the Colonized' in Sugata Bose and Kris Manjapra (eds.), *Cosmopolitan Thought Zones: South Asia and the Global Circulation of Ideas* (London: Palgrave Macmillan, 2010), pp. 97-111.

3 Rabindranath Tagore to William Rothenstein, March 30, 1930, in Mary Lago (ed.), *Imperfect Encounter: Letters of William Rothenstein and Rabindranath Tagore, 1911-1941* (Cambridge, MA: Harvard University Press, 1972), pp. 325-326 (original in the Rothenstein Papers, Houghton Library, Harvard University).

4 Rabindranath Tagore, 'The Language of Pictures' in *Rabindranath Tagore: Exhibition of Paintings* (New York: The Fifty-Sixth Street Galleries, November 1930).

5 Subhas Chandra Bose to Sarat Chandra Bose, September 17, 1912, in Sisir K. Bose and Sugata Bose (eds.), *Netaji Subhas Chandra Bose Collected Works Volume 1: An Indian Pilgrim* (Calcutta: Netaji Research Bureau and Delhi: Oxford University Press, 1997), pp. 150-152.

6 Tagore to Rothenstein, October 27, 1912, in Lago (ed.), *Imperfect Encounter*, p. 56 (original in the Rothenstein Papers, Houghton Library, Harvard University).

7 Tagore to Rothenstein, October 31, 1912, in Lago (ed.), *Imperfect Encounter*, p. 57 (original in the Rothenstein Papers, Houghton Library, Harvard University).

8 This is convincingly argued in Mou Banerjee, 'Mr. Tagore at Emerson' (unpublished seminar paper, Harvard University).

9 This is my translation of Tagore's '*E Monihar Amay Nahi Saje*'. Sugata Bose (trs.), *Tagore: the World*

Voyager (Delhi: Random House, 2012, 2013), p. 8.

10 *The Harvard Crimson*, November 22, 1913.

11 See Prabhat Kumar Mukhopadhyay, *Rabindrajibani o Rabindrasahitya-prabeshak Volume 2* (Kolkata: Visvabharati Granthalay, 1961), pp. 452-453.

12 Ibid., pp. 464-475.

13 Kris Manjapra, 'Stella Kramrisch and the Bauhaus in Calcutta' in R. Siva Kumar (ed.), *The Last Harvest: Paintings of Rabindranath Tagore* (New York: Mapin, 2011).

14 Tagore to Rothenstein, August 24, 1930, in Lago (ed.), *Imperfect Encounter*, pp. 328-329 (original in the Rothenstein Papers, Houghton Library, Harvard University).

15 Manjapra, 'Stella Kramrisch and the Bauhaus in Calcutta' and Kris Manjapra, *Age of Entanglement: Indian and German Intellectual Encounters across Empires* (Cambridge, MA: Harvard University Press, 2014).

16 Tagore to Rothenstein, March 24, 1931, in Lago (ed.), *Imperfect Encounter*, p. 334 (original in the Rothenstein Papers, Houghton Library, Harvard University).

17 This is Churu C. Chowdhuri's translation of Tagore's 'Prashna'. Krishna Bose and Sugata Bose (eds.), *Purabi: The East in its Feminine Gender* (Kolkata: Seagull, and Chicago: University of Chicago Press, 2008).

18 Tagore to Rothenstein, November 26, 1932, in Lago (ed.), *Imperfect Encounter*, pp. 344-347 (original in the Rothenstein Papers, Houghton Library, Harvard University).

19 Stella Kramrisch in in *Rabindranath Tagore: Exhibition of Paintings* (New York: The Fifty-Sixth Street Galleries, November 1930).

20 Rabindranath Tagore, *The Diary of a Westward Voyage (Pashchim Jatrir Diary)*, trs. Indu Dutt (Wesport, CT: Greenwood, 1975), p. 62.

Tagore as International Educationist: The 'Bond of Relation'

Kathleen M. O'Connell

Throughout Tagore's 150th birth anniversary celebrations in 2011, he was celebrated enthusiastically as a global humanitarian and creative artist. Though receiving some acknowledgement for his work in education, Rabindranath's pioneering role in the history of international education failed to receive adequate recognition, and he remains relatively unknown in the field. This is an area that has special relevance in today's world of clashing cultures and common dilemmas, and Tagore's contribution as a global ambassador of education deserves further elaboration and exploration. What is clear is that Rabindranath was a seminal thinker and innovator in the area, and his philosophy of inclusiveness with its sensitivity to the problems of environment, race, language, cultural differences, economic disparity, and political and gender imbalance represents one of the earliest comprehensive paradigms for international education and global collaboration. Moreover, he was (and remains) one of a handful who put his theories into constructive practice, not only at the primary and secondary levels, but as one of the earliest post-secondary models for internationalism. As he wrote many decades ago:

> When races come together, as in the present age, it should not be merely the gathering of a crowd; there must be a bond of relation, or they will collide with each other...Education must enable every child to understand and fulfil this purpose of the age, not defeat it by acquiring the habit of creating divisions and cherishing national prejudices.[1]

This essay positions Tagore's role within the pre-WWII history of international education and explores his evolving strategies for implementing international education in the context of the personal encounters and experiences of his western travels. It documents the way in which his travels to the West and interpersonal relations and growing familiarity with other global educators affected his own perceptions and were reflected in his educational model at Visva-Bharati. The relational qualities he came to emphasize in his educational experiment, such as empathy, cooperation, creativity and cosmopolitanism are explored—along with his insistence that individuals should be encountered one by one, rather than as categories or generalities.

Pioneers of International Education: World Citizenship over the Nation-State

The term 'international education' has been in use since the middle of the nineteenth century, but has come to have vague associations relating to a number of areas such as international studies, multi-cultural education, international cultural exchange and education for international business. The defining and construction of a history of international education, particularly as it predates the second world war, has been a relatively recently phenomenon. In general, the movement for international education has been linked to the humanist-progressive movement, which advocates a more democratic child-centered form of education and is indebted to such figures as Jean Jacques Rousseau, Johan Pestalozzi, Frederich Froebel and others, though the ideals of world citizenship would not have been clearly evident.

Some would trace an educational movement for internationalism back as far as John Amos Comenius (1592-1670), a Moravian bishop, who has been called 'the teacher of nations.'[2] More recently, it has been argued that international education should be examined against the 19th century background of aggressive nationalism and colonialism, and this is the educational arena in which Tagore's pioneering work has the greatest relevance. D.G. Scanlon, historian of international education, describes the nineteenth century as marked by the development of great national systems of education, which varied from despotism to the humanitarianism of a Pestalozzi, but were united in the advancement of national interests. The pioneers who began developing alternative types of education for international understanding were unique in this background of aggressive nationalism: that is, of a form of international education that privileged humanity over the nation-state. Regarding these early pioneers, Scanlon writes:

> For fundamentally all were out of step with the nineteenth century. In an era of provincial loyalties, they argued for loyalty to mankind. In an era of nationalism, they spoke of internationalism. And in an era of provincial loyalties, they argued for loyalty to mankind. And in an era of mass education for patriotism, they contended that the school was the only agency capable of advancing education across national boundaries. Little wonder that their proposals were viewed as radical, visionary, and utopian.[3]

In the early history of international education, one of most impressive examples comes from the *folkelig* movement in Denmark and the Folk Schools established in 1856 by Nikolaj F.S. Grundtvig, a poet and myriad-minded thinker like Tagore, who was instrumental in shaping his particular culture in a more inclusive manner.[4] Grundtvig argued for, and helped establish, a non-violent egalitarian educational system in Denmark for 'all the people' that would promote freedom of thought and the incorporation of the disenfranchised into the public schools. Like Rabindranath, the relevance of his educational work has not received sufficient recognition in the pre-WWII history of international education.[5]

In 1901, a bulletin was published by Franz Kemeny, a Hungarian educator, proposing that an international bureau should be initiated to encourage six areas of international education: 1) studies of educational systems in various countries; 2) international conferences for teachers; 3) international codes for the organization and structure of education; 4) the teaching of human rights based on western democratic principles; 5) Education for peace (to counter chauvinism); 6) Inter-racial education (to counteract race prejudice). However there is no evidence that his work was taken up in any practical way. Similarly, Mrs. Fannie Fern Andrews of Boston organized a conference on international education at The Hague and argued for an international office of education within the League of Nations, but, like Kemeny, she did not establish a school. The League of Nations, whose 1921 Committee on Intellectual Cooperation included Henri Bertson, Eve Curie, Gilbert Murray, Albert Einstein and Tagore's close scientist friend, Jagadis Bose, outlined its purposes in a pamphlet:

> General culture should make our pupils into men and women for whom internationalism, which we consider to be the outcome of individual nationalism, will become a reality, a second nature, a mode of their thought. We have to prepare a generation which will carry over on to the international plane ideas of brotherhood, understanding, and mutual assistance.[6]

The work of UNESCO and the New Education Fellowship in the years that followed would mark a continuation of these efforts to define and implement an idiom for international education.

Pioneers, who sought to develop a new form of education for international understanding and world citizenship through the creation of their own schools include such educators as Paul Geheeb, Aldous and Dora Huxley, Leonard and Dorothy Elmhirst, John Dewey, Maria Montessori, and, of course, Rabindranath Tagore. They stand out as some of the earliest architects of a form of secular international education that promoted a global identity rather than a nationalistic one. The German experimental school, known as the Oldenwaldeschule established by Paul Geheeb and Edith Cassirer-Geheeb in 1910, was one of the early practical experiments. The cosmopolitan ideals of his school conflicted with Nazi ideology and he had to flee Germany, later founding the Ecole d'humanite in Switzerland in 1937, which continues to the present time at the secondary level. Tagore corresponded with Geheeb and visited his school. The Beacon Hill School, a small primary experimental school that placed great emphasis upon human rights and a cosmopolitan outlook, was established by Aldous and Dora Huxley somewhat later in the 1920s.[7] Although it generated considerable interest, its history was of short duration. In 1924, a group of parents from the League of Nations and International Labour Office in conjunction with Adolphe Ferriere, and Elisabeth Rotten of the Geneva Rousseau Institute founded the International School of Geneva. The school, which has continued to be an important centre for international studies at the primary and secondary levels, considers its mandate to be the overcoming of narrow nationalistic pride and the promotion of understanding of individuals and groups of other cultures.[8] Dartington Hall, developed by Leonard and Dorothy Elmhirst, and inspired by Tagore, also provided an early model for international education and cultural exchange. Schumacher College, an offshoot on the Dartington campus, continues the tradition with its mandate of education for global sustainability. Two educators whose work has close connections with Tagore's global outreach are Maria Montessori and John Dewey.

Maria Montessori (1870-1952): Education for Commonality

Maria Montessori, a pioneer in the field of humanistic education, became the first Italian female M.D. in Italy in 1896. She became widely known for her work with illiterate children in Rome with her *Case dei Bambini*, which opened in 1905, from which she developed her own principles of learning.

Montessori's writings drew from a variety of sources including psychoanalytic insights regarding the unconscious, challenged the adult-centric perception of early childhood, and argued for a child's dignity and autonomy. An examination of some of the central concepts of her educational theory reveals many parallels with Tagore's thought.[9] Montessori's international humanism and pacifist efforts stemmed from her wide travels and close observation of children from different backgrounds and different parts of the world, and she wrote:

> Through the study of children, I have scrutinised human nature at its origin both in the East and the West and although it is forty years now since I began my work, childhood still seems to me an inexhaustible source of revelations and – let me say – of hope... Childhood has shown me that all humanity is one. All children talk, no matter what their race or their circumstances or their family, more or less at the same age; they walk, change their teeth etc. at certain fixed periods of their life. In other aspects also, especially in the psychical field, they are just as similar, just as susceptible.[10]

Tagore met Montessori during his 1926 trip to Italy, and they were both part the New Education Fellowship, an international education group seeking to break down national barriers. Tagore was supportive of Montessori's efforts, agreeing to become a patron of the Association Montessori Internationale (A.M.I.), and of the Teacher's Training Course, which she organized when she arrived in Adyar, India in 1939. Today, Montessori schools are found throughout the world, and the Montessorian World International network sees its mandate as one that promotes global communities and the values of peaceful exchange .

John Dewey (1859-1952): Education for Cooperative Learning

John Dewey was part of the progressive-humanist movement that rebelled against the traditional authoritarian educational system of his time, which was deemed out of touch with its surroundings, in this case, a democratic and industrializing America. In 1896, he and his wife, Alice Chipman Dewey, opened the Laboratory School connected with the University of Chicago, which proposed 'to train children in cooperative and mutually useful living.'[11] Although there is no evidence that Tagore and Dewey ever met personally, it is likely that Tagore would have become familiar with his experimental Laboratory School that was connected to the University of Chicago during his visit there in 1913, as well as through his later association with Leonard Elmhirst.[12] P.K. Mukhopadhyay writes that Tagore lectured at the University of Chicago in January 1913 and that 'America had given him new points about his school at Bolpur.'[13] Dewey's influence, via Elmhirst, can be found in various aspects of the rural reconstruction work at Sriniketan, particularly in the educational format of Siksha-Satra with its emphasis upon hands-on learning and social integration. In 1924, William Kilpatrick, who developed the project method after Dewey's theories, visited Santiniketan, where he observed the educational methods and had discussions with the workers there.[14]

Dewey's emphasis upon a broad democratic approach and hands on learning that was grounded in real-life activities of one's community had parallels with Tagore's educational approach. Dewey also emphasized the importance of global citizenship from the earliest days of the child's schooling.

Children would be asked to learn about the products that were part of their daily life, discovering their connection to regions across the world.[15] Dewey's influence can be found in curriculums throughout the world.

Cambridge educator J.J. Findlay – who dedicates his book *Foundations of Education* to Tagore– cites Tagore and Dewey as the 'two greatest educators of our epoch' and writes the following comparison:

> Hence while studies matter greatly, community life matters still more: the popular phrases on the lips of the reformers are in their speech too:--freedom, activity, initiative, individuality, sociality, self-government, altruism: even interest and happiness, but these are informed with fuller significance. With Tagore the environment, the *ashram*, stars and sky, friends and neighbours, are the means whereby an inner happiness is fostered.[16]

Rabindranath Tagore (1861-1941): Education for Inclusiveness[17]

Rabindranath's educational theory and practice has links to the progressive-humanist education that took shape internationally during the nineteenth century against a background of growing nationalism. Affinities can be seen between his emphases upon internationalism, social inclusiveness, human freedom and creativity and the priorities of such pioneers as Nikolaj F.S. Grundtvig, Paul Geheeb, Maria Montessori and John Dewey. Tagore's paradigm of inclusiveness draws from various sources traditional and modern, and there is evidence that he was aware of the other educational models that were emerging. As the *tapoban* had served as a prototype for Santiniketan, for Visva-Bharati, Rabindranath referred to the ancient Buddhist monastaries of Nalanda, Taxila and Vikramshila, as Indian models of hospitality, cosmopolitanism, scholarship and a harmonious relationship with the local community. The motto of Visva-Bharati drawn from the Upanisads, reflects the inclusive global scope of the undertaking: *yatra visvam bhavati ekanidam*---'Where the world meets in one nest.' There was also the generous model of cultural and social exchange which he had experienced at the Tagore joint-family mansion, Jorasanko--the 'living university'--which provided a practical example of 'the world meeting in one nest.' Here, individuals of different religious and cultural backgrounds interacted in a manner that permitted a balance between individual creativity and common concerns.

From the early days of Rabindranath's small experimental school in Santiniketan, which was inaugurated in 1901, a cosmopolitan approach was evident. The first constitutional letter interprets the Gayatri mantra, which was recited in the morning ceremony, in a manner that transcends national borders to a level of profound interconnectivity. As Tagore writes concerning his interpretation of the Gayatri: *Om bhur bhuvah svah*:

> This part of the Gayatri mantra is known as *vyahrti*.[18] *Vyahrti* means the bringing together from all directions. In the beginning of the meditation, the *bhu* realm [earth region below], the *bhuvar* realm [atmospheric intermediate region] and *svar* realm [heavenly region above], i.e., the universe, is to be

summoned within the mind. At that time I am to think that I am standing in the middle of the entire universe: *I am now no mere resident of some particular country.* Standing in the middle of the universe. one should meditate upon the excellent knowledge and power of Savita [sun-god], the master of creation. One has to think of the unimaginably vast universe which is being flashed this and every instant from Him. How is my immediate existence linked to that unlimited power which ceaselessly manifests the three regions? What path should I follow in order to meditate on Him?[19]

Educationally, he distinguished the need for various levels of education: 'education of the senses' (*indriyer siksa*); 'education of the intellect' *(jnaner siksa*). and 'cultivation of feeling' (*bodher tapasya/sadhana*) which involves an expansion of sympathy in kinship with all existence. What Tagore seems to have considered most primary was what he termed 'Education for Sympathy or Compassion', that is, an empathetic sense of interconnectedness with the surrounding world.

The theme of interconnectedness—in terms of the natural environment, with creative endeavours, society and the global village – has a special resonance throughout Tagore's writings, and his social and educational critique very precisely pinpoints the disconnect that he experienced around him: whether it be the disconnect caused by an educational methodology that alienates a child from the natural environment, or the disconnect caused by a foreign-based educational system taught in English. He also delineated the disconnect between urban elites and uneducated villagers, the barriers between women and the educational system, and the alienation and misunderstanding that was being caused at all levels by aggressive nationalism. Having defined these areas, he began devising practical and innovative educational strategies to overcome them. He saw the arts, for example, as an area that encouraged cross-cultural collaboration, and mutual service as a vehicle for social cohesiveness. With his own increased travels and experience, Tagore became aware of the dangers of globalized technological exploitation and the challenges of environmental sustainability, and he strove through his teaching and writings to increase awareness of the threats that faced future societies.

Tagore's Foreign Travels and the Inaguration of Visva-Bharati

Tagore's international travels and the personal relationships he formed during that period, greatly contributed to his educational thought and practical experiment. In his 1912-13 visit to England and North America, he met with artists and educators, visiting institutions that would affect his ideas on education, art and internationalism. Simultaneously, his own ideas were having a far-reaching effect on those he met, and his manuscript of *Gitanjali* began to have a circulation that would culminate in the Nobel prize. During his first visit to North America in February, 1913, he received an invitation to give three lectures at Harvard by James Houghton Woods, Professor of Indian philosophy. Tagore's lectures were from the *Sadhana* series, but his educational work was also becoming known. Ellery Sedgwick, who was the editor of the *Atlantic Monthly* from 1908-38, met Tagore during his Harvard lectures and requested him to write an article on Santiniketan, which was published in July 1913. During the same trip, he spoke on 'Race Relations' at the annual Congress of Religious Liberals in Rochester. In a later trip to Chicago, Rabindranath visited a school, which was likely

Dewey's Laboratory School, which he described in a letter back to one of his Santiniketan teachers.

One of Rabindranath's first statements regarding his intention to establish a centre for a type of international education that would counter aggressive nationalism came in a 1916 letter written from Los Angeles, California to his son, Rathindranath:

> I have it in mind to make Santiniketan the connecting thread between India and the world. I have to found a world centre for the study of humanity there. The days of petty nationalism are numbered – let the first step towards universal union occur in the fields of Bolpur. I want to make that place somewhere beyond the limits of nation and geography—the first flag of victorious universal humanism will be planted there. To rid the world of the suffocating coils of national pride will be the task of my remaining years.[20]

Tagore put forth a blueprint for an educational centre modelled on sustainability and cultural collaboration in his essay 'Centre of Indian Culture'. He envisioned a centre that would be in constant touch with the totality of life: economic, environmental, intellectual, aesthetic and social--with aesthetic education having a prominent seat of honour.[21]

1921: Visva-Bharati as a Centre for Global Cooperation [22]

I have in mind not merely a University—that is only one of the aspects of our Visva Bharati,—but I hope this is going to be a great meeting place for individuals from all countries who believe in our spiritual unity and who have suffered from the lack of it, who want to make atonement and come into human touch with their neighbours... This we have tried to do in Visva-Bharati. Our endeavour has been to include this ideal of unity in all the activities in our institution, some educational, some that comprise different kinds of artistic expression, some in the shape of service to our neighbours by way of helping the reconstruction of village life. As I wanted this institution to be inter-racial, I invited great minds from the West. They cordially responded, and some have come permanently to join hands with us and build a place where men of all nations and countries may find their true home, without molestation from the prosperous who are always afraid of idealism or from the politically powerful who are always suspicious of humans who have the freedom of spirit.'[23]

Tagore's 1920-21 visits to Europe and North America brought him into contact with various artists and intellectuals, resulting in an exchange of ideas that would have an influence on the shaping of Visva-Bharati. In Paris he met Sylvain Levi, a noted Indologist from the Sorbonne, who supported his ideas concerning the unifying influence of Hindu and Buddhist culture throughout Asia. Levi indicated his willingness to come to India and help develop the Asian level of Visva-Bharati.[24] After visits to Belgium, Denmark and the Netherlands, where Tagore was enthusiastically received, he set sail for New York with William Pearson. Since his previous visit in 1916, there had been considerable technological advance in North America, and the postwar atmosphere was less responsive to Tagore's idealism than it had been on previous visits. However, valuable friendships were made with such individuals as Helen Keller and Jane Addams – and with Leonard Elmhirst, who would later head the Institute for Rural Reconstruction at Sriniketan.

After five months he returned to Europe where there was more support for Visva-Bharati ideals, particularly in his contacts with Romain Rolland, Scottish biologist and sociologist Patrick Geddes, and art historian Stella Kramrisch. His reception in Germany was unprecedented with huge crowds turning out in Berlin, Munich, and Darmstadt, where Count Keyserling declared a 'Tagore Week' at his School of Wisdom. The final stops before returning to India in July 1921 were Vienna and Prague, where he began a friendship with Indologists Moriz Winternitz[25] and Vincent Lesny, who would also become visiting professors at Visva-Bharati.[26]

In terms of education, Rabindranath repeated his concerns for a holistic learning process in touch with nature, a process that seeks full expansion of the personality through the arts, and cultivates social and cultural collaboration. His address, 'An Eastern University,' contains the plan for Visva-Bharati which was set forth in 'A Centre of Indian Culture,'[27] with the specific addition of a far Eastern dimension - Chinese, Japanese and Tibetan — 'for, in the past, India did not remain isolated within her own boundaries.'[28] The meeting-ground of cultures must be a type of learning centre where conflicting interests are minimized 'where we can work together in a common pursuit of truth, share together our common heritage, and realise that artists in all parts of the world have created forms of beauty, scientists discovered secrets of the universe, philosophers solved the problems of existence, saints made the truth of the spiritual world organic in their own lives, not merely for some particular race to which they belonged, but for all mankind.'[29]

The Constitution, which reflected the influence of Tagore's travels, designated Visva-Bharati as an Indian, Eastern and Global cultural centre whose goals were:

a) To study the mind of Man in its realisation of different aspects of truth from diverse points of view.

b) To bring into more intimate relation with one another through patient study and research, the different cultures of the East on the basis of their underlying unity.

c) To approach the West from the standpoint of such a unity of the life and thought of Asia.

d) To seek to realise in a common fellowship of study, the meeting of East and West and thus ultimately to strengthen the fundamental conditions of world peace through the free communication of ideas between the two hemispheres.

e) And with such ideals in view to provide at Santiniketan a centre of culture where research into the study of the religion, literature, history, science and art of Hindu, Buddhist, Jain, Zoroastrian, Islamic, Sikh, Christian and other civilizations may be pursued along with the culture of the West, with that simplicity of externals which is necessary for true spiritual realisation, in amity, good-fellowship and co-operation between the thinkers and scholars of both Eastern and Western countries, free from all antagonisms of race, nationality, creed or caste and in the name of the One Supreme Being who is Shantam, Shivam, Advaitam.[30]

Rabindranath began building up a network of Indian and international scholars. Among those from Bengal were Kshitimohan Sen, a scholar of Pali and folk culture; Sanskritist Bidhu Shekhar Sastri, who took charge of Sanskrit, Pali and Prakrit; linguist Haricharan Bandyopadhyaya, who was compiling his Bengali dictionary; Vidhusekhara Bhattacharya, who taught the Vedas, Puranas and Buddhist scriptures; and others.[31] Those outside Bengal included: Dharmandhar Rajguru Mahasthabir, a Sri Lankan Buddhist monk who taught Buddhist culture; Kapilesvar Misra from Mithila who taught Sanskrit grammar; Maharashtrian Bhimrao Shastri, who taught Sanskrit; Gurdial Mallik, a Punjabi who taught Sufism; and Jain monks, Jinvijayaraj Muni and Sukhlaji. In 1927, Islamic studies were given a boost by the establishment of the Nizam Chair of Islamic Studies made possible by a donation of a lakh of rupees from the Nizam of Hyderabad.[32]

East-West Links

In 1924, Nandalal Bose, Kshitimohan Sen, Kalidas Nag, Leonard Elmhirst and others accompanied Tagore on a four-month tour of Burma, Malaya, China and Japan.[33] The *Modern Review* coverage of the Chinese tour mentions discussions with Chinese artists and scholars, such as Hu Shih and Wang-chi-Lin. Such discussions centered on the 'various thought-currents of modern China as well as on the possibility of the compilation of the history and philosophy of Buddhism through the collaboration of Indian and Chinese scholars.' They gave lectures on Indian art, presented the publications of the Indian Art Society and made plans for a student exchange program.[34] Their visit led to the establishment of an Asiatic Society in Shanghai and to an opening-up between Chinese and Japanese scholars.[35] The trip to Japan and a subsequent one in 1929, led to the formation of a Tagore Society in Tokyo and further cultural exchange with Visva-Bharati.

In 1927, Tagore visited Indonesia, Java and Bali, accompanied by comparative philologist Suniti Kumar Chatterjee, Dutch Sanskritist and musician, A. Bake and others, who studied the pervasive influence of Hinduism and Buddhism in those areas.[36] Rabindranath was particularly taken with the dance forms and various individuals and agencies presented books and art pieces for Visva-Bharati. Surendranath Kar brought back the techniques of batik which later became a special feature of Santiniketan.

An important link between India and China was made through Chinese Buddhist scholar Tan Yat-Shan who came to Santiniketan from China in 1928, and then returned to raise funds for the foundation of the Sino-Indian Cultural Society. He became the first Director of Cheena Bhavan, inaugurated in April, 1937. Others on the staff included: Tibetan scholar Lama Angi Wangoli; Persian scholar Aga Puri Daud; Japanese Ju-jitsu artist Nokuzo Takagaki who started Ju-jitsu classes in 1929; Prof. Lim from Rangoon who taught Chinese literature; and Javanese dancer Soekoro; as well as numerous other visiting scholars and artists. An extensive library was set up for Eastern writings,[37] and arts and crafts from Japan,[38] China, Malaya, Thailand, Java and Sumatra were introduced. Asian students who came to study at Santiniketan became cultural ambassadors.

As the Non-Cooperation movement was taking shape on the Indian side, Rabindranath invited creative individuals from around the world. The extended-family model represented a shift from European models whose chief goal was intellectual. Tagore's model emphasized the development of human sympathy through community life and mutual service in the presence of nature. At the centre of the international community was the shared vision of Tagore and C.F. Andrews.[39] Andrews held many positions at Visva-Bharati – both academic and administrative – until his death in 1940. In his memory, an institute 'Deenabandhu Bhavan' was established to promote a 'fuller contact and understanding between the East and the civilizations of the West in their cultural and humanistic aspect.'[40]

By 1924, the academic curriculum included modern and classical languages, logic, philosophy, political economy, sociology and science. The Visva-Bharati staff and students came from different parts of India and the world. Tagore himself taught English literature along with Andrews and Pearson, with Rathindranath Tagore covering biology. A sampling of scholars who taught Western subjects included Hirjibhai Pestonji Maurice, a Parsi from Bombay who taught French and English; Narsinghbhai Patel, a Gujarati from East Africa who taught German; Jehangir Bakil, an Oxford graduate who taught English and philosophy and was Principal of Shiksha Bhavan. Vincent Lesny (of the Charles University) taught German, and M. Ferdinand Benoit came from Switzerland to teach French. Sir Patrick Geddes visited Santiniketan in 1922 and created a plan for the layout of Santiniketan and Sriniketan and his son, Arthur, worked at Sriniketan. Among those joining in the late 1930s were Alex Aronson of Haifa, who came as a lecturer in English, and Marjorie Sykes, a British Quaker who served in various capacities at Santiniketan and Sriniketan and later worked with Gandhi at Wardha.

European specialists in Eastern studies included Winternitz, Prof. Sten Konow from Oslo University and Bogdanov, a Russian scholar of Persian and Pali. Julius Germanus of Hungary became the Nizam Professor of Islamic Studies during 1929. During Tagore's 1925 European trip, he invited Italian Indologists Carlo Formichi and Giusepe Tucci to join the staff.

European artists and musicians included Austrian art historian Stella Kramrisch, French artist Andree Karpeles,[41] Viennese sculptress Lisa Von Pott, Swedish craftspersons Miss Jeanson and Aina Cederblom and many others. As Santiniketan evolved to include an ever-widening circle, Tagore's concern with societal and global issues increased. The Centre for Rural Reconstruction is Sriniketan, inaugurated in 1922, extended Tagore's vision of inclusiveness to include the rural areas surrounding Santiniketan. The 1929 Visva-Bharati prospectus states that:

> College students are expected to become familiar with the working of existing institutions and new movements inaugurated in the different countries of the world for the amelioration of the social conditions of the masses. They are also required to undertake a study of international organisations so that their outlook may become better adjusted to the needs of peace...The aim of this education is to ensure that the students should, in thought, emotion and action, attain truth and achieve the fullest development in all the various manifestations of the human spirit.[42]

Conclusion

This article has provided a brief overview of some of the pioneering efforts in the history of pre-WWII international education and explored Tagore's role as a global ambassador of education and the manner in which his foreign travels affected his own educational paradigm. It has argued that his vision of inclusiveness and interconnectedness—in terms of the natural environment, with art, society and the global village – represents one of the most comprehensive paradigms for international education. As well, it has documented the evolution of Visva-Bharati, one of the earliest post-secondary models of international collaboration that privileges world citizenship over a narrow nationalist identity. The objectives of Rabindranath were summed up in a letter (September 16, 1934) to Gilbert Murray, who was instrumental in establishing the League of Nations International Committee of Intellectual Cooperation:

> I agree in complete agreement with you in believing that at no other period of history has mankind as a whole been more alive to the need of human co-operation, more conscious of the inevitable and inescapable moral links which hold together the fabric of human civilization... Willingly therefore I harness myself, in my advanced age, to the arduous responsibility of creating in our Educational Colony in Santiniketan a spirit of genuine international collaboration based on a definite pursuit of knowledge, a pursuit carried on in an atmosphere of friendly community life, harmonized with Nature, and offering freedom of individual self-expression.[43]

Notes

1 Rabindranath Tagore, 'My School' (lecture in Tokyo, Japan, June 1924), *Modern Review*, May 1925, in *English Writings of Tagore*, vol. IV, edited by Nityapriya Ghosh (New Delhi: Sahitya Akademi, 2007), 522.

2 Robert Sylvester, 'Mapping international education: A historical survey 1893-1944,' *Journal of Research in International Education* vol 1 (2002) 90-125.

3 D.G. Scanlon, ed., *International Education; A Documentary History* (New York: Columbia Universisty Bureau of Publications, Teachers College, 1960) 3-4.

4 Tagore would likely have heard about Grundtvig's educational ideas from a Danish woman Hiordis Siggard (Tagore renamed her Haimanti Devi), who came to Santiniketan in the 1920s and married his secretary Amiya Chakravarty.

5 For a discussion of Grundtvig's contribution, see Steven M. Borish, *Danish Social Movements in a Time of Global Destabilization* (Vegle, Denmark:Kroghs Forlag A/S, 1996), pp.119-132.

6 WFEA(World Federation of Education Association), 1929: 94, quoted in Sylvester, 'Mapping International Education,' *Journal of Research in International Education* vol 1 (2002), p. 104.

7 See: Deborah Gorham (2005) 'Dora and Bertrand Russell and Beacon Hill School,' *Russell: the Journal*

of Bertrand Russell Studies: Vol. 25: Iss. 1, Article 4. http://digitalcommons.mcmaster.ca/russelljournal/vol25/iss1/4

8 Sylvester, 'Mapping International Education, p. 103.

9 The first indication of Tagore's awareness of Montessori's work comes in a Bengali article published in the *Tattvabodini Patrika* in 1913, when he was the editor.

10 *Maria Montessori: a Centenary Anthology 1870-1970* (Amsterdam: Association Montessori Internationale, 1970), p. 19. One finds similar echoes in Tagore's poem, 'On the seashore of endless time children play.'

11 A.E. Meyer, *An Educational History of the Western World* (New York: McGraw Hill Gook Company, 1965), p. 471.

12 After leaving the University of Chicago, Dewey was appointed professor of philosophy at Columbia University in 1905, where he taught in the philosophy department as well as in the teacher's college. Dorothy and Leonard Elmhirst were personal friends of Dewey, and likely would have shared his educational ideas with Tagore.

13 P.K. Mukherji, *Life of Tagore* (Calcutta: Indian Book Co., 1973), pp. 146-7.

14 See G.D. Khanolkar, *The Lute and the Plough*, p. 340.

15 See Martha Nussbaum, *Not for Profit* (Princeton: Princeton University Press, 2010), p. 85.

16 J.J. Findlay, *The Foundations of Education* (London: University of London Press, 1927), Vol. II., pp. 239-40.

17 The sections on Tagore's educational work draw from my study *Rabindranath Tagore: The Poet as Educator* (Kolkata:Visva-Bharati, 2012).

18 '*Vyahrti*' may be translated as 'utterance'.

19 Translated by myself from a twenty page letter written in November 1902, by Tagore to Kunjalal Ghosh, a teacher and administrator at Santiniketan. According to Tagore's biographer, P.K. Mukhopadhyay, the letter served as the first constitution of the school and remained in effect for much the first decade. This is substantiated by the fact that when Kshitimohan Sen came to the school in 1908 and requested a written statement on the school, Tagore gave him the letter. The letter, included in a Visva-Bharati pamphlet entitled *Santiniketan Brahmacharyasram,* is titled 'First Constitution/Rules and Regulations' ('*Karya Pranali*').

20 Letter to Ratindranath Tagore, quoted in *Selected Letters of Tagore,* edited by Krishna Dutta and Andrew Robinson (Cambridge: Cambridge University Press, 1997), 179.

21 Rabindranath Tagore, 'The Centre of Indian Culture', *Towards Universal Man*, (New York: Asia Publishing House, 1961), pp. 225-226.

22 This section draws from my study of Visva-Bharati in *Rabindranath Tagore: The Poet as Educator* (Kolkata:

Visva-Bharati, 2002)

23 'My School', *English Writings of Tagore*, vol. IV, 522-3.

24 Sylvain Levi's scholarship concentrated on the influence of Indian culture throughout Asia. Levi wrote: 'India gave her mythology to her neighbours who went to teach it to the whole world. Mother of law and philosophy, she gave to three-quarters of Asia a god, a religion, a doctrine of art.' Quoted in Wilhelm Halbfass, *India and Europe* (Albany: SUNY, 1988), p. 190.

25 Winternitz, whose thought also supported Rabindranath's concept of an Asian university, had spent ten years at Oxford helping Max Muller prepare the second edition of the *Rigveda-Samhita*. He catalogued the Vedic manuscripts for the Oxford Bodleian Library and prepared the collection of South Indian manuscripts for the London Royal Asiatic Society. As well he worked on the *Sacred Books of the East* series and authored *The History of Indian Literature*. See Manilal Patel, 'Professor Moriz Winternitz,' *The Visva-Bharati Quarterly* (Feb., 1937), pp. 91-96.

26 Both Indologists would later teach at Visva-Bharati.

27 The plan called for a co-ordinated study of the Vedic, Puranic, Buddhist, Jain, Islamic, Sikh and Zoroastrian cultures with Western culture finally placed beside the others.

28 RT, 'An Eastern University,' *Creative Unity* (London: Macmillan & Co., 1922), p. 195.

29 Ibid., pp. 171-2.

30 From the first Visva-Bharati prospectus, 1922.

31 Kshitis Roy, *Rabindranath Tagore; A Life Story* (New Delhi: Ministry of Information and Broadcasting, 1980), p. 66.

32 Ibid.

33 'Rabindranath Tagore's Visva-Bharati Mission,' *Modern Review*, Sept. 1924, pp. 287-301.

34 'Rabindranath Tagore's Visva-Bharati Mission,' *Modern Review*, Sept. 1924, pp. 287-301.

35 Leonard Elmhirst reports in his 'Personal Memories of Tagore,' that when the Chinese scholars challenged Tagore's interpretation of Japan, he persuaded three scholars to accompany the group to Japan. When they returned to Peiping, they held an exhibition of the art and crafts – ancient and modern – which had been presented to them in Japan. *Rabindranath Tagore; A Centenary Volume, 1861-1961* (New Delhi: Sahitya Akademi, 1961), p. 22.

36 Suniti Kumar Chatterjee 'The President's Tour in Indonesia and Further India,' *The VBQ* (April 1928), pp. 98-102.

37 The library in 1938, contained 100,000 volumes including rare volumes which had been contributed by the Sino-Indian Cultural Society in Nanking.

38 Tagore introduced the Japanese tea ceremony to Santiniketan. Later a department of Japanese studies was established under the direction of Prof. Shinya Kasugai of Bukkyo University, Kyoto.

39 William Pearson was similarly affected by Santiniketan, but his premature death prevented him from taking a role at Visva-Bharati.

40 *Visva-Bharati* (Booklet) Santiniketan, January 1950, p. 10, Quoted in Mukherjee, p. 203. The *Visva-Bharati Bulletin,* No. 4C (March 1948), describes it as 'Dinabandhu Bhavan (Andrews Memorial Hall for Christian and Western Studies),' but the listing of Dinabandhu Bhavan is omitted in the later syllabi.

41 She founded a section of Kala Bhavan called 'Bichitra' in 1923, at the request of Abanindranath Tagore.

42 'Santiniketan and the Educational Institutions,' *Visva-Bharati Bulletin*, No. 12 (Feb., 1929).

43 *International Education; A Documentary History*, 106.

Tagore Ocampo: A Turning Point in India-Argentina Relationship

Gustavo Canzobre

A hundred years from today
who are you, sitting, reading a poem of mine,
under curiosity's sway -
a hundred years from today?

There lived then a poet, ebullient of spirit,
his heart steeped in song,
who wanted to open his words like so many flowers
with so much passion-
one day a hundred years back.

A hundred years from today
who is the new poet
whose songs flow through your homes?
To him I convey this springtime's gladsome greetings.
May my vernal song find its echo for a moment
in your spring day, in the throbbing of your hearts,
in the buzzing of your bees, in the rustling of your leaves
a hundred years from today.[1]

If you are reading these lines written by Gurudev Tagore in *Chitra* in 1896, this means Satyajit Ray´s words were prophetically true: Tagore´s mortal remains perished, but he left behind a heritage which no fire could consume.

Our paper will focus on how this legacy reached Argentina through Victoria Ocampo, and not only changed her personal life, but also the Argentine attitude towards India.

Anyone who tries to do some research on the relationship between India and Argentina cannot, shortly after starting, fail to come across the description of the fascinating, exotic and even chance encounter that takes place on the banks of the Rio de la Plata in the early 20th century, if the word *chance* can be suitably used to refer to the unknown meaning given to life events.

This seemingly odd relationship that Rabindranath Tagore and Victoria Ocampo began in 1924, about which a lot has already been written and published[2], radically changed the way Argentina sees India. As a matter of fact, that attitude is still alive today, and it explains how Argentina and India have come to be what they are to each other.

When Victoria Ocampo read *Gitanjali* for the first time, from Andre Gide`s French translation, she was so shocked that it was one of the capital moments in her life. As she herself wrote later for the centenary volume published by Sahitya Akademi in 1961: 'I remember the moment and the exact spot where this took place. I was leaning against a white marble fireplace in a room upholstered in light gray silk. The house no longer exists. Neither do those I was afraid of hurting, or those who were hurting me. Nor does the poet who was bringing me the gift of tears, as not even the closest friend would have been able to do. The images which now live only in my memory will cease to exist together with it, as easily, as irrevocably as all that has preceded them into nothingness. But the *Gitanjali* over which I was weeping will remain.'[3]

Ocampo´s statement would be certainly true, not only about *Gitanjali*, but also towards all Tagore and Indian thought and their influence over Argentine cultural life along the 20th century. When in May 1953 Victoria Ocampo is imprisoned by the Peronist government, one among all the voices that is heard in the world crying out loud for her release was that of Pandit Nehru, Prime Minister of India. His intervention was decisive in twisting the decision of General Perón at a particularly belligerent time of his government. Nehru's demand serves as a clear example of how remarkably close these two far-off nations had become by that time.

Argentina and the Eastern Cultures Before Tagore - Ocampo

Since its foundation as an independent nation and culture, Argentina has been interested in the East, in general, and in India, in particular. Although often not taken into account, that interest, and its subsequent influence, does not begin in the 20th century, but goes back to the first half of the 19th century. Argentina's history shows the picture of a culture that has never remained indifferent to what happens in the rest of the world although it has always been under certain and determinant influences.

The main cultural streams that marked the process in which Argentina´s national identity was built are, on the one hand, its own native American tradition, and on the other and equally important, the European tradition. This cosmopolitan identity was transmitted first by the political and economic conquerors in the 19th century (Spain and Great Britain) and later, by the flow of immigration mainly from Italy, Spain, Germany and the Middle Eastern countries (Arabian and Jewish descendants) in the second phase between 1870 and 1930. All this makes the Argentine identity a melting-pot of cultures and races right from its very beginning, an identity clearly different from any other South American country, such as Bolivia or Perú, where their native roots are the main character of their national identity.

This is the classical explanation of the way the Argentine national identity has been built. We agree with Axel Gasquet on his assumption[4] that the influence the East will have on the Argentine elite from the very birth of the nation cannot be ignored *'Of the initial conceptual and ideological use of European Orientalism, inherited from the Enlightenment in the generation of 1837, which has served to establish the aesthetic of the Pampas as a basic topic of Argentine literature and also to define the contours of native barbarism through Eastern barbarian's statements, we see that there is a political, even social, concern for the East among the members of the Generation of '80'.*[5] More significant is to know that already in 1869 the journalist Lucio Vicente López, son of the historian Vicente Fidel López, publishes in *'La Revista de Buenos Aires' ('The Magazine of Buenos Aires')* an extensive review of '*The Ramayana*' using the French and Italian versions that had just come out as source material. The travel chronicles of Eduardo Wilde in the late 19th century inaugurate this new relationship with the East.

Nevertheless, Argentina's lack of desire to conquer gives its intellectuals the opportunity to not only move away from the biased European desire for closeness but also start appreciating the East in a different way. Thus, '*... from the very beginning, the South American orientalism played a self-reflective role: it looked for a support for the development of a new national political program or for the configuration of a local aesthetic in vague and exotic latitudes.*'[6]

The modernist American literary movement led by Rubén Darío, though Lugones was its most important representative in Argentina, sets up a new stage of Eastern cultural penetration, even if it was somewhat ambivalent. On the one hand, the importance these writers give the East in their works allows it to gain a growing space in culture. However, the East is superficially appraised, its assessment often being limited to the mere exoticism that the impact of the Arabian Nights stories had on the literati. What also contributes to this belief is the generalization made of the entire East and, in all aspects of culture, of the perceptions of the Muslim social structures that the Argentine generation created while visiting the Arab countries.[7]

This scenario witnesses the birth of the new century and the cultural eagerness aroused in 1910, the Centenary of Argentina. Both attest to the increasing interest in India, in particular, that Argentina starts developing. The crisis unleashed after the First World War, which challenges the thus-far unquestionable values of Western culture, awakens an existential craving for the search of new horizons of the Argentine intelligentsia. And India will hold a central place.

The milestones listed below confirm this awakening: some of them are clearly described by Gasquet; the others are our original identifications.

1. In 1896 Emilio Roqué publishes the first edition of the *Bhagavad Gita* in Argentina.

2. In 1907 the young Danish Nicolas Kier buys Emilio Persico's bookstore and transforms it into the Theosophical Bookstore, where those fond of India can come into contact with the Orientalist literature of the time, mainly with the esoteric kind. The bookstore becomes a meeting point for the elite interested in the East and Kier founds the publishing house that bears his name and that will be one of the main promoters of Orientalist literature in Argentina throughout the 20th century.

3. Works by Omar Khayan and Khalil Gibran as well as the oriental literary classics, especially from the Middle East, are made available to all the public thanks to other publishers, like the popular Tor, *Nosotros* Magazine, and the Mercatali printer on the first quarter of the 20th century.

4. The growth of a circle of young intellectuals who will look for humanistic alternatives to the increasing progress of positivism is promoted at the University of La Plata by Joaquín V. González, the well-known Argentine politician, also known for his theosophical interests and his affiliation to the Argentine Masonic Lodge where he held very important degrees. He was also followed by Alejandro Korn, the philosopher, whose undisclosed Buddhist affiliation remained unknown for a long time, since he didn´t want it to be public.

Among the men in this circle is the writer Carlos Muzzio Sáenz Peña, who then founds and, for many years, directs *El Mundo*, a well-known cultural newspaper. Peña will be the first one to make Tagore known to important Argentine audiences through the editions of *Poems* in 1915, *Fruit Gathering* in 1917, and his celebrated edition of *The Gardener* in the Nobel collection in 1924, the same year when Peña met the poet in Buenos Aires. They had both also published Kabir's poems in 1915, using Tagore's English translations. This means that in the same year the first Spanish translations by Camprubi would be published in Spain, these two Argentines were doing the same in Argentina. Apart from that, Vicente Fatone, the famous Philosophy professor and one of the architects of the new appreciation of India that closely followed the Tagore-Ocampo meeting – as we will remark later - would also emerge from this fruitful group.

EL POETA
a su llegada a Buenos Aires, con Carlos Muzio Sáenz Peña

5. The Theosophical Society, an 1875 global philo-Hindu philosophical movement that quickly attracted and captured much of the Argentine intelligentsia as it had done with modernists Rubén Darío and Amado Nervo, founds its Argentine branch in Buenos Aires in 1893. Everybody, from Leopoldo Lugones to Alfredo Palacios, including the Joaquín V. González and José Ingenieros among others, shows rising interest in enhancing their knowledge of India. The theosophical movement seeks to face the reigning monolithic rationality and European scientism with a mystical inquiry, but rationally formulated, that easily vanishes before the lure of 'Western esotericism' and its desire for the paranormal phenomenon, a parasite that accompanies the theosophical movement and often prevents it from achieving more metaphysical depth.

6. In 1909 the first Spanish translation of *The Gospel of Sri Ramakrishna*, as translated from the English version by Swami Abhedananda of the Swami Ramakrishna Order of India, is published in Buenos Aires. By 1922 the translations of *Karma Yoga*, *Jnana Yoga* and *Raja Yoga* written by Swami Vivekananda have already been published. Swami Vivekananda becomes Ramakrishna's best known disciple in the West as a result of his trip to the United States and his famous speech at the World's Parliament of Religions held in Chicago in September 1893.

This is the beginning of the 20s. There are three important facts that should be taken into account: first, a friendly, deepening and sincere interest in Indian culture in Argentina; second, a distrust and frustration with the Western myth of eternal progress severely questioned by the war and its aftermath. Lastly, cursory knowledge of India, of its ancient philosophy and wisdom due to the remnants of the exotic oriental appreciation of the late 19th century together with superficial and phenomenal mass communication led by theosophical and occultist circles which use the generic name of Orientalism to refer to both the Supreme Philosophy of the Upanishads and the Vedanta and also parapsychological investigations of so-called mediums, fakirs and seers, information that could be found in the libraries of these well-known characters as the publications were available in Kier's bookstore. Even today the same situation may repeat itself at non-specialized stores in the West.

However, none of the above mentioned information denies that this interest was genuine. A good example is the fact that, in the early 20s, groups gathering around the teachings of Sri

Ramakrishna and led by Mr. Ricardo Vivie and Mr. López Villamil formed the first Vedanta Society of South America in Buenos Aires in the early 20s, clearly showing that Ancient India together with its deep culture, science, literature and art was ready to flow over the Argentine Pampas.

The Turning Point: The Tagore-Ocampo Encounter

This was the Argentine—India panorama when Tagore arrived in Argentina. As stated above, the society of Buenos Aires was not indifferent to India, in general, and to Tagore, in particular. This is what Rafael Ruiz López, journalist from the *Radio Cultura* weekly magazine, writes a few days after Tagore arrived in the city: '*a man so great, a heart so perfect, an imagination so clear, an intelligence so luminous and capable of bathing in soft light the most unfathomable mysteries!*'[8] Much more balanced, *La Nación,* the most traditional Argentine newspaper, wrote the following the first Sunday after Tagore´s arrival: *'No act of his does not interest us, nor an aspect of his thought does not preoccupy us'* [9]

'I am restless, I`m athirst for the distant' Tagore had said about himself, and that certainly explains his constant trips, all along his life, all around the world. He used to think of himself as a citizen of the world. In addition, Tagore often travelled worldwide for fundraising purposes for his university Visva Bharati and all the pedagogical institutions he ran in Santiniketan near Kolkata. It is amazing to find that in Tagore's soul there is a peculiar combination which defines true idealistic human beings: both the love for contemplation and solitude, and the quest for depth, hand in hand with an endless capacity for working to make that ideal quite real. Just like our bodies, life depends on inhalation and exhalation; the soul's health requires action and contemplation, both deep and intense at the same time.

While Tagore was travelling around China and Japan, he was invited to the Peruvian celebrations of independence that were to be held in December 1924, on account of the centenary of the battle of Ayacucho in which Perú was finally freed from the Spanish rule. Only his idealistic spirit can explain his being able to leave for America just after coming back from such a trip. His friend Leonard Elmhirst was in charge of all the travelling arrangements for his trip to South America. Shortly after arriving from China in September 1924, Tagore left India for Europe where he met Elmhirst, who had offered to become his honorary secretary for his new trip. They both sailed from Cherbourg in France aboard the *Andes* ship, which would take them to Perú through the east coast of South America.

The ship was expected to stop in Rio de Janeiro and Montevideo and also in Buenos Aires so that they could travel to Santiago de Chile by railway and, then, to Perú. But Tagore had no plans to stay in those countries. Besides, although he was aware of the existence of Brazil and Argentina, he did not expect to see what, in fact, he did find when he arrived in the *Rio de la Plata*. His perception of Argentina and South America, in general, based on the works by the South American naturalist W.H. Hudson, was very naive: he imagined a rustic scenario, filled by large meadows and fields where the 'gauchos', riding their horses, would deal with the Pampas. Buenos Aires was, in fact, quite different. Although he seemed a bit lost in his first days in the city, in the end what awaited him there did not disappoint him at all.[10]

At that time Buenos Aires was the main center of Latin American artistic, philosophical and literary culture. As soon as Buenos Aires became aware that the Poet would be arriving on his way to Perú, all the cultural circles got ready to meet him. A special commission, led by the writer Ricardo Rojas, who at that time was the Chancellor of the University of Buenos Aires, was sent to Rio de Janeiro to arrange the arrival and welcome the poet. All the newspapers were writing about Tagore in those days. The main one, *La Nación*, whose literary Sunday supplement was one of the leading voices of the Spanish speaking culture all over the Hispanic world, sent a Journalist to meet Tagore in Montevideo. Only three days after his arrival in Buenos Aires, did *La Nación* proudly announce that Tagore would regularly contribute to the paper.

As a matter of fact, that interview, as wisely pointed out by Miss Dyson, clearly shows what Tagore´s thoughts and expectations were at that moment. What the journalist says about meeting Tagore on the ship, after working really hard to get Elmhirst to let the interview take place, is extremely revealing: *'If it could be said of any poet that he manifested his poetry in his face, it would not, without a doubt, be as true of anybody else as it would be of Rabindranath Tagore. The visions of night, of peace, of silence, the magical lyrical flowers which he has ceaselessly created since his childhood seem to surround this image lying on the pillow in the corner of the cabin'.*

What is fascinating too is the interest and knowledge the journalist showed Tagore as soon as the interview moved on and they dealt with the challenges of Western civilization after the First World War, the European struggle between the material quest and the need for preserving spiritual values, his quest for an encounter between East and West; Gandhi´s struggle and their differences in points of view. All this told Tagore of the profound interest in him and his quest, an interest he would find in his unexpected stay in Buenos Aires.

Tagore had been ill since the moment they got to Rio de Janeiro and still felt unwell on the voyage to the Rio de la Plata. Hundreds of people were waiting for the *Andes* when it finally got to the port of Buenos Aires in the evening of November 6th in 1924. As stated in the papers, very few personalities had caught the attention of the '*porteños*'[11] as Tagore did. On arriving, he was feeling better but tired. In spite of that, and of Elmhirst's efforts, he didn´t avoid the journalists with whom he shared his plan to stay in Buenos Aires for a few weeks while on his way to Perú.

From the port, he was driven to the most elegant hotel in the city, even by today's standards, the Plaza Hotel facing San Martin Square located in one of the most exclusive areas downtown, just a few blocks from Maipú 994, the emblematic building where Jorge Luis Borges would live from 1944 on.

Tagore with the Argentinian poet Victoria Ocampo : "He is as near to me as my life."

And here is where Victoria Ocampo appeared: '*He had caught a nasty influenza on board, had become exhausted, and the doctors in Buenos Aires would not let him proceed to Perú, as the Andean train journey was considered to be dangerous for his heart.*'[12] That was the conclusion intimated to him the next day.

The young Argentine writer Victoria Ocampo, at that time thirty four, accompanied by her friend Adelia Acevedo, was trying to contact the poet that very day. They met Elmhirst in the hall of the hotel, while he was still concerned about the doctor´s conclusion and about the need to take Tagore to the countryside to rest before leaving the country. This was Victoria's opportunity. On an impulse she offered to accommodate the Poet in a country house in the suburbs of the city. She had no certainty how that could actually be done, but her proposal sounded quite firm to Elmhirst. In fact, in a couple of hours, and even after her parents' refusal to provide the poet with accommodation in the Villa Ocampo, she was back at the hotel with good news. She had managed to borrow another villa for a week. It was called *Miralrío* (Facing the River) and was located a few blocks away from her parents.

This newly-started relationship was for the poet one of the richest experiences in life. Ocampo was not the first foreign woman that would fall under his spell: first, in 1878, he had met Ana Turkhud while staying in Bombay before his first trip to England. Years later, in England, the Austrian art specialist, Stella Kramrisch, after meeting Tagore in London, decided to move to India, arriving in Santiniketan in 1921 and staying in India close to Tagore's life and work for decades.[13]

But Vijaya's presence in his life (this is the name by which Gurudev used to call Victoria, translating her name from Spanish into Sanskrit) was particularly important for his own life, and, of course, for the Argentine relationship with India. In 1939, two years before passing away, but almost fifteen years after this two month stay in Buenos Aires, Gurudev would say: 'Time and again, my mind flies back to that verandah in San Isidro. I still recall quite vividly the exquisite festival of red and blue flowers glittering in the morning sun. And the endless display of colors atop that great river. Those unforgettable days, and her tender and compassionate care have been enshrined in my poems: they may well be among the best I have written.'[14]

It´s well known that he addressed his book Purabi to Ocampo. There we find *Atithi, The Guest,* showing his full feelings, and also the way Victoria addressed him:

> With what infinite sweetness you fill
>
> The days of my sojourn, Woman, and with
>
> What ease you make the traveller from
>
> A distant land as one of your own- with
>
> the ease of those strange stars of the evening
>
> Sky who with a calm and bening smile greeted
>
> Me from Heaven. As I stood at the quiet
>
> window
>
> All alone and gazed upon the southern sky,
>
> from high above there came into my soul
>
> In unison, a message of light. And I heard
>
> A solemn voice -'You are no stranger,
>
> we know
>
> You from when the earth took you in its arms
>
> From the arms of the dark; you are a guest
>
> Of Ours, a guest of light for ever '

Blessed one, you gazed upon my face even as
Those stars and said in the same strain,
'You are no stranger, I know you fully well,'
And though I do not know the language you
Speak, I heard you sing, 'The poet is a guest
Of love and a guest of mine for ever' [15]

Being close to Tagore, Dilip Kumar Roy recalls what Tagore once told him: he mentioned that he felt these affective encounters with both Indian and non-Indian women were a blessing in his life, and that he was always so satisfied with them that he could never ignore any intimate feeling that arose after those encounters: *'It doesn't matter what kind of love they evoke: those feelings make new flowers blossom in the jungle of our own hearts. These flowers don't need any physical presence to be kept alive, or even a concrete act of intimacy. They keep growing, even in absence, even in solitude, for a long, long, time. But in Victoria's case, I know they did depend on her presence.'* [16]

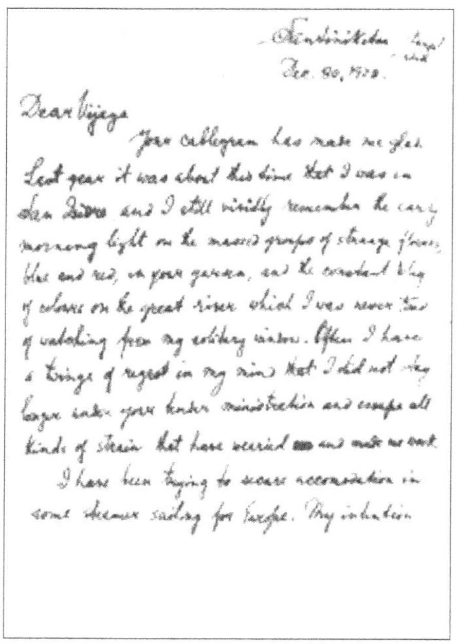
Santiniketan
Dec. 30, 1925.

Dear Vijaya
Your cablegram has made me glad. Last year it was about this time that I was in San Isidro and I still vividly remember the early morning light on the massed groups of strange flowers, blue and red, in your garden, and the constant play of colours on the great river which I was never tired of watching from my solitary window. Often I have a twinge of regret in my mind that I did not stay longer under your tender ministration and escape all kinds of strain that have wearied [illegible] and made me weak.
I have been trying to secure accommodation in some steamer sailing for Europe. My intention

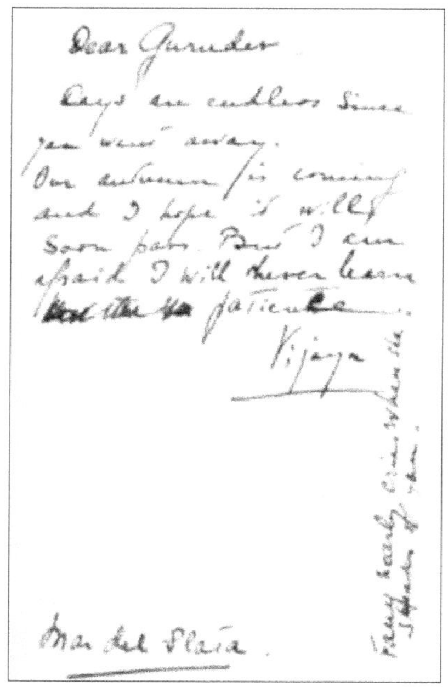
Dear Gurudev
Days are endless since you went away.
Our autumn is coming and I hope it will soon pass. But I am afraid I will never learn [illegible] patience....
Vijaya

Mar del Plata

Nothing could have foretold what, in the end, happened: both Tagore and Victoria Ocampo belonged to two different cultural and social realities, and she was a very young woman when they met, while Tagore was almost 63. But Indian wisdom states that what has to happen will, in the end, happen. And in this case, it was once again true. It was also quite clear in her encouraging Gurudev to paint (and her support of his first exhibition in Paris in 1930): the influence went from India to Argentina, and also from there back to India.

Their close relationship is one of the most fertile encounters of two great personalities of universal culture that has ever taken place. They kept in contact, one with the other, during the life time of the poet, and many letters, quoted and analyzed in Dyson`s book, are a valuable testimony to this special relationship. Due to her contact with Gurudev, and her leading role in the cultural life of my country, these values began to get a place among us.

'She was interested in the universe. She appreciated and thanked the infinite variety of human souls, and the circunstance that each of them is unique.' What Borges wrote on Victoria Ocampo, could have also referred to Tagore and, if so, we wouldn`t have known to which of the two he referred to.

Both Ocampo and Tagore devoted their lives, their time and wealth, including all their personal belongings, to try to give their countires all the treasures they could find worldwide, without losing their own national tradition. Victoria, at the end of her life, after having spent all her money on the Sur project, donated her two houses to UNESCO, with the purpose of continuing there her cultural dream. The main one is still being run by an UNESCO office according to her will.

They were both neither colonialist nor nationalist: they fought against sectarian foms of nationalisms, dreaming for their people and their countries a life crowned by happiness and brotherhood. They both believed in what today we call International Cooperation. Neither of them was a politician, but they couldn´t avoid their political gestures.

Both of them refused formal education. They never got a degree, though we could hardly find, in their surroundings and time, somebody as aware of man and culture as they were. Both never lost their faith in humanity, and worked hard to win back the lost human heritage.

I find very accurate Dyson's statement: if somebody could properly call Tagore *'Gurudev'*, that was Victoria Ocampo. She was deeply impressed not only by his appeal and artistic talent, but by his entire global vision. The closing words on the volume she wrote for Tagore's 100 aniversary in 1961 are clear enough:

> I know, however, that although I am not qualified to decipher my life's design, enigmatic the same as the design of all lives, I can contemplate it without despairing or feeling almost no melancholy. And I owe part of this wisdom, (which is not mine but an intuition I receive as if I were an erial) to two men born in a faraway country and who belong to a civilization apparently different from mine: the Mahtma and Gurudev (Gandhi and Tagore).
>
> Nothing stops us, except our human limitations. Love gives the heart the same feeling of space, of infinitude. But the heart is more powerful than the eyes.
>
> Tagore tells us in Reminiscences that there came a moment in his life when 'the apparent triviality of the finite and the apparent emptiness of the finite' vanished, . . . and the joy of reaching the infinite through the finite is felt only when the abysm that separates them is filled up with love.

William Blake was thinking about the same idea when he wrote 'everything on earth is the word of God'. This is the reality the Upanishads speak about, that everything on earth is the word of God. Blake believed that natural events were symbolic messages from the unknown powers. Tagore must have shared that belief. However, it is not easy to decode those messages.

What I, as a Western and American , owe men like Tagore and Gandhi is like a restitution of property that was in my inheritance, but I had taken no notice of.

Both Tagore and Gandhi half opened the doors to perception which separate us from reality if they remain permanently and hermetically closed. If we cleaned them, says Blake, everything would look to us the way it is: infinite. There is only history of the soul.[17]

Tagore and the Spanish Speaking World

Tagore's approach to Spanish speaking countries was not new. It has already been stated that his books were first published by Joaquín V. González and C. Sáenz Peña in Argentina in 1915. On that very same year, on July 31st, *The Crescent Moon* was published in Spain, translated by Zenobia Camprubí. She had met the Spanish poet Juan Ramón Jiménez three years before at the Students' Residence in Madrid where the famous poet was working. After Tagore won the Nobel Prize in Literature, she became really interested in him, and read his works in English at that time. She found his poetry very close to Jiménez's style. A year later, now married, Camprubí, a poet herself, and Jiménez would begin a great job translating the main poetry and dramatic works of Tagore from English. Zenobia was the translator and Jimenez revised and stamped his own poetic style on the translation, which was very fashionable at that time. That would make their translations quite special, since it is not usual for a famous poet to be translated by another famous poet. A North-American critic even wrote: 'his translations are more Tagore than Tagore himself'.

With the passage of time, Tagore became a close friend of both Juan Ramon Jimenez and his wife. According to a Spanish researcher,[18] he would even have travelled to Spain in 1921. Among Juan Ramon´s papers he found a very detailed schedule of the never-made trip, and of several acts to be held in Tagore's honor. The trip was unexpectedly cancelled for no known reason. In 1921 Tagore stopped travelling to attend the opening of the university, Visva

Bharati, in Santiniketan. That could explain why he never went to Spain.

The translations by Camprubí-Jiménez made Tagore's work known to all Spanish speaking readers. Among them was the Spanish philosopher José Ortega y Gasset. Although they translated nearly 30 works of the Bengali poet: poems, plays and aphorisms, and were the main promoters of his work in Latin America, they never had the chance to meet him in person, as Victoria Ocampo did.

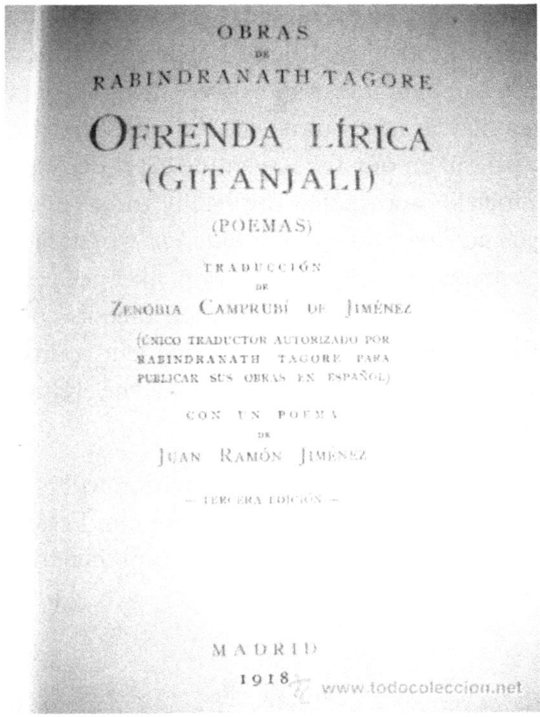
OBRAS
DE
RABINDRANATH TAGORE
OFRENDA LÍRICA
(GITANJALI)
(POEMAS)
TRADUCCIÓN
DE
ZENOBIA CAMPRUBÍ DE JIMÉNEZ
(ÚNICO TRADUCTOR AUTORIZADO POR RABINDRANATH TAGORE PARA PUBLICAR SUS OBRAS EN ESPAÑOL)
CON UN POEMA
DE
JUAN RAMÓN JIMÉNEZ
— TERCERA EDICIÓN —
MADRID
1918

Argentina - India After Tagore-Ocampo

One of the best fruits of the Tagore–Ocampo relationship is the magazine *Sur*, founded by Victoria Ocampo in 1931, that certainly shows influence of *Viswa Bharati Quarterly*, published by Tagore. *Sur* (meaning South in spanish) was not only a magazine, but also a complete cultural project in which Ocampo developed her cultural and political views. In *Sur* she promoted a sincere dialogue between cultures: Latin and North America; America and Europe; the West and the East. Paul Valery; Albert Camus; Vladimir Nabokov; José Ortega y Gasset; Romain Rolland; Mahatma Gandhi; Jorge Luis Borges; Rabindranath Tagore, Pedro Henríquez Ureña, became regular

Grupo Sur, 1931. Francisco Romero, Eduardo Bullrich, Guillermo de Torre, Pedro Henríquez Ureña, Eduardo Mallea, Norah Borges, Oliverio Girondo, Ramón Gómez de la Serna, Ernest Ansermet, Nenona Padilla y María Rosa Oliver.

contributors in the decades *Sur* was a shining star in America. And, besides, the way Victoria Ocampo introduced these important characters coming from the East shows the new place the Eastern cultures and India were being given: they would be considered as equals among equals, a way that clearly showed she did not share the idea of the European cultural superiority that used to prevail among the Argentine elites decades before. In this sense, Victoria Ocampo and Tagore's influence anticipated the reality to come in the 21st century. Both Tagore's and Gandhi's ideas, even acknowledging their different way of facing the challenges of life, show the West the complete spectrum of what India is really like: not just mystics unaware of what happens around, but wise men working hard to make their dreams come true while being endlessly patient; a global vision of life as a whole and interdependent reality in which philosophy, science, and politics work all together towards a common goal: the welfare of Humankind. The real face of India starts to be shown at that time.

All the so-called *Sur* Generation will promote this change in viewpoints: the still young Jorge Luis Borges with his affection for the Eastern and mystical aesthetics; Hector Murena with his philosophical inquiries that took him then to a definitely mystical inquiry; Enrique Pezzoni with his valuable translations.

In *Sur* magazine there was a special place for the new ideals of a non-violent movement led by Mahatma Gandhi, translated into politics by

Jawaharlal Nehru and into the spiritual and personal practice by Vinoba Bhave in India and Lanza del Vasto in the West. These new ideas spread all over the Spanish speaking countries through Sur´s editions: Nehru, Vinoba and

Lanza del Vasto´s books haven´t been reprinted in Spanish by any other publishing house till now. Besides, out of all the political approaches coming from the East, this non-violent movement was the one that fitted the democratic tradition of the Argentine intellectuals, and that can partially explain why Argentina seems to be closer to India than to China, although the latter has done more than the former to be present among Argentine culture along the last decades.

Other literati, though not from the *Sur* group, were also under this budding influence: near the end of his life, the famous writer Ricardo Guiraldes (1886-1927), always close to India after his theosophist passion, developed a mystical sense of life together with his wife, Adelina del Carril, who, in 1932, published his posthumous and spiritual work, *El Sendero* (The Path). Once widowed, she returned to India and stayed there until 1950 when she came back to Argentina with her adoptive Indian son, Mr. Ramachandra Gowda, who established himself in Argentina till his death few years ago.

Also Arturo Capdevila, the famous Argentine poet (1889-1967), was always close to the Indian way of life. In 1948 he published *'Los versos del bosque'* ('Poems from the Aranyakas'): this unfamiliar book in Argentina is a vivid proof of how the presence of India can be felt in Argentina.

India and Argentine Universities and Cultural Institutions

Many facts show how the Tagore-Ocampo relationship made the deeper realities of India known in Argentina, over the old occultist fashion attached to India in the 19th century. We will just mention the most important for us:

1. In 1931 a group of Argentine devotees led by Mr. Joseph Schmidt wrote to Belur Math, head of the Ramakrishna Order, asking about the possibility of sending an Indian Monk to Argentina. They were so insistent that on December 1932 Swami Vijoyananda arrived in Argentina, and a few months later founded the First South American Ramakrishna Ashram in Buenos Aires. From then on, it`s been a source of vedantic teaching for all South America, with that sense of brotherhood Swami Vivekananda developed in his disciples. Vijoyananda, till his death in 1973, was a restless monk who lectured along the South American countries, appearing weekly on the radio and on the main cultural centres, teaching in simple words the traditional Indian view of life adapted to modern times and western countries. Adelina del Carril worked with him through the 50s for the development of his work. All these teachings, along with important books published by the Ashram publishing house, are still alive in its followers.

2. As previously mentioned, Victoria Ocampo made Lanza del Vasto known in South America. She was the one who invited him in his first visit to Buenos Aires, at her own home. In fact that would be the first of seven trips that, along 20 years, would take Lanza del Vasto to Argentina. He formed a group of his Gandhian Community named 'El Arca' and taught the philosophy of non-violence to great audiences. His best disciple was one of the members of the Sur group, Adolfo de Obieta, not only well known because he was Macedonio Fernandez`s son (one of the greatest writers and thinkers of Argentina along the 20th century), but also because he was as

fond as Victoria Ocampo of Indian philosophy. At the age of 84, on 1996, he received the Jamnalal Bajaj award in India for his preaching of Gandhian values outside India.

3. In 1949, India`s political presence in Argentina turned from a simple office to an Embassy. A year later, the former Argentine consulate present in Calcutta since the 20s, moved to Delhi as the Argentine Embassy.

4. As quoted, the appearance of Vicente Fatone has to be highlighted as one of the best products of the new relationship Argentina was developing with India. Fatone, one of the most lucid but forgotten humanists of Argentine culture, left the Engineering High School because he found it too empirical for his metaphysical quest. As already mentioned above, he attended Alejandro Korn´s Philosophy classes at the University, and also joined the students' group of Joaquin V. Gonzalez. Already a Philosophy teacher at the University of Buenos Aires in 1926, he visited India for the first time in 1936/37 after winning a scholarship from the National Culture Commission. He stayed there for a year, studying with Surendranath Dasgupta and with other top Indian masters in Calcutta. When he returned to India 20 years later, first as Argentinian delegate at a UNESCO conference, and then, unexpectedly, as Argentine Ambassador to India, he used to lecture at the main Indian Universities, and was often asked to explain the clear and precise sense of traditional Indian Philosophy. Motilal Banarsidas, the famous and traditional publishing house, published his book *The Philosophy of Nagarjuna* in 1981 and 1991.

While teaching at the University of Buenos Aires, Fatone promoted the introduction of Oriental Studies. As a result of that, five years after his death, in June 1968, the 'Eastern Philosophy Studies Section' was created under the wise direction of Professor Armando Asti Vera in the Philosophy Department. The most eminent specialists in Buenos Aires gathered at that Center and produced publications that were remembered for years.

But Fatone wasn't just an intellectual or a philosopher; he rejected both terms. As with the classic sages, what was human was close to him and his interests. '*Philosophy and Poetry*; *Epic Mysticism*; *Architecture and Dance*, are the titles of only some of his books. They exemplify the variety of his interests. His belief that knowledge is useless if it does not transform the whole person and that there is no knowledge without transformation clearly shows that he is far away from the knowledge that comes only out of books. However, his library, made up of more than 10.000 volumes and acquired through an endless number of occupations throughout his life, was one of the best mystical libraries ever found in Buenos Aires. After his death, more than 6.000 volumes were donated to the Institute of the University of Buenos Aires housed in a big room in the Department of Medieval Philosophy. In the 70s and 80s, those interested in mystical traditions, both from the East and the West, could have access to these books which were not readily available in any other place in Argentina.

With Fatone and all the scholars that followed his path, India definitely became a well-known subject in Argentine culture. What should be pointed out is his vision, excellence and integrity, which made him the most important but usually unknown Argentine indologist in the first half of the 20th century. Just a few months after his death, Victor Massuh, another great figure in this new Argentine approach to a deeper knowledge of the East, wrote the remarkable 'Vicente Fatone, the man and the philosopher' in *Sur*'s edition # 286, as a tribute to the great master. Many of Fatone's direct and indirect disciples have been trying to keep the light of the Real India alive among us.

5. In 1968, another important event also takes place that shows how the India-Argentina relationship had been growing in the last decades: Indira Gandhi's visit to Argentina. Among all her activity here, she handed Victoria Ocampo the honorary doctorate from the Viswa Bharati University. Her photograph by Victoria Ocampo is still one of the main icons at one of the guest rooms at Villa Ocampo's Museum in San Isidro. Both Tagore and Ocampo promoted the dialogue between the East and the West on an equal footing, with mutual respect of their differences, going beyond the idea of Western superiority accepted by many of their contemporaries, or the anti western ideas held by the radicals. The International Meeting called 'Dialogue for the Cultures' took place there on 1968, clearly showing this spirit, with great world personalities taking part in it.

6. Ismael Quiles, the Jesuit father also present at the meeting between Indira and Victoria, performed a great job at the Universidad del Salvador in Buenos Aires, one of the main private universities in South America. He was a relevant personality in Argentine philosophy, and as vice-chancellor, in 1961, he created the Eastern Studies Center, which six years later, when he became Chancellor, was transformed into the School of Eastern Studies. He directed the School until his death in 1993. It was the first in Argentina and is still today the only university that offers a course in Indian studies where students earn a Bachelor's Degree in Eastern Studies. He also started a well-known yoga training school at the university.
7. In the seventies, the Philosophy Teacher, Ada Albrecht, after getting instruction in India with her Himalayan masters, founded the Hastinapur Foundation inspired by Indian spiritual universalism and Tagore pedagogical practices. She devoted her life to teaching the upanishadic legacy received from them: faith in the oneness of God, and service to mankind.

Many other facts could be quoted. Lots of Argentine people, passionate about India, worked hard during the 20th century lecturing, travelling, publishing books, sharing their experiences.

Like a delicate, good, red malbec wine, as time passes, it`s aged flavour becomes deeper: that`s how Gurudev`s ideas and ideals taste.

This long story is merely a brief summary of the way the relationship between Argentina and India changed during the 20th century, starting with that 'chance' encounter between the young poetess and the white-bearded Master that took place at the Plaza Hotel in Buenos Aires in the summer of 1924.

Our permanent purpose is, not only to salute his memory but to commit ourselves to keeping his legacy alive.

List of Illustrations

Notes

1 'The year 1400', *Chitra* (1896). Translation from Bengali by Ketaki K. Dyson, published in *www.Parabaas.com* (2001)

2 Still a must-read classic: Dyson, Ketaki K. (1988) *'In your blossoming flower-garden: Rabindranath Tagore and Victoria Ocampo'*, Sahitya Academi. Unfortunately still not translated to Spanish.

3 Translated from the Spanish version: Ocampo, Victoria (1961). *'Tagore en las barrancas de San Isidro'*, Ediciones Fundación Sur.

4 Gasquet, Axel. (2007). *'Oriente al Sur. El orientalismo literario argentino'*. Eudeba.

5 Gasquet, Axel. *'El Orientalismo Argentino: 1900-1940'. Working Paper 22.* Latin American Studies Centre, University of Maryland - College Park, 2008.

6 Op. cit, Axel Gasquet (2007)

7 A similar phenomena is seen today in western media, with the assumption that all terrorists are characterized as Muslims.

8 Quoted by K. Dyson, *op cit*, page 82

9 Quoted by K. Dyson, *op cit*, page 3

10 This in fact happened to Tagore not only in relation to Argentina, but also to several other countries he visited.

11 Argentine word for the people living in Buenos Aires, since they are from a port city, the port of Buenos Aires being the main Argentine trade center in all the country's history.

12 Dyson, K.K. *'On the Trail of Rabindranath Tagore and Victoria Ocampo'*, 2001. Special Rabindranth Tagore section from *Parabaas.*

13 For the time being, she became one of the most important specialists. As first professor of Indian Art at the University of Calcutta, she laid the foundations for the systematic study of Indian Art in India. Her works are masterpieces for indologists: *'Principles of Indian Art'*; *'Indian Sculpture'* and, mainly, *'The Hindu Temple'*.

14 Quoted by Dr. Rajat Chanda, *'Rabindranath Tagore in South America: some perspectives'*. English translation by Monish R. Chaterjee. Published in Monish R. Chaterjee´s web page: http://academic.udayton.edu/monishchatterjee/tagore/index.html

15 Krishna Bose and Sugata Bose. (Ed). *Rabindranath Tagore: 'Purabi: the East in its Feminine Gender'*, Translatated by Charu C. Chowdhuri. (Kolkata: Seagull Books, 2007).

16 Shyama Prasad Ganguly, *'Ocampo Tagore: entre la visión de lo uno y lo otro'*. Minutes of the 13th Congress of the International Association of Hispanists, 1998. Published by Editorial Castallia, España, 2000

17 Translated from the Spanish version: Ocampo, Victoria, *'Tagore en las barrancas de San Isidro'*, Ediciones Fundación Sur, 1961.

18 Francisco Garfias (1983). *'Rabindranath Tagore en Español'.* Plaza and Janés.

Tagore and China

Dong Youchen

Tagore was not only a poet, novelist, dramatist, composer, painter, educator and community leader, but he was also a great friend of the Chinese people.

I. Tagore's Visit to China

In 1920, Cai Yuanpei (1868-1940) invited Tagore to visit China on behalf of China's educational circles. At that time, Tagore was busy with fundraising for Visva-Bharati and his visit to the U.S., so the trip was not realized. Yet, he'd always wanted to visit China.

During the winter of 1922 winter, the Peking Lecture Association led by Liang Qichao (1873-1929) invited Tagore to visit China. He was very happy to receive the invitation, and made up his mind to visit China in August 1923. However, Tagore postponed the visit till the next spring, as he was not feeling well at the time.

On March 21st, 1924, Tagore left India by ship, and travelled to China. The ship which Tagore and his group traveled by reached Hong Kong on April 8th. At that time, Sun Yat-sen, who lived in Guangzhou, was not well, so he was not able to visit Tagore. Therefore, he sent someone to call on Tagore in Hong Kong, along with a letter of invitation, hoping that Tagore could come to Guangzhou. The letter contained the following:

> Dear. Mr. Tagore,
>
> When I heard of your visit to China, I would be honored to see you, if possible. It's our tradition to respect Confucian. The reason why I am so happy to hear about your visit is not only that you have contributed a lot to Indian literature, but also that you spare no efforts to seek for happiness for all mankind and the great achievements of spiritual culture, which is so rare and commendable. I specially invite you to visit me, and I hope that you can come for a visit.
>
> Yours sincerely,
>
> Sun Yat-sen

When he heard that Tagore planned to visit Shanghai first and then head for Beijing, Sun Yat-sen asked someone to tell the poet, 'The life center of China is Beijing. The work of Indian delegation can start from the North. If it's possible, I would waste no time to go there and visit you.' Yet, Sun Yat-sen was so ill that he could not realize his dream of meeting with Tagore.

Tagore's delegation reached Shanghai at 09:15 am on April 12th. At that time, representatives from Shanghai's cultural circles and a lot of celebrities went to welcome Tagore at Huangpujiang wharf.

The major goal of Tagore's visit was to renew friendly contact between India and China. He said at the tea parties held in honor of him by various academic communities in Shanghai, 'My dear friends, the reason of my visit is to request you to ask for the reopening of our channel of communication. I believe that such a road still exists. Although it has been sent to oblivion by wild grass and thorns, traces are still visible. I hope that you do not treat me as a mere visitor but rather a brother whom you meet again after a long separation.'

From 14th to 16th of April, accompanied by Xu Zhimo, Wang Tongzhao and so on, Tagore and his delegation went to Hangzhou for a visit. They saw the beautiful scenery at the West Lake, visited Xiling Society of Seal Arts, had a taste of Longjing Tea, and gave a speech at Zhejiang Education Convention. They returned to Shanghai on April 17th. On the 18th, various communities in Shanghai held a welcome party for them at the Conference Hall of the Commercial Press, where Tagore gave a speech. In the morning of April 19th, Tagore and his delegation took a ship against the current and reached Nanjing on April 20th. They visited Ming Xiaoling Mausoleum in the morning, gave a speech at 2 pm at Southeast University, and headed for the North by train at 9 pm. They reached Jinan at 05:40 am on April 22nd, and gave a speech at Shandong Conference Hall in the afternoon. At 05:30 on the 23rd, they left Jinan by train, and reached Beijing at 7 pm.

From April 23rd to May 20th, Tagore and his delegation stayed in Beijing. On the noon of April 25th, the British and American Association held a welcome party for them. At 3pm, the Peking Lecture Association hosted a tea party to welcome Tagore and his group at the Heart-East Study of Beihai Park, where Tagore made an impromptu speech. On the afternoon of April 26th, Tagore went to enjoy the lilac at Fayuan Temple of Beijing. They visited the Royal Garden at the Forbidden City on April 27th. At noon, Emperor Puyi held a reception for them. In the evening, a welcome banquet in honor of them was held by the Navy Gala Society at Jinyu Hutong, where Tagore spoke about his creative ideas.

At 3 pm on the 28th, Tagore met with the university students in Beijing at the Temple of Agriculture, where he made a speech. Thousands of people were present. On 29th, Tagore and his delegation went to Guizhou Hall at Yingtao Xiejie to visit a calligraphy exhibition, attend the tea party held in his honor by Chinese artists, and meet with Xu Beihong (1895-1953), Qi Baishi (1864-1957), Chen Banding (1877-1970), Yao Mangfu (1876-1930) and so on. He talked to them about friendship, and exchanged creative ideas. In the evening, they headed for Tsinghua University.

On May 1st, Tagore made a long speech at Tsinghua University. On the afternoon of May 5th, the Buddhist Association threw a welcome party for Tagore. On the afternoon of May 6th, the Beijing Federation of English Teachers hosted a lecture by Tagore.

May 7th was Tagore's 64th birthday. On the evening of the 8th, a birthday celebration party was held in his honor at the Union Hospital, where the host dressed him up in traditional Chinese attire. Liang Qichao first delivered gracious blessings, and then gave him a Chinese name 'Zhu Zhendan.' Later, he explained the meaning of the name: 'Zhu' comes from 'Tianzhu' by which name the ancient Chinese called India. 'Zhendan' is what Indian people call China, the transliteration of 'Cina,' and 'dan' signifies that the Sun rises from the horizon, which connects with Tagore's Bengali name 'Rabi.' Then, Tagore went up on stage to give his thanks. He said that he was very happy to have a name which symbolized the India-China friendship. At the celebration party, the beautiful Lin Huiyin (1904.6.10 - 1955.4.1) performed Tagore's famous drama *The Flower Girl* in English. She played the leading role of the Flower Girl, and the handsome young poet, Xu Zhimo, acted as the God of Love. The two young people mastered the language of English and spoke the language without any accent. Tagore couldn't be more satisfied with their performance. Before leaving Beijing, Tagore gave Lin Huiyin a short poem:

> Blue Sky,
>
> Looks down at the green woods,
>
> Between them,
>
> Breathes a gust of cool breeze.[1]

This short poem seems to portray the beauty of nature, but it in fact implies a very touching love story. During Tagore's visit to China, Xu Zhimo was his interpreter, for which reason the two of them became lifetime friends. Xu not only talked with the poet about literary creativity and goal of life, but he also recounted the hardship he felt about love, showing the deep, bitter-sweet love he felt for Lin Huiyin. Tagore also seemed to feel that the two young people were made for each other. When he politely conveyed Xu's love for her, he learned that Lin had already been engaged to Liang Qichao's son, Liang Sicheng (1901- 1972), about which even Tagore felt that there was nothing he could do. In his poem, Tagore compared Xu to 'the blue sky,' and likened Lin to 'the green woods.' Both of them were beautiful and pure, but they could only be like the sky and the woods, looking at each other from afar for all eternity, unable to have a happy ending. Tagore compared himself to the kind-hearted 'cool breeze.' He could not be the Moon Lord, but only to sigh for the sky and the woods.

The birthday celebration left a wonderful impression on Tagore. Up till his later years, he could still bear the memory in his mind. For example, he dictated a poem on February 21, 1941, where he affectionately thought of the birthday celebration in China:

I visited China,
The host I'd never known before
Wrote on the Bindi of my forehead,
'You are our bosom friend.'
The strange veil falls down unknowingly,
The ever-lasting people rise in my heart,

Unexpected closeness

Opens the door to happiness.
I give myself a Chinese name.
I put on my Chinese attire.
I deeply realize:
Where there are friends,
There will be rebirth and the miracle of life.[2]

From May 9th, Tagore gave a series of three speeches. On May 17th, Zhang Xiangwen, Shen Junru and some others from Beijing Buddhist Assembly visited Tagore, and they talked about the idea of establishing an India-China Academic Society. On May 19th, various religious associations from home and abroad held welcome parties for him. In the evening, Mei Lanfang performed *The Goddess Luo* at Kaiming Theater (later renamed Minzhu Theater after the Chinese Revolution), to welcome Tagore and his delegation. After the performance was over, Tagore himself went backstage to congratulate Mei Lanfang, saying, 'I am very happy to watch the performance. I would like to talk about some of my thoughts on the show tomorrow face to face.'

On the noon of May 20th, Liang Qichao, Mei Lanfang, Yao Mangfu (1876-1930) and some others gave a banquet to Tagore to see him off. During the lunch party, Tagore said that on the day when he watched *The Goddess Luo*, he praised Mei Lanfang's performance, and at the same time he offered some advice on the setting of the island of angels for *Chuan Shang Zhi Hui*. He suggested to Mei, 'The colors for the setting had better be red, yellow, black, purple and other such dark colors, so as to create strange peaks, stones, magical grass and flowers which human eyes have never behold before, and also to delineate gold and silver frames to set off mythological atmosphere.' Mei was very happy to take his advice.

At the table, Liang Qichao asked Tagore, 'You must have some beautiful lines in mind for this visit. Would you be willing to share them?'

Tagore replied, 'After watching *The Goddess Luo*, I am thinking about a short poem which I would like to give to Mei.' With this, he mused for a few seconds, and then wrote a short Bengali poem on a fan, and gave it to Mei Lanfang.

> My dear friend, you've used
>
> The veil of language unknown to me,
>
> To cover your looks;
>
> Which is like looking at the same
>
> Wispy clouds and rainbows,
>
> And the mountain peaks surrounded by mist and fog [3]

At 11 pm on May 20th, Tagore and his delegation left Beijing for Taiyuan. They reached Taiyuan at 6:40 pm on the 21st, where they stayed for two days. He made a speech at Wenyinghu Park, visited Jin Memorial Temple, and attended the welcome party held by Shanxi educational authorities. They reached Hankou on May 25th, and gave a speech on May 26th. They returned to Shanghai on May 28th. At 10 pm on May 29th, Tagore and his group embarked on the ship at Huishan Wharf in Shanghai. At 8 am on May 30th, the steamship set sail. Tagore stayed in China for almost 50 days, during which time he visited seven big cities.

Tagore's visit to China in 1924 had active influence on China's literary circles. Before and after his visit, the introduction and translation of Tagores works reached high tide in Chinas cultural circles. His visit strengthened the traditional friendship between India and China, and it also made new way for India-China cultural exchange.

Tagore got to know some people from China's cultural circles, and formed a strong friendship with them. During the 50 days of Tagore's stay in China, Xu Zhimo accompanied him all the time, interpreting for him. The two of them became great friends regardless of their difference in age. After Tagore finished his visit to China, Xu also accompanied him to Japan and saw him off all the way to Hong Kong, only to part reluctantly from him there. After he returned to India, Tagore compiled and published his speeches in China as *Talks in China*. He wrote on the title page, 'To my friend Xu Zhimo. With his help, I've introduced myself to the great Chinese people.' Constant correspondence between them strengthened their friendship. In 1928, Xu realized his dream of visiting India, and met with Tagore whom he thought of day and night.

In late March of 1929, Tagore passed Shanghai on his way to Japan and North America. From March 19th to 20th, he stayed at Xu's place for two days. It was Tagore's second visit to China. On his way home after his trip abroad, Tagore stayed at Xus for three days, from July 13th to 15th. This was Tagore's third China visit. Under the constant care of Mr. and Mrs. Xu, Tagore had a good rest at their home, and spent an enjoyable time there. Therefore, before leaving, Sudhidranaath Datta, the poet accompanying Tagore on this trips abroad, copied a short poem written by Tagore on a piece of paper, signed on it, and gave it to Xu. The poem is as follows:

My dear friend, I stopped over for my trips.
Time is wasted; Sakura withers and falls.
Luckily, all over the mountains and plains is Azalea,
Showing your consoling smiles.[4]

This short poem expressed Tagore's mood for his trips abroad this time. He lost his passport in Canada by accident, which caused him a lot of trouble. He was also insulted at USCIS, which upset him a lot. In Japan, he was given a cold shoulder, which depressed him greatly. That was why he wrote, 'I stopped over for my trips./ Time is wasted.' The Japanese authorities were very cold to him, and that was why he said 'Sakura withers and falls.' However, the moment he arrived in Shanghai, the handsome Xu welcomed him with a big smile. The poet received warm hospitality and great care from the Xus, which was like putting him among the flower beds of Azalea, for which he felt joyful in body and mind. Lu Xiaoman, Xu's wife, later recalled, 'Tagore was so natural and kind. One time, he fondled my hair kindly and called me 'my child." The poet treated the Xus as his own son and daughter. One time, he took them to a dinner party for the local Indian community. When introducing them to his fellow Indians, he called them his own son and daughter-in-law, for which Xu and Lu felt very warm and gratified.

During his stay in Shanghai, Tagore was received by Song Qingling who gave him a set of models of Beijing Opera Facial Masks.

Liang Qichao (1893 – 1988) was a key member of the Beijing Lecture Association, and also a key figure in Tagore's reception in Beijing. When welcoming Tagore at Beijing Normal University, he made a speech titled *The Family Relation between Indian and Chinese Cultures*. Liang said, 'Tagore also said that he couldn't give us any presents this time, only to offer his total sincerity to China on behalf of Indian people. I said that only this point had already triumphed over any expensive presents. We open our hearts to welcome you as our big brother and we give you more to take home with you… I hope that we can renew our friendship which to Tagore will not be ephemeral. Our big brother shoulders a great deal of burden for mankind. We will enjoy a long time together for cooperation and friendship.' (*Morning Paper, Supplement*, May 3rd, 1924)

On May 8th, Liang Qichao gave a warm speech at Tagore's 64th birthday celebration party in Beijing. He also gave Tagore a bloodstone signet engraved with 'Zhu Zhendan.' Liang told Tagore, 'I hope that the old friendship between India and China can be rekindled by Zhu Zhendan.'[5]

Hu Shi, the renowned intellectual who knew both the East and the West at the time also got to know Tagore. He praised Tagore's tremendous character, and criticized the uncivilized behavior of passing out leaflets against Tagore when he made speeches. To console Tagore, he wrote one of his own poems, *Dedication*, into a banner and gave it to the poet as a birthday gift.

> He walks out of the strong wind,
>
> And moves uphill to the peak.
>
> On the hill is only peace and beauty,
>
> Without the oppressing wind and rain.
>
> ...Behold! He walks down the hill,
>
> To the place covered by thick cloud.
>
> Forget about rain and hailstone!
>
> If they can suffer, then so can I.[6]

Tagore also kept contact with Xu Beihong, Mei Lanfang and Liang Shuming, with whom he enjoyed heart-to-heart conversations.

Since Dec. 1939, at Tagore's invitation, Xu Beihong started visiting India and his international university. In 1940, he did some sketches and color paintings for Tagore. The watercolor painting where Tagore is silver-haired and wears a pair of glasses comes from Xu. Besides, at Tagore's request, he also compiled and edited a collection of works for the poet.

Chinese modern writer Xu Dishan (1893.2.14—1941.8.4) also had deep respect for Tagore. He called the poet his 'all-time respectful bosom friend and elder.' In 1924, when Tagore visited China, Xu Dishan was studying at Oxford University in the UK, so he got no chance to meet him. Therefore, when he returned home from the UK in 1926, he specially made a detour to India for a visit to Tagore. He visited Tagore's international university, and paid a visit to the poet at his home. Tagore gave some of his photos and a mascot – a white elephant – to Xu Dishan as something to remember him by. In 1934 spring, when he went to India to study and research on Sanskrit, Xu would like to pay the poet another visit. Unluckily, the poet was not at home, so he didnt have the chance to meet him, for which he felt very regretful. On April 20th, 1934, he mentioned in a letter to his wife, Zhou Qisong, 'I recall that Tagore wears long wavy hair to his shoulders, and he has a silver beard. He talks cheerfully and humorously. He is elegant, with good manners. He has made a strong impression on me up till this day. He suggested that I write and edit a Sanskrit dictionary which is applicable to the Chinese people so it can not only be used for India-China academic exchange, but also be used for promoting India-China friendship. I will get right to it once I get back home. The draft of the dictionary is in my study at Peking University. I was planning to visit Tagore, telling him that I have taken his instructions to compile a Sanskrit dictionary, which he would be thrilled to hear about. However, I have heard that he is not at home right now but giving lectures somewhere else. I do not know where he has gone. Therefore, I am not able to pay him a visit... although we are so near to each other, but I still cannot visit him, which is a great pity. How people are irregularly meeting and parting with each other!'[7] Xu Dishan half jokingly said to his wife, Zhou Qisong, 'Tagore is my bosom elder, while you are my bosom wife. I am so happy to have the two of you in my life.'[8] As is seen, Xu Dishan and Tagore connect with each other.

When talking about Tagore's contact with Chinese intellectuals from the cultural circles, we cannot keep silent about Master Tan Yunshan (1898 – 1983). As early as 1924 when Tagore visited China, Tan especially cared about news reports about Tagore's visit to China. He sought to read his works and started to admire the poet.

Only in 1927, when Tagore visited Singapore did Tan get the chance to meet with Tagore whom he had always admired. The two of them instantly felt like old friends at their first meeting. From then on, they became good friends in spite of their age difference. Tagore invited Tan to teach at his international university, to which the latter quickly agreed.

In September 1928, Tan said goodbye to his newly-wedded wife Chen Naiwei, and went to teach at Tagore's international university by himself. He opened a Chinese course on the one hand, teaching Chinese language and culture, and researched on Buddhism, so they could have the chance to be together from morning till evening, to forge a profound friendship. Soon, the idea of establishing an India-China Cultural Association struck Tan, so he told Tagore about his idea. The poet agreed willingly because Tagore had had such an idea since a long time ago. Thanks to the active efforts from Tagore and Tan Yunshan, the India-China Cultural Association was founded in May 1934. The address was set up at Visva-Bharati, the international university where Tagore was president, and Nehru honorary president.

In November 1934, Tan returned to China. He reported the establishment of the India-China Cultural Association, and gave Cai Yuanpei a letter from Tagore. The poet wrote in the letter, 'I am happy to invite the India-China Cultural Association to use my university as its activity center in India. I hope that my friends can warmly welcome the founding of this Association, and that they can lavishly help my friend, Prof. Tan Yushan, to carry out his plan and set up a permanent group so as to promote India-China cultural exchange.'

Thanks to the efforts and planning of Cai Yuanpei, Tagore and Tan Yunshan, in May 1935, the India-China Cultural Association (China) was founded in Nanjin, where Cai was selected to be President of the Council, Dai Jiyao President of Board of Supervisors. Soon afterwards, the India-China Cultural Association (China) decided to donate a great many books, altogether more than 100,000 books, to the Association. In August 1935, Tagore received a remittance of INR 31,712 from the India-China Cultural Association (Nanjing) to build the compound in China. From September 1935, plans to construct a College of China were put into action. The construction took one year. The College of China was completed in the spring of 1937.

In April 14, 1937, the opening ceremony for the establishment of the Chinese College at Visva-Bharati was held in Santiniketan, hosted by Tagore, who gave a passionate speech. Tan Yunshan made a long speech at the opening ceremony. Cai Yuanpei, Dai Chuanxian and Chen Datong jointly sent a message to Tagore, expressing their warm congratulations. Chiang Kai-shek also sent a congratulatory message. Mr. Tan Yunshan was appointed President of Chinese College. From then on, he'd been continuing to work with Tagore till the death of the latter on Aug. 7th, 1941. After Tagore's death, Tan Yunshan continued to work for the international school for 26 years till he retired in 1967. He devoted all his life and talents to the China-India cultural exchange. Therefore, he was widely respected and loved by the Indian people, and he was also highly praised by the leaders of New China.

Tan Yunshan passed away at the Chinese Buddhist Temple in Bodhgaya on Feb. 12th, 1983. Indian Prime Minister Indira Gandhi wrote in his letter of condolence :

> 'The death of Prof. Tan Yunshan greatly saddens me. He was a great scholar and cultural person. Both Tagore and my father admired him a lot. He was of one mind with Santiniketan, and he made a great contribution to improving the understanding between both civilizations.'[9]

II. Tagore's Influence on Modern Chinese Literature

Tagore had a great influence upon modern Chinese literature. His works, especially his poems, not only greatly influenced the creative ideas and the writings of a series of young Chinese writers during the 1920s and 1930s, but also played an active role in the birth and development of Chinese new-style poems, which can be clearly seen from the works of such writers as Guo Moruo (1892—1978), Xie Bingxin (1901—1999), Zheng Zhenduo (1898—1958) and so on.

Guo Moruo once said, 'I am afraid that I have been the first one that has ever come into contact with Tagore.' (*Modern Times* Bi-weekly, 1st Issue, Aug. 6th, 1939) In Oct. 1923, when he recalled the first time that he had ever read Tagore's works, he wrote, 'I got to Tagore's name in 1914 when I first went to Japan. That was the time when Tagore's name became popular in Japan. In September, I went to prep school and lived with a junior college student who was a relative of mine. One day, he brought back home several English mimeographs which he told me were some poems from an Indian poet. I saw the title read as "baby's way", "sleep stealer", "clouds and waves." I opened and read it, and I was surprised.'

First, it was very easy to understand. Second, it was in prose style. Third, it was fresh and meaningful. From then on, Tagore's name had left a very deep impression on me. Later, I would very much like to buy and read his works, but it was not very easy to get my hands on his books in Tokyo because it sold like hot cakes. When I got myself a copy of his *The Crescent Moon*, it was a year later. At that time, I was not in Tokyo any more. I had already gone to Okayama Higher School for undergraduate studies. When I got *The Crescent Moon* and saw the elegant binding and several illustrations, I was like a little child who was overjoyed to get an illustrated magazine.... I remember that in the fall of 1916, I suddenly came across such books in Okayama Library as *Gitanjali*, *The Gardner*, *The King of Black Chamber* and so on. I felt like I'd explored the 'life of my life,' or 'the spring of my life.' Every day after school, I would run to a very dark reading room, sit in a little corner, hold the book in my hands and read it. Sometimes, I wept and remembered with gratitude. A sense of bitter sweetness ran through my body. I enjoyed the happiness of nirvana.'[10] Here, Guo Moruo talked about the influence of Tagore's works on him when he felt lost, as well as his own feelings when he was reading Tagore's works. Later, he also said:'I was first influenced by Tagore and some others when it comes to the experience of writing poems. He focused on dilution.' His 'focus on dilution' was the style of Guos new poems during that time period, which means that he upheld the dilution of grace. For instance, his new poem the *Temptation of Death* is a very compelling example.

(1)

I have a small knife,

Leaning against the window and smiling at me.

She chucked, 'Moruo, don't worry!

Come to me quickly and kiss my lips,

So I can help remove your troubles.

(2)

The green seawater outside the window,

Keep calling me as well,

She called to me, 'Moruo, don't worry!

Come quickly to my embrace,

So I can help remove your troubles.

His long poem *Goddess* was also influenced by the style of Tagore's poems in English translation.

Xie Bingxin was another writer who was greatly influenced by Tagore. As early as in her adolescence, she was especially fond of Tagore's poems. In 1920, she wrote a short essay titled *To Indian Poet Tagore*. She wrote in the essay, 'Tagore, thank you for curing the melancholy of my talent with your lively and beautiful poems. Thank you for comforting the loneliness of my heart with your excellent philosophy.'[11]

Later, when she talked about her poem writing, Bingxin wrote, 'When I wrote *Stars* and *Spring Water*, I wasn't writing about poems. I was only influenced by *Stray Birds*. I merely gathered a lot of 'loose thoughts' into a collection.' 'I was just reading Tagore's *Stray Birds*, and imitating his style. That's all.' Those poems were 'rather short, full of poetic illusions and philosophy.'[12]

Aside from Guo Moruo and Xie Bingxin, Zhu Ziqing (1898—1948), Zheng Zhenduo, Wang Tongzhao, Zong Baihua, Ye Shaojun (a.k.a. Ye Shengtao, 1894—1988), Guo Shaoyu (1893—1984) and a series of other writers and poets all started to create such short poems. At that time, the supplements of such famous newspapers as the Morning Paper, *the Lamp of Learning* and *the Awakening*, as well as such monthly magazines as *Fiction Monthly Magazine*, *Poetry* and so on all published short poems. In this way, short poems became a very fashionable literary genre on China's news papers and magazines at the time.

Not only the form and style of Tagore's literary works, but also the content of his poems deeply influenced Chinese modern literature. For example, Tagore's children's poems had a great influence on Bingxin, Zheng Zhenduo and so on. His novels on children's lives and family issues, as well as the theme of 'love' permeating his novels, all had a very great impression on China's modern writers. For instance, *Qiu Feng Qiu Yu Chou Sha Ren* by Xie Bingxin, *A Feng* by Ye Shengtao, *Smile* by Wang Tongzhao and so on were all influenced by Tagore's works.

III. Tagore's Support of the Just Cause of the Chinese People

During his lifetime, Tagore had always had friendly feelings toward China and its people. Ever since his childhood, he had cared about China. In 1888, when he was 20 years old, he wrote an article titled the *Trading of Death*. In the article, the author severely criticized the English colonialists for forcing the selling of opium to China and harming the physical and psychological health of the Chinese people. He deeply sympathized with the Chinese people. He wrote in the article, 'This way of trading and hoarding wealth can only be called trade in a polite tone. It is nothing short of freebooting.'

When he visited China in 1924, Tagore praised the huge contributions of China's ancient civilization to mankind, cherished the memory of the traditional friendship between the two peoples, and advocated recovering and strengthening friendly contacts between the two countries. He said to the Chinese audience, 'You've definitely made amazing progress in the past eras. You have several great inventions, and your civilization has been borrowed and imitated by other nations.' He also said, 'The reason why I can win your hearts is that we are so close with one another. It is also because that I have a belief. When your country stands up and shows your own spirituality, you and even the whole Asia will have a great prospect..... I recalled the day when India called you brothers and showed great respect for you. I hope that such a relationship can still exist and is still buried deep in our minds, that is, the bottom of our Oriental people's heart.'[13]

He also said to the Chinese people, 'My dear friends, the reason why I've come this time is to request you to reopen the channel of communication. I hope that it is still there. . . In the event of this visit, China comes close to India, and India comes close to China – not just for the sake of political or commercial reasons, but for the utilitarian love of mankind. Except these two, he wanted nothing more, for which I am deeply honored.' (*Complete Works of Tagore*, V. 20, p. 29)

The most moving thing is Tagore's support to China's Anti-Japanese War. At first, Tagore had a good impression of Japan. He once praised the Japanese people's love of beauty, their hospitality and their patience, and he even hoped that Japan could fight for the Orient. However, after he went there in 1916, he just saw 'a nation that revealed its ugliness stark-nakedly.' He saw 'a nation that would like to gain benefits with undivided attention, and expand their political power - severe, exclusive, suspicious and impersonal.' Therefore, he changed his attitude and started to criticize 'Japanese nationalism.'

On Sept. 18th, 1931, Japan started the 'Sept. 18th Incident,' and later occupied the Northeast of China. In 1936, Tagore read from the newspaper that the Japanese invaders went to a temple, praying for victory. He was so angry that he wrote a poem full of a sense of justice. He wrote in the poem,

Bell strikes chime,

Smoke curls up,

Prayers rise up to the sky:

'The merciful Buddha,

Bless us with victories.'

They will use bayonets to

Start the startling

And heartrending bellow,

Cut off the link of love from thousands of households,

And insert the Flag of the Rising Sun into

The ruins of leveled villages.

They will destroy the palace of knowledge,

And crush the altar of 'Beauty.'

For this, they will receive

The blessings of the merciful Buddha.

The war drums beat,

The bugles ring,

The world stands trembling.

From Jul. 7th, 1937, the Japanese army started full-scale invasion into China. At that time, Tagore was really infuriated. On Sept. 21st, 1937, he wrote in his reply to his message of condolences to Cai Yuanpei, 'My people and I completely sympathize with your country.' In Oct. 1st, 1937, he published an article in the *Dew Market Daily*, condemning Japanese aircrafts bombing China's peaceful cities and supporting India in its boycotting Japanese products.

Japanese militarists continued to expand the invasion into China, and the vastness of China's land fell into their claws. Tagore was saddened by their invasion with a heavy heart. As was recalled by Prof. Tan Yunshan from the Chinese College of the international university at the time, to support Chinas anti-Japanese war and relieve its refugees, Tagore started a donation activity at the university, and he himself donated 500 rupees. In April 1938, he wrote *Letter to the Chinese People*, showing his respect and offering his blessings to the Chinese people who rose to fight against the invasion.

In July 1938, a Japanese poet called Yone Noguchi wrote a letter to Tagore, exonerating Japan from its policy of aggression against China. He said that the war 'was an inevitable method to build a great new world on the Asiatic continent.' On Sept. 1, Tagore refuted the Japanese poet's idea in his reply. Yone remained unwilling to give up after receiving the letter. Once again, he wrote to Tagore, trying to justify Japan's policy of invasion. He stubbornly emphasized that Japan's actions in China were to 'save China,' and that it was for the greater good of 'common prosperity of the whole Asia.' Tagore replied: 'Unless you can convince the Chinese people that your army bombed their cities and enabled their women and children to become homeless beggars... Unless you can convince those who have sacrificed their lives that what they've suffered is just a merciful treatment and that such a merciful treatment will save 'their nation' in the end, there is no need for you to lobby us to believe the noble purpose of your country.' At the end of his letter, he wrote, 'I sincerely hope that the people of your country whom I love will achieve no success but rather regret.'[14] Tagore's righteous stand was very rare and unforgettable. Tagore was not only a great figure whom we admired, but also a great friend to the Chinese people!

In 1956, when visiting Tagore's international university in India, Premier Zhou En-lai said, 'Tagore was not only a talented poet who made excellent contributions to world's literature, but he was also an outstanding representative of the great Indian people who hate darkness and fight for light. The Chinese people can never forget Tagore's love for them. The Chinese people can also never forget Tagore's support to them for their hard fight for national independence. Up till now, the Chinese people can still recall with a warm heart Tagore's visit to China in 1924.'[15]

Notes

1. *The Complete Works of Tagore*, V. 5, (Hebei Education Press, 1st E. 2001), p.181.
2. *The Complete Works of Tagore*, V. 8, (Hebei Education Press, 1st E., 2001), pp. 179–180.
3. Mei Lanfan: *Recalling Tagore*, see *People's Literature,* Issue May, 1961.

4. Bai Kaiyuan: *Tagore's Three Poems to Chinese Cultural Celebrities, see Collected Research Papers on Indian Literature*, Issue 6, P.135—136. Shanghai Translation Publishing House, 1st E., Nov. 2003.
5. Yang Tianhong: Dream of the New People, (Liang Qichao, Sichuan People's Publishing House, 1955), p. 329.
6. Bai Jiyan: *Biography of Hu Shi*, People's Publishing House, p. 216.
7. *Valuable Collection of Chinese Modern Literature – Xu Dishan*, P. 320-321.
8. Ibid., p.321.
9. *Tan Yunshan and China-India Culural Exchange*, (The Chinese University Press, 1988), p. 295.
10. *The Works of Moruo*, V.10, 'Tagore Comes to Visit Me in China'
11. *Collected Works of Bingxin* – Notes.
12. *On How I Write 'Stars' and 'Spring Water.'*
13. *Complete Works of Tagore*, Vol. 20, p.26.
14. *Poet – Sage Tagore and Sino-Japanese War.*
15. *Xinhua Semi-monthly,* Issue 6, 1956.

Tagore's Travels to China

Sreemati Chakrabarti

China Studies in India began with Rabindranath Tagore. In that sense, he was also the founder of area studies in this country much before it became a fashion in American universities. In 1929 Tagore invited Prof. Tan Yunshan, who he had met in Malaysia, to come to Santiniketan to set up a programme for the study of Chinese culture and civilization. This was realized in 1937 with the establishment of *Cheena Bhawan* on the campus of Vishwa- Bharati. This is indeed remarkable as no other area studies centre got the privilege of becoming an exclusive department in that university. This was Tagore's way of showing reverence to China, India's civilizational twin, to use the words of an eminent scholar. It was in Cheena Bhawan, Visva-Bharati, that the seeds of Sino-Indian Studies were sown. Tagore made relentless efforts to initiate China studies at his institution. This was despite the fact that Tagore's first visit to China in April-May 1924 was not a pleasant one. Nonetheless, he always held China in high esteem. During this tour, Tagore said in one of his lectures to his Chinese audience: 'Your civilization has been nurtured in its social life upon faith in the soul. You are the most long-lived race, because you had centuries of wisdom nourished by your faith in goodness, not in mere strength. This has given you a great past'. (Cited by N. Rao, Foreword, Tagore and China)

For Tagore and his family, China was never an unknown, alien land. Since childhood he had learnt a lot about the country as both his father, Debendranath and grandfather, Dwarkanath had visited China. His father also had great interest in Chinese philosophy. From a young age Rabindrnath had acquired considerable knowledge about Chinese culture. His admiration for Chinese culture and philosophy was complemented by his profound sympathy for the Chinese people suffering under unchecked colonial exploitation. In 1881, when he was barely 20 years old, he wrote an essay in Bengali condemning the opium trade which the British had imposed on China. Titled 'Chiney Maraner Byabasay' meaning 'the commerce of killing people in China' he attacked the British strategy of using India (where the opium was grown) to devastate China. (Amartya Sen, Tagore and China p.6)

In Tagore's lifetime, East Asia as a whole showed the two opposite trends of the then existing world order. On the one hand, there was Japan, rising economically and militarily, and on the other, were China and Korea, the former a target of Japanese imperialism and the latter completely subdued by the Japanese into a colony. Tagore had always had a deep admiration for these countries and believed that bonding between East Asian countries and India was vital to world peace. Moreover, Tagore was also greatly attracted to the civilizational past of all these countries, which led to certain traits in their cultures, many of which, he believed, they shared with India.

Being the first Asian to win the Nobel Prize, Tagore was also held in high esteem by the peoples of China, Japan and Korea. He was seen as a symbol of Asian pride whose philosophy, poetic genius and vision were recognized the world over. As his fame spread across the Asian continent with the translation of *Gitanjali* and other writings, there was curiosity and enthusiasm among the intellectuals to hear him and interact with him. Invitations poured in from Japan and China to visit, which he did. However, despite his keen desire to go to Korea he could not, as circumstances were not conducive for a visit. However, during his first visit to Japan, he addressed a gathering of Korean students and enthralled them with his ideas on the value of culture and civilization. He even read out a short poem of his to these students where he referred to Korea as the 'Lamp of Asia'. It is believed that this poem gave massive inspiration and encouragement to the national movement in Korea that was struggling against Japanese imperialism. Up to this day, Koreans mention this often in their discourses on India.

Tagore's winning the Nobel Prize won him many admirers in China. Intellectuals who knew English had read his works and obviously were impressed. In the second decade of the twentieth century, a cultural movement of massive dimension engulfed China. Known later as the May Fourth Movement, it had many facets. It was an anti-imperialist movement, a movement for women's liberation, a literary revolution, a youth movement and so forth. It attacked Confucianism and along with it Chinese tradition in almost all its aspects. The radical Leftists within this movement formed the Chinese Communist Party in 1921. The May Fourth activists, though greatly opposed to Western and Japanese imperialism, were proponents of western science and culture. They considered Chinese traditional culture to be an impediment to China's modernization and development. It is in this context that one must see and evaluate Tagore's visit in April-May 1924.

However, before Tagore's 1924 visit which was his first, there was great enthusiasm among the intellectuals who were looking forward to meet the Nobel-laureate. Scholars like Guo Moruo, Hu Shi and Xu Zhimo expressed their admiration for Tagore through their writings. Years before the visit several Chinese scholars, including the Leftists, translated many of Tagore's writings in the established journals and magazines. Chen Duxiu, professor at Beijing University, who later became one of the founders of the Chinese Communist Party, translated portions from Tagore's *Gitanjali* for the prestigious magazine *Xin Qingnian* or 'New Youth'. In a letter written just before this visit Xu Zhimo wrote to Tagore: 'Nearly all leading magazines here have published articles on you and special numbers are issued in honour of your visit. Most of your works in English have appeared in Chinese translation, some are in more than one version. Never before has a single writer, eastern or western, excited so much interest in the heart of our young nation, and few, not even our ancient sages perhaps, have gifted us with such a vivid and immense inspiration as you have done' (Xu zhimo 2005, vol.6:378 cited by Tan Chung, Tagore and China)

Obviously, the above facts in no way indicate that there was any reason to think that Tagore's views would be opposed by several Chinese intellectuals. One interpretation of this opposition to

Tagore's speeches in China is that the official invitation to visit China came from a traditionalist group called the Beijing Lecture Association, headed by Liang Qichao. This group was opposed by the young radicals and communists and that led to a rejection of Tagore. Another reason why Tagore's visit led to some unease in the minds of China's progressive intellectuals was his meeting with the deposed emperor of China Pu Yi. The ex-emperor still lived in the Forbidden City from where he reigned China as a small child from 1908 till 1911 when he was deposed by the Republican Revolution (Hay, p.160). It is quite likely that China's young radicals did not appreciate this as monarchy was being seen as part of an anti-people tradition. A third factor was that in the Hankow-Wuchang area, a Buddhist organization hosted and organized his lecture. Perhaps this too was seen by the radicals as Tagore's inclination towards conservatism.

Professor Stephen N. Hay of the University of California, Santa Barbara has, in a pioneering work published in 1970, given a detailed account of Tagore's 1924 trip to China. Tagore arrived in China on April 12, 1924 when his ship reached Shanghai. Two days later he delivered his first speech, the English version of which has been lost. One has to rely on the Chinese translation. Tagore began with a modest tone. He said 'I have come to China not with the attitude of a tourist... or as a missionary bringing a gospel, but only as one seeking wisdom, like a pilgrim wishing to pay homage to the ancient culture of China, in an attitude of reverence and love' (Hay, p.152). As Hay puts it: 'Despite this modest prologue, Tagore did have a message for China. What he had seen in one of her largest industrial centers made him fear for her future'. He said:

> I am a poet, not a politician nor a diplomat and can say what I feel most sincerely in my heart. I feel that China is now going the same way as India. I love culture, I love life. I cannot bear to see Chinese culture endangered day by day. Therefore I sincerely warn you: know that happiness is the growth of the power of the soul. Know that it is absolutely worthless to sacrifice all spiritual beauty to obtain the so-called material civilization of the West!

He ended by saying 'bring the world to idealism, to humanism, and destroy materialism' (Ibid.). Tagore saw himself as the 'representative of Asia'. However, he did point out that his purpose in coming to China was not to create hostility to western civilization, but to strengthen the unity of Eastern civilization as a necessary step toward achieving a complementary relationship between the two. (Hay, pp.156-57)

In a period of about seven weeks, Tagore had lectured in Shanghai, China's largest city, Beijing which was its nominal capital, and in five provincial capitals: Hangzhou, Nanjing, Jinan, Taiyuan and Wuchang. He had meetings with scholars and students, actors and artists, military officials and politicians, poets, religious leaders and even an ex-emperor. He was received well by the general public almost everywhere. In the provincial capitals, most students seemed receptive to his ideas though he was not sure how much of his English they understood or how well his translators had rendered his words into Chinese. Dissident voices were being heard and their 'chorus of disapproval grew louder as the poet's tour of China progressed. (Hay, p.182)

Being a true internationalist in every sense of the term, Tagore was staunchly opposed to the nationalism which was displayed both by the states as well as mass movements of his time. Japanese nationalism was expressed through militarism, which Tagore's strongly condemned as destructive. (He did not live to see that it was ultimately even self-destructive!). In China too Tagore's first visit ended up being controversial as what he said in his speeches was not quite palatable to the activists of the May Fourth Movement, which was then the mainstream anti-imperialist and nationalist movement. Tagore opposed the infatuation for modernization in the name of science and democracy, the twin goals of the May Fourth Movement. In this period, radical Leftist ideas were dominant in China and hence there was a near complete rejection of tradition. Even though Tagore was not a traditionalist in any way, he did not accept that westernization was the solution to all ills of Asia. The World War (1914-18) and the uneasy peace after it weighed heavily on Tagore's mind. This came out in his speeches and writings.

However, much later there came the realization among scholars in China that Tagore was not out of sync with the times, nor was he preaching anything that could be detrimental to their society. Tagore's lectures in China have been published but it is difficult to summarize what he said. As an example, I give a quote from one of his speeches. He said: 'It would be degradation on our part, and an insult to our ancestors if we forget our own moral wealth of wisdom which is of far greater value than a system that produces endless material and physical power that is always on the warpath.' (Sisir Das, *Across the Himalayan Gap*)

Wei Liming, a scholar who now teaches at Beijing University, has done very useful research on Tagore's visit to China. This has been published in the form of an essay in a recent publication titled *Tagore and China*. The following account is partly paraphrased from her writing.

China's great nationalist leader, Dr. Sun Yat-sen, had become an ardent admirer of Tagore much before his visit. He had sent a warm invitation to Tagore but regretted that they would not have a chance to meet as Dr. Sun would not be able to come to Beijing due to ill-health. Dr. Sun always believed that China's revitalization has to be from an Asian perspective. In his opinion, Tagore had dedicated his entire life to social upheaval with art and literature as his weapons, advocating the revitalization of traditional Indian culture, enhancing national self-pride, and promoting the national liberation movement. When Tagore was visiting China, Dr. Sun thought that he was doing so with the aim of 'promoting the rebirth of the traditional cultures of the East as well as the unity of Asian nations'. In his words, Tagore was opposed to the 'absolute reliance on European civilization and whole-hog Westernization at the expense of one's own native civilization and values', condemned the West's aggression, exploitation and devastation of China at will, and advocated the coming together of the people of Japan, China and India under the banner of the rebirth of Oriental thought.' Sun hoped that Tagore's visit would show China the correct road ahead to come out of crisis and peril. Tagore, however, did not respond to this letter. Moreover, he met the deposed Manchu emperor, Pu Yi as well as some warlords, thus sending wrong signals to many Chinese revolutionaries. (pp. 20-21)

After he had delivered a few lectures, opposition began to mount. Chen Duxiu, who had earlier translated sections of *Gitanjali* became a staunch opponent. He and Qu Qiubai, another prominent leader of the newly-formed CCP, attacked Tagore through their journals and magazines. They also sent people to the venues where Tagore was lecturing, to distribute leaflets condemning Tagore's views. During one such lecture, a leaflet reached Tagore's hands. His interpreter and close friend, Xu Zhimo was too embarrassed to translate it and so his companion's got it translated by another person. Tagore was shocked to know about the contents of the leaflet and decided to cut short the visit and return to India. Later the gist of the criticism was published in a Calcutta magazine.

In his farewell speech in Shanghai, Tagore said that it was difficult to bring the different human races close together because life has become so conventional. He added: 'Let me confess this fact, that I have my faith in higher ideals. I believe that through them we can best serve the higher purposes of life. At the same time, I have a great feeling of delicacy in giving utterances to them, because of certain modern obstacles which make it almost a disreputable thing to be frank and free in the expression of ideas.' (Das, 1999: 75-76). Tagore realized that his otherwise perfect visit had become unpleasant.

Chinese intellectuals of the radical left criticized Tagore as they feared that it could have an adverse effect on the mass movement that they were struggling to organize. To be fair to the critics of Tagore, one cannot be oblivious to the fact that those were the years when China was going through a massive external and internal crisis when all progressive elements were determined to end once for all not just warlordism but, along with it, any remnants of feudal and traditional culture. While it is true to say that the Chinese revolutionaries misunderstood Tagore, it is also true that Tagore despite his greatness and genius failed to read the mood of the then Chinese people. He did not prepare adequately for the visit and his Chinese friends failed to communicate to him the existing mood in China. It was a clash of an idealism and vision on one side, with on-the-ground political activism on the other. Tagore's lofty ideas and message were for the entire world and applicable for all times to come. The Chinese intellectuals he was addressing were engrossed in their own typically Chinese problems and issues of that era.

According to some present-day China scholars, one possible reason why Tagore was misunderstood by the Chinese was due to improper and inadequate communication. Tagore did not know Chinese and in his time western Sinology had not advanced much. His understanding of the contemporary situation in China, therefore, could have been limited. Similarly, his own 'lofty ideas and profound emotions' expressed mostly in Bengali or English were often untranslatable in Chinese. Perhaps one can infer from this that his serious effort to set up a Chinese language department in Visva-Bharati was due to this conviction that Indians need to understand China through their language.

Indian author, Pankaj Mishra, has made a deeper analysis of the Chinese response to Tagore. According to him, at one lecture, Tagore faced hecklers who greeted him with, 'Go back, slave from a lost country'. He points to the fact that since the dawn of the colonial era in India, large numbers of

Indians were involved in the opium trade. Indians were recruited as soldiers to fight the Chinese in wars, and as policemen and security guards in British Concessions to harass the ordinary Chinese. For more than seventy years before Tagore's trip, the Chinese had an image of Indians as a defeated and slavish race. Chinese intellectuals, including the 19th century reformers, too had this view and saw India as a 'lost nation'. Perhaps this negative image loomed large in the minds of the Chinese who saw in Tagore an anti-modernist figure.

Tagore's second and last visit to China was a private one. Not much is known about it. At present, there are a large number of people in the East Asian region who are interested in scholarship on Tagore and realize that a reconsideration of Tagore's philosophy is desirable. Wang Bangwei, eminent India scholar at Beijing University, said recently, 'We in the past, had respected Tagore but also misunderstood him… the misunderstanding was due to complex causes. Now it is the time to reopen the issues and understand them afresh'.

Interestingly, in today's modernizing China, we see a society looking for a philosophy and ideology to deal with tensions of modern living. Had that not been the case, Confucianism, which many thought was consigned to history, would not have bounced back nor would concepts of a harmonious society and a harmonious world been propagated by the political elite. It is hoped that the growing number of Tagore scholars in China will come out with new perspectives on how Tagore viewed China and how the Chinese viewed him.

References

1. Sisir K. Das, 'The Controversial Guest: Tagore's 1924 Visit of China' in Tan Chung (ed.) Across the Himalayan Gap (New Delhi: Indira Gandhi National Centre for Arts, 1999).
2. Tan Chung, Amiya Dev, et.al (ed.) Tagore and China (New Delhi: Haranand, 2011).
3. Stephen N. Hay, Ideas of East and West: Tagore and His Critics in Japan, China and India (Cambridge, Mass., Harvard University Press, 1970).

Apocalypse of Tagore: Nature and Man, Japan and India

Yukifumi Takeuchi[1]

Abstract

It was the 150th anniversary of Rabindranath Tagore's birth last year. This year, 2013, is the 60th anniversary of the establishment of India-Japan diplomatic relations. We, the Japanese people, commemorate these occasions not only for celebrating historical memories, but also to recollect meaningful messages for contemporary Japan, India, Asia and the world.

Since Tagore was very interested in Japan, he visited the country five times. In particular, he visited Tokyo for the second time in 1924. It was the year after the Great Kanto Earthquake that devastated the Tokyo Metropolitan Area. The number of victims and missing exceeded 105,000. Seeing the Japanese people working and rebuilding the devastated city, Tagore wrote that he was impressed by their strong dedication and tireless effort for reconstruction. He also wrote that it was difficult to prevent natural disasters but man-made calamity was far worse.

Last year, we, the Japanese, experienced the Great East Japan M9 Earthquake and lost almost 20,000 lives. Since Tagore observed the Great Kanto disaster in Japan, nearly 90 years have passed and, since then, human society has made great technological innovations. But before such natural disasters, we human beings are small and weak, as Tagore had said. But his praise of the Japanese people's efforts toward reconstruction encourages people living in the damaged areas of Fukushima or other Tsunami-affected areas. The fact that India dispatched an emergency rescue team is also highly appreciated.

Although Tagore criticized Japanese imperialism very sharply since his first visit to Japan in 1916, on the other hand, he was also critical of the attitude of the major powers. Tagore's thoughts were taken up by other Indian leaders such as Jawaharlal Nehru and, in keeping with this trend, India decided to be a non-signatory to the San Francisco Peace Treaty. In its place, India agreed to have a bilateral Peace Treaty with Japan in 1952 and we have celebrated the 60th anniversary of that Treaty in 2012.

Taking advantage of such memorable events, we should deepen our spirit of exchange as Asian countries and partners for a better future.

Speech to Japanese people

To the people of Japan:

My friends, the warmth of your welcome deeply touches my heart. I feel extremely glad that I have been able to meet you once again and I offer you on behalf of my country a deep sympathy for the disaster which only later you have met with. I must tell you how deep was our sorrow when we heard of it. We felt this sorrow the more because it was Japan which had aroused us to the consciousness of life all through Asia. For that we shall ever be thankful to Japan and that we all accept this disaster as our own.

This was the only reason which brought me to Japan. This time I have no desire to speak to you or to prolong my stay in this country for any special purpose of my own. I always feel it unfortunate, my friends, that you should know me as a speaker, which I am not. My mind only speaks through silence and solitude, when I dive deep into the depths of my being and find a voice hidden there. I never had any training as a speaker. But it has always been my fate to be known as a speaker when I travel outside my own country. I know today that some of you will go away with the impression that you have seen me, a poet of reputation, and have known me because on one of my short visits you have been in touch with me on the platform. Nothing can be further from the truth. You may come to know me more truly after I am no longer on the earth but not now.

There are men strong of arm and strong of purpose who build institutions stone upon stone and brick upon and who finish their building before the eyes of public, but I do not belong to their brotherhood. I come like a sower of seeds; I just scatter them on the soil and do not have time to see them germinate. I may even go away with the misgiving that these will not come to their fulfillment. But this is my mission, and a greater part of my work must be carried on in obscurity.[2]

Earthquake in 1923

This speech, titled 'To the People of Japan', was delivered by Tagore at Asahi Shimbun Hall in Osaka, on June 3, 1924. It was the year after the Great Kanto Earthquake happened on September 1st, 1923. Tagore spoke in a very tender tone.

Tagore visited Japan for the first time in 1916. Seven years after his first visit, there was the Great Kanto Earthquake whose magnitude was 7.9. The number of people affected by the earthquake and the tsunami was more than 1.9 million and the number of victims was approximately 105,000. After he heard about this disaster, he decided to visit Japan again. He came the following year and visited the areas devastated by the earthquake such as Tokyo and Yokohama. There, he witnessed the dedication of the people in their reconstruction work.

At the time, Tagore, a great author, honoured as the first Asian Nobel Laureate in Literature, was over sixty years old. Witnessing the scene of people working to reconstruct their lost houses, he lamented that he was just an outsider and unable to help the victims directly except for reporting about the situation.

He compared himself, reporting messages from the frontline of the stricken areas, to a sower of seeds in a field, but he considered his mission very seriously.

Thereafter, Tagore sent a letter to one of his Japanese friends, Eiichi Shibusawa, said to have been a pioneer of modern Japanese entrepreneurs and 'a Father of Japanese Capitalism'. Tagore wrote saying, 'It is encouraging that Japanese people do not stop their reconstruction efforts. We cannot prevent natural disasters, but it is much worse to have a man-made calamity. I guess that the Japanese people are really angry about the disaster. But it is hopeful that they are apparently very even-tempered, vigorous and optimistic.'[3]

We can still take strength from how greatly he was impressed with the Japanese people in dealing with the disaster. That view would equally apply even with the great disaster that happened in 2011, and Tagore's message encourages us living in the present day.

Lessons of Disaster

The Great Kanto Earthquake of 1923, which damaged the larger metropolitan area of Tokyo, was the biggest disaster in Japan's history.

Sixty-two years later, in 1995, the Kobe-Hanshin Great Earthquake occurred. With this epicentral earthquake of magnitude 7.3, more than 6400 people died.

And then, on March 11th, 2011, the Tohoku Earthquake and Tsunami occurred. With a magnitude 9.0, it was of the biggest in the recorded history of Japan. The stricken area was so over 500 km from north to south. Then, just 30 minutes later, the tsunami reached the coastal area. The height of the biggest tsunami wave was over 40 metres. It devastated and swallowed people, houses, cars and railways.

The number of evacuees was more than 400,000. The victims and missing exceeded 19,000. The total loss was approximately $30 billion. The images reported from the damaged area greatly shocked the people of India.

Of the repeated earthquakes in Japan, the Great Kanto Earthquake in 1923 had a significant historical importance. Its number of victims was five times larger than the East Japan Great Earthquake of 2011.

Tagore's observation on the disaster was very insightful. Although people were angry, devastated and saddened by the disaster, they controlled their emotions and worked relentlessly for reconstruction. Tagore was profoundly impressed in 1923.

Such extraordinary efforts by the people were again observed and reported from the devastated area in 2011, in particular through several articles in the foreign media.

The Indian newspaper, 'Business Line' reported just after the disaster, quoting Indian witnesses who were visiting the Honda factory located in the area on March 11th.[4] These Indians were a delegation of auto parts suppliers doing business with Honda. They coincidently visited the Honda R&D Center of Tochigi Prefecture on that day.

They said, 'Even amid the disastrous situation of ceilings and walls falling down, all order was well maintained;' 'Contingency plans were perfect and prepared well. They responded to the earthquake like a precision mechanism;' 'Soon after the shaking subsided, all the workers got together to the planned evacuation site very quickly and fulfilled their own roles very precisely.'

'After we moved to the town from the R&D Center, there are no confusion. They kept quiet and order at all times.'

The Russian media, Tass also reported surprisingly, 'In other countries' during a disaster, trouble, violence and looting happens quite often. But those were not reported at all in Japan', 'When all the transportation was suspended in the Metropolitan areas of Tokyo after the earthquake, people helped each other. Shops and restaurants provided people free drinks and allowed them to use toilets as they walked back home through the night.'

Among the US media, the Wall Street Journal reported in its editorial entitled 'Invulnerable Japan' the very next day. It said, 'we have to point out how stoically the Japanese people prepare for survival, even if they face an attack and damage from the Mother Nature,' and appreciated Japan's disaster prevention system and earthquake resistant construction.

Man-made Calamity

On the other hand, Tagore's view that we cannot prevent natural disasters but that it is much worse to have a man-made calamity was really significant.

In 1923, owing to the damage of the Great Kanto Earthquake, newspaper companies were stricken, burnt out and stopped printing papers. It accelerated the human disaster. At that time, the newspaper and the telegraph service were vital media and communication tools. In such a situation, the suspension of newspapers had a major detrimental impact. Rumors and speculations that some Koreans, a minority of Japanese society, had incited riots and poisoned ground-water wells prevailed. It stimulated racial friction and there were victims. The number of victims was 233 according to government sources, but put at 2613 – 6661 according to the Korean media.[5]

With this historical lesson, the introduction of new media, namely radio broadcasting, started two years after the disaster. In the end of 1923, the Ministry of Communication formulated the rules for broadcasting in Tokyo, Nagoya and Osaka and it started, for the first time in Japan, in 1925.

On the other hand, many buildings and houses were also burnt and collapsed. The low standard of wooden buildings caused the fire and spread the disaster over a wider area. It was then that the government laid down rules about building construction that subsequently developed into the

Building Standard Law.

The government also encouraged local prefectural governments to make preventive plans for disasters and draw up regulations for periodic drills in all offices, schools and local residential areas. All these were the lessons learnt from the earthquake.

In this manner, the experience of disasters was reflected in the social, public and institutional infrastructure of Japan. Whenever people experience major disasters, not only in Japan but in many countries, historical wisdom develops into social infrastructural reform and improvement. It is the human way to learn from our own history.

By the way, what were the man-made calamities of The Tohoku Earthquake and Tsunami?

Although the magnitude of this disaster was greater than the one that happened in 1923, people were evacuated as fast as possible because of their experience and their disaster preparedness education and drills. As a result, the number of victims was less than one fifth as compared to 1923. Moreover, in terms of constructions, incombustible materials were more prevalent and caused less fire destruction of buildings.

However, what happened with the nuclear power plant accident in Fukushima was unprecedented.

After the reactors were damaged and shut down by the earthquake and tsunami, the government and the Tokyo Electricity Power Company were late in reacting to the situation of a possible melt down. It resulted in a belated notification of evacuation to the residents.

There have been no victims directly as a result of the nuclear accident so far. However, many people, who could have avoided exposure to radiation if the authorities had made a proper judgement, were exposed to radiation. Those people worry about the radioactive accumulative impact on their health and those who have young children still stay out of the area.

It has also had a big influence on agricultural production. Some production is regulated for sales because of the higher radiation contamination found. But some sales have been affected by baseless rumors and images of contamination. Among prefectures hit by the disaster, Fukushima is called as the land of 'Four-Fold Handicap' as it includes the earthquake, tsunami, nuclear accident and agricultural production damage.

If we define man-made calamity as a disaster caused by human carelessness and default, the Fukushima nuclear accident, caused by natural disaster obviously accelerated into a man-made calamity and stands as the most serious accident in the history of nuclear technology.

Japan has developed its own traditional manufacturing culture to world-class, front-runner state both in industry and export. Its technology, know-how, management excellence and quality-control were utilized, not only for mass industrial production, but also for the operation of mass transportation and distribution by trains and trucks. They had also been applied to the management of nuclear plants.

However, given the energy mix of Japan, which has scarce raw energy resources, the government and power industries, considering nuclear energy inevitable for several years, have been creating a myth about 'Safe Nuclear Technology'. As a result, Japan has developed into the third biggest nuclear plant state in the world with fifty-four plants.

When the serious accidents happened in Three-mile Island in 1979 and in Chernobyl in 1986, the Japanese nuclear energy community launched a campaign to say that such accidents only happened overseas and that Japanese technologies were safe. Regional preparedness education in case of a nuclear accident or disaster and training drills were only held as a show, saying, 'No worries about accidents but it is necessary because we said we would do it.'

After Fukushima's accident, all of Japan's nuclear plants have been suspended, one by one, for safety inspection and, in May 2012, all their operations were stopped. An anti-nuclear energy movement has risen and is very active nationwide. The number of people who support the idea that Japan should choose the policy of ending all nuclear energy dependence and enlarge renewable energy, like the Germans have done, has been growing.

Message of 'Mukthadhara (The Water Fall)'

Then, did Tagore predict that the age of large-scale science and technology including that of nuclear energy would come in the future?

It is well-known that Tagore had a meeting and a dialogue with legendary scientist, Einstein, at the latter's cottage in Berlin in 1930. Their argument about truth was a reflection of the contrast between poet and scientist, between east and west. However, it was only at an abstract level because of the limitations of the era in which people could not foresee future science development.

Fifteen years later, nuclear technology developed for military use emerged as a huge issue confronting the future of human beings. Nuclear bombs were dropped on Hiroshima and Nagasaki in 1945.

Although nuclear emergence surprised the world, US President Eisenhower turned its direction toward peaceful uses in his famous speech 'Atoms for Peace.'

Tagore passed away on August 7th, 1941 and hence did not know about nuclear weapons or nuclear power plants.

But we can observe his prediction in his works that the the huge scale of science and technology would be a major obstacle for human beings and nature owing to their environmental implications.

Tagore, in his play 'Mukthadhara - The Waterfall', mentioned natural destruction by man and voiced his criticism more than 90 years ago.[6]

It is the story about the Kingdom Uttarkut ruled by a dictatorship and the people living in the valley downstream. The King tries to build a dam to keep the flow of the river into the beautiful water fall and to make the downstream valley dry up in order to rule over the people. Prince Ojibito pained by this, destroys the dam. The water flow resumes and green is restored in the valley, but the waterfall swallows the prince.

Among the characters, the significant person was Bibuti, who symbolized science, technology and a machinery civilization. He designed and operated huge mechanical creations as an engineer. Although he has no interest in national benefit or the well-being of the Kingdom, he has extraordinary compassion for his mechanical creations that become mechanical monsters.

It has a scene, where Bibuti tells a mother who lost her son owing to the dam construction. 'Mothers cursed me. But my machine won. When we win against god's curse, people don't heed the god at all.'

It was a measure that Bibuti always brought. He tried to find a solution through calculation and technology. And some people praised Bibuti's method. Tagore pointed out to and criticized people who had the tendency to follow and surrender to political power in human society.

This story was enacted in New Delhi in 2011 for commemorating the 150th anniversary of Tagore's birth. Bhanu Bharti, the director of the production 'Tamasha Na Hua' (There was no Fun) said, 'The re-enactment of the 1922 play, which raises vital issues about man, machine and the conflict between humanity, development and nature in a contemporary light, is a tribute to the bard from Bengal.'

'The issues that the characters argue are more complex than those articulated by Tagore because machines are taking over in a big way in all spheres of our lives. Machines have become so important to human existence that it has bred an addiction for them,' Bharti said.

The dam on the waterfall symbolises this widespread mechanisation, the playwright said. The play also touches on the other side of the dam-versus-man debate that 'big dams are important to development and existence', he said.

Views of Professor Azuma

Professor Kazuo Azuma, a famous scholar who specialized in Tagore study and died in 2011, pointed out in his book 'Tagore---His Poems, Thoughts and Life'.[7]

'Tagore said that every single leaf, flower, fruit and stone is necessary for life, thoughts and art. Now we argue about the threat to the earth caused by natural destruction. Tagore continued to talk about the necessity of leaves, fresh air and water. He kept appealing that ignorance and the wanton destruction of nature means the death of man through disconnection of the unity of man and nature. He hit an alarm for contemporary issues more than 100 years ago.

Tagore's appeal that ignorance and the destruction of nature means the death of mankind has a strong impact on the Japanese people of today. In the past, during the 1970s and the 1980s, although Japan had a higher rate of economic growth, air pollution and the contamination of rivers were prevalent nationwide. At present, people are disconnected from nature in Fukushima, and their agricultural products are in danger. They cannot live in their own home towns and villages any more.

On the other hand, Tagore's poignant connection between machinery, civilization and political power was exactly like the structure for nuclear energy in Japan and in other countries around the world.

Emerging countries, such as India, which has twenty reactors now and likely to have more in the future, are not an exception. Japan's incident is not just a fire on the other side of the river.

Tagore's philosophy that man should unite with nature motivated him to develop villages. However, he didn't necessarily deny the co-existence with a machinery civilization.

Mahatma Gandhi did a campaign with the Charkha, the spinning wheel, to appeal for self-sufficiency and to symbolize the struggle against the British and its machinery civilization for India's Independence.

In contrast, although Tagore accepted such a means as the Charkha, he introduced the latest spinning machines from Japan and the US.

He appreciated the effect of machinery civilization to some extent and adopted the use for diversified production and harmony with modern human life.

The ideal of Tagore was to pursue the real independence of people with a diversified human culture.

Criticism against Japanese Imperialism

Tagore visited Japan for the first time in 1916, when he was 56 years old. He made a speech entitled 'The Spirit of Japan' at Keio University.[8] In this speech, his ideas were more clearly set out on science and civilization.

He said, 'Science, when it oversteps its limits and occupies the whole region of life, has its fascination. It looks so powerful because of its superficiality, as does a hippopotamus, which is very little else but physical. Science speaks of the struggle for existence. Man truly exists in the ideal of perfection, whose depth and height are not yet measured'.

He also extended his ideas on comparison between Western and Japanese civilization. 'All particular civilization is the interpretation of particular human experience. Europe seems to have felt emphatically the conflict of things in the universe, which can only be brought under control by conquest. Therefore, she is ever ready for fight, and the best attention is occupied in organizing forces'.

'But Japan has felt, in her world, the touch of some presence, which has evoked in her soul a feeling of reverent adoration. She does not boast of her mastery of nature, but to her she brings, with infinite care and joy, her offerings of love. Her relationship with the world is the deeper relationship of heart'.

Tagore regarded highly Japanese ideas and values towards nature, in contrast with Western ones. We, the Japanese, appreciate his profound views toward our civilization.

However, we need to know his purpose for visiting Japan. That was to bring his message and criticism against the First World War to the Japanese people.

Now let us have a look at the history of Japan during the end of the 19th century and beginning of the 20th century. Japan won the first Sino-Japan War that took place in 1894 and also won the Russo-Japanese War of 1904. Owing to these victories, Japan was regarded as a new Asian power and respected by other Asian countries.

But, on the other hand, Japan expanded its territory from China and Russia and competed with European imperialism. Japan's territorial ambitions were feared as an emerging threat to European countries.

In 1914, two years before Tagore visited Japan, the First World War broke out. Japan, based on the agreement of the Anglo-Japanese Alliance, made a proclamation of war against Germany and took part in the war as an Alliance member. The Japanese military attacked and occupied Qingdao, a port city Germany controlled in Eastern China, and also occupied the Micronesian islands that Germany controlled.

Japan sent her vessels to the Indian Ocean and the Mediterranean Sea to protect the sea lanes of the alliance. When the Russian Revolution occurred in 1917, Japan also joined the Siberian Intervention in 1919. After the First World War ended, Japan participated in the Paris Peace Conference and became a member of the Permanent Security Council of the League of Nations.

Receiving the Nobel Prize, Tagore further assured himself of the belief in the significance of the human community, which he had embraced since his early days. He became confident that people in the world could mutually understand and respect each other. This belief made him 'a Missionary of Peace.' The direct motivation was the First World War.[9]

Although the war was triggered in Europe, Tagore was deeply concerned. He wanted to appeal against its stupidity to the world outside of Europe, in particular, Japan and the US. He thought that the countries of Europe were filled with hatred toward each other but that Japan and the US still kept their calm and right thinking. He wanted to visit these two countries to talk with the people there.

In Japan, Tagore discovered the delicate sense of beauty, courtesy, simplicity and diligence of lifestyle and strongly praised them. Even beforehand he had a basic recognition of Japanese culture through more than a decade of friendship and exchange with Japanese artists and intellectuals such as Tenshin Okakura, Taikan Yokoyama and Shunso Hishida.

During his stay in Japan, when Tagore, however, realized only about half of his expectations, he must have been somewhat disappointed. In particular, he noticed the imperialistic tendencies in Japanese society and the excessive imitation of Western culture. His speeches at Keio University and Tokyo University represented his views very frankly.

Japan, from his standpoint, rushing into modernization and Westernization, had forgotten its own traditions and morals. He criticized Japanese imperialism and imitation of Western culture harshly because he loved the Japanese sense of beauty and politeness based on its own traditions. In particular, Tagore could not accept the Japanese ambition to invade China, the most important Asian neighbor for Japan, in order to expand her territory.

He said at Keio University, 'I cannot think that the imitation of the outward aspects of the West, which is becoming more and more evident in modern Japan, is essential to her strength or stability. It is burdening her true nature and causing weakness, which will be felt more deeply as time goes on.'

He also pointed out, 'I do not for a moment suggest that Japan should be unmindful of acquiring weapons of self-protection. But this should never be allowed to go beyond her instinct of self-preservation. She must know that the real power is not the weapons themselves, at the cost of his own soul, for then it is he who is in even greater danger than his enemies.'

On the other hand, the then-Japanese establishment, which liked to enjoy the hegemony of the Far East, had welcomed Tagore as a Nobel Prize Winner as a tactic for supporting a rise in Asian nationalism against European colonialism. For this reason, they felt betrayed when they learned of Tagore's pacifistic thoughts.

A British missionary, who attended Tagore's visit to Japan, then reported that the majority of Japanese society perceived that Tagore's pacifistic ideas were dangerous for them during the wartime. He also wrote that the atmosphere of welcome for Tagore's visit faded quickly and Tagore was isolated in Japan. His purpose for visiting Japan was not realized.

Contemporary Asia and Tagore

Tagore continued his dialogue with Japanese intellectuals, including Yonejiro Noguchi, a Japanese poet and author, about civilization and the Asian future. Through many letters with Noguchi, they exchanged their views. Noguchi wrote, 'If you think of the Japanese war in China as a criminal result of Japan's imitation of the Western colonialism, you are wrong. It is an inevitable method to realize the great world in Asia.' Tagore wrote, 'Chinese civilization has excellent cultural resources and will be never conquered. Japan and China should wipe out the memory of their past sad history and hold hands with each other in the future. Both countries should restore the true Asian love for humanity.'

Until his death in August of 1941, Tagore tried to continue his exchange with these people.

He criticized Japanese and European imperialism while listening to the sounds of soldiers' boots. However, he did not lose his faith in humanity and kept alive his expectation about the future of Asian civilization.

He did not witness the outbreak of the Pacific War between Japan and the US in December 1941, nor Japan's defeat after the atomic bomb attacks against Hiroshima and Nagasaki.

In India, his spirit was passed on to the younger generations and Independence was realized in 1947. Although Tagore criticized the then Japanese imperialism very sharply since his first visit to Japan, on the other hand, he was also critical of the attitude of the other major powers. Tagore's thoughts as such were inherited by other Indian leaders such as Jawaharlal Nehru and resulted in the Indian diplomatic policy of being a non-signatory to the San Francisco Peace Treaty. In place of this decision, India agreed with Japan to have a bilateral Peace Treaty in 1952 and we have celebrated the 60th anniversary of this Treaty in 2012.

Taking into account such memorable incidents, we should deepen our spirit of exchange as Asian countries and partners for a better future.

In particular, East Asian countries face difficulties in the regional situation more than seventy years after the death of Tagore.

Japan has territorial issues with China, South Korea and Russia. These island issues have a history of several decades but were not major issues during the Cold War Era. Since the Cold War ended, national integration ideologies of these countries have shifted the focus to their own nationalism. With these trends, territorial issues come to the centre of nationalism and become a symbol of the integration of nations.

We can explain and exemplify with the Senkaku-islands case with China. In 1972, Japan established diplomatic relations with China and the Peace and Friendship Treaty was entered into by the two countries. In this process, both countries, prioritizing economic cooperation, agreed that territorial issues should be shelved for the time being. But, nowadays, a sense of rivalry with Japan motivates Chinese nationalism and it focuses on the islands' territorial issues more drastically. The historical education in China criticizing Japan's imperialism has been strengthened and promoted. Dissatisfaction about income gaps and social inequality among the urban young has been increasing and apparently reflected in anti-Japan demonstrations.

On the other hand, in Japan, there also have been serious movements of inward-looking nationalism. It is accelerated by the sense of frustration with the current economic slump and with the increasing Chinese economic and military power. A famous right-wing politician and the former Tokyo governor, Shintaro Ishihara, declared the intention to purchase Senkaku Islands to strengthen actual control. Finally, Japan's government decided to purchase the Islands in order to prevent fanatical nationalists making use of such territorial issues. But even this step has stirred China and the tension between the two countries has increased. Some hawkish ideologues, including

Ishihara, also mentioned the possibility of Japan's nuclear armament to deter China.[10]

It is not necessary to describe the ordinary Japanese people's perception toward such extreme ideas. But how one is to deal with narrow-minded nationalism of our own and of other countries is not easy to answer.

More than seventy years ago, Tagore wrote that all the countries should have their own nationalism while respecting other countries' culture and nationalism. Along with this thought, all of us must promote and develop our own culture and traditions, while respecting and learning from other cultures and traditions. It is a long path, but we need to follow it.

Notes

1. Visiting scholar of Edwin O. Reischauer Center, SAIS, Johns Hopkins University in Washington DC and Visiting Professor of South Asian Institute, Gifu Women's University in Japan. The author wishes to thank Dr. Nivedita Ray, Tsutomu Kawai, Rika Sasaki and Jennifer Takeuchi for their assistance.
2. Rabindranath Tagore, *Talks in Japan*, Shizen, 2007, pp. 77-78.
3. Tagore150Japan: http://www.jaip.org/tagore150japan-e/110628/kawai.htm
4. Business Line, India, March 13th, 2011.
5. Sakuzo Yoshino, *Gendai-shi Shiryo, vol.6,* pp357
6. Rabindranath Tagore, *Tagore Chosaku-shu,vol.6*, Daisan-bunmeisha, pp. 189-244.
7. Kazuo Azuma, *Tagore – His Poems, Thoughts and Life,* Reitaku University, 2004, pp 79.
8. Rabindranath Tagore, *Talks in Japan*, Shizen, 2007, pp. 55-71.
9. Tatsuo Morimoto, *Gandhi and Tagore*, Daisan-bunmeisha, 1995, pp. 54.
10. ABC News, *Japanese politician argues for nuclear weapons*, Nov.21,2012 http://www.abc.net.au/news/2012-11-21/japanese-politician-argues-for-nuclear-future/4383362

Rabindranath Tagore in Thailand: His Visit and His Diplomatic Roles[1]

Sawitree Charoenpong

Rabindranath Tagore's Visit to Siam[2]: A Part of His Ninth Foreign Tour

Rabindranath Tagore visited Siam during 8–16 October 1927 as part of his ninth foreign tour which included visits to several Southeast Asian countries and extended to three and a half months. He sailed from Madras on 12 July 1927, his first destination being the countries on the Malay Peninsula, including Singapore (20–26 July), Malacca (27–30 July), Kuala Lampur (30 July–7 August), Ipoh (8–11 August), Taiping (12–13 August) and Penang (14–16 August). He then proceeded to Java, arriving in Batavia on 22 August. Tagore stayed in Java and Bali for more than one month until 30 September. Then he sailed to Bangkok via Penang. The tour ended with his arrival in Calcutta on 27 October 1927.[3]

While Tagore was staying in Singapore, Professor Suniti Kumar Chatterji, a member of his entourage, approached the Siamese Consul General in Singapore, Praya Pradibaddha Bhubal, regarding Rabindranath Tagore's visit to Siam. However, it was not within the Consul's power to make arrangements for them. Therefore, the Siamese Consul General in Singapore wrote a letter on 27 July 1927 to inform Prince Traidos Prabandh, Minister of Foreign Affairs of Siam:[4]

> Your Highness,
>
> I have the honour to inform Your Highness that yesterday Professor Suniti Kumar Chatterji came to see me on behalf of Dr. Rabindranath Tagore who is now on a visit to Malaya. He informed me that Dr. Rabindranath Tagore and a party of four intend to pay a visit to Siam, leaving Penang by train on or about the 10th of August and wished to stay for about ten days, and they would like to see Ayuthia and Lopburi. During their visit to Singapore they were entertained by the Governor and stayed at the Government House. It seems to me that he came to approach me in order to see what facilities could be accorded to them by way of accommodation in Bangkok. I replied that it was not within my power to make any arrangement for them. The most that I could do would be to communicate with the Authorities in Bangkok, through Your Highness, concerning their intending visit to Siam.
>
> I beg to enclose a list of party and also newspaper-cuttings regarding their stay in Singapore.

There are some interesting points in this letter. Firstly, it shows that Tagore was not invited by the Siamese government, as is sometimes claimed. Anyhow, one newspaper in Siam, *Lak Muang*, reported on Monday, 17 October, 1927, that Tagore was visiting Siam 'at the invitation of Indian and Chinese residents here.[5] Secondly, Tagore had to have knowledge about Siamese history and Indo-Siamese cultural relations to a certain extent because he requested to see Ayuthia and Lopburi. Ayuthia or Ayutthaya was the old capital of Siam from the fourteenth to the eighteenth century before moving to Thonburi and Bangkok respectively. Its name was the same as Rama's town in Ramayana because Siamese revered their kings as 'Rama'. Lopburi was an ancient town which had been dominated by 'Khom' during the sixth to the eleventh century and had been under the influence of Khom's art and culture. It is not necessary to define in this article who the 'Khom' people are because it is still debated. However, it has been accepted that the center of the Khom Kingdom was in present day Cambodia and that the Khom people were influenced by Indian civilization, both Hinduism and Mahayana Buddhism. For this reason, Khom built many impressive temples in Lopburi. Moreover, Lopburi used to be the second capital for a while during the Ayutthaya period too. This is the reason why Tagore was interested in Lopburi.

Apart from Tagore, the list of his entourage included:

1. Professor Suniti Kumar Chatterji, M.A. (Calcutta), D.Lit. (London), Dhaira Professor of Indian Linguistics at the University of Calcutta, Member of the Senate (Covering Body) of the University of Calcutta, Member of the Covering Body of Visva Bharati (Rabindranath Tagore's University).
2. E. Ariam Williams, M.Ed. (Leeds), Professor at Visva Bharati.
3. Surendranath Kar, Vice Principal, Kala Bhavana (Art School) of Visva Bharati.
4. Dhirendranath Deva-Barman, Kala Bhavana (Art School) of Visva Bharati.[6]

Although Tagore was not invited by the Siamese government, he was treated as a 'guest of the government' [7] and a 'distinguished visitor'. Therefore, the accommodations arranged for him and his party were at the Phyathai Palace Hotel,[8] which had been the palace of the late king, King Vajiravudh (King Rama VI, 1910–1925).

Purposes of the Visit to Siam

Apart from promoting the syntheses of different cultures and raising funds for his university, Visva Bharati—the common purposes of his visit to other countries—Tagore also hoped that a Chair of Buddhism could be established in Visva Bharati with the support of the king of Siam, a land where Buddhism was deep-rooted. As Tagore said in an interview on Monday, 10 October 1927,

> My aim is the unity of man… to bring about the internationalisation of education. We have different chairs at our university, but we have no chair of Buddhism, and it is essential that one should be established to fit into the scheme which we have undertaken. We greatly need such a chair to give

students proper guidance.

You may ask why it is that I have sought to develop my scheme in India. The answer is that India in her long history has been a cultural centre… It may be said then without exaggeration that India has attracted all races and cultures to her bosom. India, therefore, is the centre for such an institution that aims at the unity of man, a unity of cultures, Eastern and Western, so necessary if the various civilisations are to have peace.

I, therefore, have come to Siam to seek your co-operation and your help in the desire to establish a chair in Buddhism, and also, if possible, to bring a message in a wider perspective…[9]

He also emphasized this main purpose of his visit again the day before leaving Siam for Penang:

I hope that my visit will take the form of a permanent memorial by His Majesty's Government establishing a Chair of Buddhism in the name of His Majesty the King of Siam.[10]

Therefore, seeking the support from the king of Siam, King Prajadhipok (King Rama VII, 1925–1934) to establish a Chair of Buddhism in Visva Bharati was the particular purpose of Tagore's visit to Siam. However, there was no documentary evidence indicating what action the King took on this proposal.

Rabindranath Tagore in Siam

It seems that Tagore placed a great deal of importance on the programme's arrangements. Professor E. Ariam Williams, Tagore's secretary, was sent to Bangkok in advance in September 1927 to arrange the itinerary. The most important Siamese scholar whom Professor E. Ariam Williams requested to meet was Prince Dumrong Rajanubhap or Prince Dumrong, the King's uncle, who has also been credited as the 'Father of Siamese History'.[11]

Although Prince Dumrong was the first person whom Professor E. Ariam Williams requested to meet to discuss Tagore's programme in Siam, Prince Dhani Niwat,[12] the Minister of Education, was ultimately the most important host for this visit.[13] Phra Rajadharm Nides, the Secretary to the Minister of Education, was assigned to act as liaison to Tagore.

The first event for Tagore in Siam was a crowded reception at Hualumpong Railway Station, reported in the *Siam Observer* on Monday, 10 October 1927:

A scene without a parallel was witnessed at Hualumpong Terminus on Saturday evening when Dr. Rabindranath Tagore arrived by the International Express from Penang. The train…was due to arrive at round about 7-30, but long before that hour well over 500 people had gathered on the platform. A thing, remarkable to those who have lived in India, or to those who have given study to Indian religious, social and political questions, was that the gathering included people of different faiths who ordinarily are not too ready to mix together. But a bond of unity was provided in the desire to pay tribute to one of the few really great men of the world.

The train arrived almost on the stroke of 7-30. There was rush along the platform to the saloon... As the aged Poet alighted the excitement swelled and cheers were raised.

People forced their way in a seething mass round him, and it was with the utmost difficulty that a passage was made. Every yard of the way had to be fought by the advance guard...Dr.Tagore looked extremely tired...This is not to be wondered at. It was four months since he left India...He was satisfied with the arrangements made during journey and was delighted with the travelling conditions in the Siamese railways. Everything had been done for his comfort, he said, and nothing neglected by Siamese officials to make the journey as pleasant as possible.

It took half an hour to pass from the saloon to the entrance of the station.

A new Dodge car (of M/s. A.R. Salebhai & Co.), beautifully decorated and illuminated, was waiting in readiness to escort the Poet to Phya Thai Palace. Tagore was accompanied in the car by A. E. Nana...[14]

The itinerary set for Tagore during his one week in Siam follows:[15]

8 October

7:30 p.m.: arrives at Bangkok Railway Station by southern line train from Penang.

9 October

10 a.m.: meets Prince Dhani Niwat, Minister of Education.

6 p.m.: meets Sankaraja, the Buddhist Supreme Patriarch.

10 October

10 a.m.: meets Prince Boripatara, Prince of Nakorn Sawan, the King's elder brother.

11 a.m.: lays wreath before the royal remains at the Dusit Maha Prasad and visits the Grand Palace and the Temple of the Emerald Buddha.

3:30 p.m.: meets Prince Traidos Prabandh, Minister of Foreign Affairs.

5 p.m.: has afternoon tea with Prince Dumrong, the King's uncle.

6 p.m.: meets Prince Bhanurangsi, the King's uncle.

11 October

10 a.m.: visits Vajiravudh Library, the National Museum, and two important Buddhist temples: Wat Pra Jetupon (the Temple of the Reclining Buddha), and Wat Mahathat.

4 p.m.: meets Prince Kittiyakara, Prince of Chantaburi, the King's elder brother.

5 p.m.: joins the Hindu reception at the Phyathai Palace Hotel.

8 p.m.: has dinner with the Minister of Foreign Affairs.

12 October

10 a.m.: visits Vjiravudh College and Wat Benjamabopitara (the Marble Temple).

2:30 p.m.: travels along the Chaopraya River and nearby canals, including Wat Arun (the Temple of Dawn).

13 October

9 a.m.: meets Prince Narissaranuwattiwong (Prince Naris), the King's uncle, visits Wat Suthat (the Temple of the Big Buddha) and the Hindu shrine nearby, Ananta Samakhom Throne Hall, and the pavilion for white elephants.

1 p.m.: has lunch with the Minister of Education and delivers speech to an audience at Chulalongkorn University.

5 p.m.: joins the Chinese reception at Pei-ing Chinese school.

9.45 p.m.: led by Prince Dhani Niwat to the presence of the King at Dusit Palace.

10 p.m.: delivers speech to the King, the Queen, some members of the royal family and high-ranking officials.

14 October

9 a.m.: calls on two high dignitaries of the Buddhist Church of Siam at two important Buddhist temples: Wat Bowoniwet and Wat Tepsirin, including a Pali language school at Wat Tepsirin.

3 p.m.: visits 'Poh Chang' College of Arts and Crafts.

4:30 p.m.: delivers speech to an audience at the National Museum.

7.30 p.m.: has dinner with a German Minister.

15 October

6:28 a.m.: visits Lopburi and Ayutthaya by train.

In the evening: visits the Visnu temple at Windmill Road and delivers a short address in Hindustani.

16 October

In the morning: leaves Bangkok for Penang by train.

Tagore's itinerary reflects that his main purpose in visiting Siam was cultural. This itinerary can be divided into three types of activity: meeting with members of the elite and scholars of Thai society, visiting historic places and places of cultural and educational importance, and attending receptions and delivering speeches.

It is noteworthy that all of Tagore's meetings with members of the elite and scholars of Thai society were with individuals well-known for their outstanding cultural contributions. Prince Dumrong, for example, was the 'Father of Thai History' and President of the Royal Institute. Prince Boripatara and Prince Bhanurangsi were experts in Thai classical music, although their styles were grounded in different schools. Prince Naris was versatile in every kind of Thai art, especially Thai traditional architecture, and Prince Kittiyakara was a scholar of Pali.

Yet, while they had all been responsible for producing outstanding cultural work, they were also key individuals with important roles in the politics of their time because they were all members of the Supreme Council of State.[16] This means that, during his visit, Tagore actually met both outstanding cultural figures and key Thai politicians.

With respect to the speeches he delivered, he gave five lectures on different subjects at different occasions:

1. 'India's Roles in the World' at the Indian reception;
2. 'Child Education' at Chulalongkorn University;
3. 'Chinese Birth' at the Chinese reception;
4. 'Asia's Continental Culture' in the presence of the King and the Queen, some members of the royal family and high-ranking officials at Dusit Palace; and
5. 'Ideals of National Education' at the National Museum.[17]

The itinerary and activities demonstrated that, although Tagore stayed in Siam for only one week, he met Siamese from several backgrounds and various social strata. Moreover, both the visitors and the hosts had good opportunities to exchange knowledge of culture and history and to develop a better understanding of each other.

Rabindranath Tagore's Impressions of Siam

Tagore wrote two poems which reflect his good impressions of Siam. One, entitled 'To Siam,' was written two days after arriving in Bangkok and read before the King. This poem runs as follows:

> When the thunder-voiced Prayer of the Three Refuges
>
> rang from sky to sky across deserts and hills and distant shores,
>
> the awakened countries poured their rejoicings
>
> in great deeds, and noble temples,

in the rapture of self-dedication,
in mighty words,
in the breaking of the bond of self.

At an unheeded, unconscious moment,
that prayer, wafted by some sudden wandering breeze,
touched thy heart, O Siam, lived in thy life
and shaded it with a branching wealth of well-being.

A centre to thy revolving centuries,
an end to thy endeavours, which is Freedom of Spirit—
it helped to bind thy people in a common bond of hope,
to strengthen them with the power of a single-pointed devotion
to one Dharma, one Sangha, and one immortal Teacher.

Let those words, potent with an inexhaustible creative urge,
ever direct thee to the adventures of new ages,
light up new truths with their own radiant meaning,
and in one single garland string all the gems of knowledge newly gathered.

I come to-day to the living temple that is one with thee, —
to the altar of united hearts
in which is seated on his lotus seat Lord Buddha,
whose silence is peace, whose voice consolation.

I come from a land where the Master's words

lie dumb in desultory ruins, in the desolate dust,
where oblivious ages smudged the meaning of the letters
written on the pages of pillared stones,
the records of a triumphant devotion.

I come, a pilgrim, at thy gate, O Siam,
to offer my verse to the endless glory of India
sheltered in thy home, away from her own deserted shrine,
to bathe in the living stream that flows in thy heart,
whose water descends from the snowy height of a sacred time
on which arose, from the depth of my country's being,
the Sun of Love and Righteousness.[18]

This poem shows that Tagore appreciated the fact that Siam had preserved certain aspects of Indian culture very well, especially Buddhism, which had not played a significant role in India since the beginning of the Mughal Empire, some seven centuries earlier. Thus, he compared India to a 'deserted shrine,' whereas Siam was still a 'shelter' for 'the endless glory of India.' This poem was also later translated into Thai by Phra Rajadharm Nides.

Another poem, 'Farewell to Siam,' offers a message of farewell, as follows:

The signet ring of a primaeval friendship
had secretly sealed thy name, O Siam, on my mind, in its unconscious depth.
This is why I felt I had ever known thee,
the moment I stood at thy presence,
and why my traveller's hasty hours
were constantly filled with the golden memory of an ancient love,
and centuries' silent music overflowed
the brink of the seven short days
that surprised me with the touch

of an immemorial kinship
in thy words, worship and aspiration,
in thy numberless offerings to Beauty's shrine
fashioned by thy own hands,
in thy fragrant altars
with their candles lighted
and incense breathing peace.
To-day at this sad time of parting
I stand at thy courtyard,
gaze at thy kind eyes,
and whose ever fresh flowers had blossomed ages ago.[19]

Taken together, these two poems reveal that Tagore was impressed by at least two aspects of Siamese culture that derived from India and had flourished in Siam since ancient times: religion and language.

Before leaving Bangkok, Tagore asked Phra Rajadharm Nides to convey his thanks to Prince Dhani Niwat and all concerned for the kind reception accorded him. His message ran:

> The Princes I have met are the most cultured in the East. My conception of Siam is that she still has a living Kingship, Righteousness and Religion. I shall always cherish the memory of my most pleasant stay. I shall treasure the privilege of giving a lecture before the First Eastern Monarch. The programme drawn up by Prince Dhani has been marvelously accomplished to the letter, and I must say I have been greatly impressed by the warm hospitality accorded me by the Thai people as a whole.[20]

Even after Tagore had returned to India, he still kept in touch with Siam, especially with the royal court. We have evidence that he sent a postcard to King Prajadhipok in 1929. The contents of this card are as follows: [21]

Rabindranath Tagore's Emblem

Their Royal Highness

The King and the Queen

Of Siam Rabindranath Tagore

My salutation is to him

Who knows me imperfect

and loves me.

Santiniketan

Founders' Day

7th Paush, 22nd December 1929 Rabindranath Tagore

While Tagore conveyed his impression of Siam to the Siamese people through the two aforementioned poems, there exists, later, at least one poem from a member of the Thai elite that expresses a response to Tagore's visit. It was written under the pseudonym *Prem Chaya*, by Prince Prem Burachatra who was the Thai Ambassador to India from 1967 to 1972. The complete poem follows:

To the learned master, whose vision gave

Reality to this poor poet's pen,

Are dedicated these few lines of verse

Lauding the Sage of Santiniketan

Who, from the far land where the mind was free

Though the body was in chains, spoke to us

With vibrant voice of hope and dignity
Of the glories of India in our home.
Then one day he came and moved in our midst,
Giving us words of wisdom and knowledge;
Extolling the Faith born in his own land,
And transplanted here in the days of yore.
He understood the Buddha's Middle Way;
He did not want a world replete with hate.
He saw on Siam's soil the fulfilment
Of the Great Teacher's whole enlightened dream;
Where the fob of ignorance had lifted,
Where the barriers of hatred were down,
And where the tongue and mind were not enslaved.
Here he breathed deep the air of our common
Heritage, forming a glorious bond
'Twixt India and Siam, which shall endure
While freedom, knowledge, wisdom, faith and love,
And the Buddha's Truth, sway the hearts of men.[22]

Rabindranath Tagore's Diplomatic Roles

Although Tagore's visit to Siam was not a visit of a head of state, it had some significant impacts on Indo-Siamese diplomatic relations. Firstly, Tagore could be considered as the first unofficial envoy of India to Siam. It is well-known that, since ancient times, Siam has had cultural and economic relations with India and inherited various aspects of Indian culture, especially religion and language. Most importantly, Siam has been a land of deep-rooted Buddhism, which originated in India. However, Indo-Siamese relationship in the past was conducted through the medium of Indian merchants and priests whose names have neither been recorded in historical documents nor become well-known in the general public. Up to the beginning of the twentieth century, Siam had enjoyed a cordial relationship with India for a long time, but no Indian, famous enough to be recognized by both countries, had ever visited Siam. Tagore is supposed to have been the first great

Indian to take this role. In other words, we can say that Tagore is the most important Indian figure who opened a new phase of the relationship between Siam and India in modern times.

Secondly, although Tagore was not trained to be a diplomat, he was well-versed in diplomatic etiquette. During his visit to Siam, there were some points that reflected his qualification as a good diplomat. First of all, he had some basic knowledge and knowledge of the history of Siam. In Siam, there were three main pillars of the kingdom—nation, religion and king. Considering Tagore's itinerary, he paid visit to all of these main institutions. Regarding history, Tagore's request to visit Ayutthaya and Lopburi reflects that he had substantial knowledge of the history of the country he visited. Another interesting point is that he was very familiar with Thai customs. During the time of his visit, the royal court was in mourning for one of the princess consorts of King Chulalongkorn (King Rama V, 1868–1910). Accordingly, Tagore was always cloaked in a black robe throughout his stay in Siam. On the occasion of his meeting with the King, it was reported in the *Siam Observer* on Friday, 14 October 1927:

> Tagore was dressed in a black robe with a white silk scarf embroidered with gold. The rest of the party were also in black Hindoo costume owing to Court mourning.[23]

One of Tagore's photographs which he took with Prince Dumrong also showed him in a black robe. Therefore, his itinerary, his historical knowledge of Siam and how he dressed are examples that confirm Tagore's extra-ordinary sense of a good diplomat throughout his visit to Siam.

Thirdly, Tagore's visit had significant influence in shaping the relationship between Siam and India from then on. It would be fair to say that this visit has established a unique relationship between these two countries in comparison with bilateral relations with other countries. This uniqueness arose from cultural relations, especially Buddhism and languages. Based on the perspective reflected in the two poems mentioned above, 'To Siam' and 'Farewell to Siam', Tagore was very much impressed by these aspects of Indian heritage which had flourished in Siam since ancient times.

With regard to Buddhism, reflected from the poem 'To Siam,' Tagore accepted that although Siam inherited Buddhism from India, Siam was also a contributor to Buddhism as its important preserver because this religion was eclipsed in India for a long time. On the contrary, it has flourished and has become a fundamental root of Siamese society. Therefore, in the context of Buddhism, Siam and India were both contributors to each other. This is the reason why Tagore tried to seek the support from the King of Siam in the establishment of a chair of Buddhism in Visva Bharati University. Moreover, during the time of Tagore's visit, Siam was the only free country of which the King was Buddhist. Until now, Thailand still takes the role as an eminent land of Buddhism and makes significant contributions to the revival of Buddhism in India—especially in supporting the restoration of the four most significant Buddhist historical sites where the Buddha was born, attained enlightenment, presented his first sermon, and achieved nirvana.

Pandit Raghunath Sharma[24] recounted that the King and Prince Dhani Niwat requested Tagore to send an Indian scholar to stay permanently in Siam because they wanted to bring these two nations closer to each other in brotherly attachment through the works of this scholar. Language, another aspect of cultural relation between Siam and India since ancient times, was chosen as the medium of this rapprochement. Consequently, after returning to India, Tagore located a scholar who was very keen on languages and sent him to Siam as an unofficial Indian Cultural Ambassador. This scholar was Swami Satyanandpuri, who proceeded to Siam in 1932.[25]

Swami Satyanandpuri was born in Bengal on 3 March 1902. His real name was Prafulla Kumar Sen. He got his first M.A. in Philosophy from Calcutta University and a second M.A. in Sanskrit from Benares Hindu University with the most outstanding marks. Then he worked as a lecturer at Calcutta University. Later he was ordained as an ascetic (*sanyasi*) and received the name Swami Satyanandpuri. From the second day of his arrival in Bangkok, the Ministry of Education appointed M.L.Pin MalaKul [26] to teach him the Thai language. Through his excellent Sanskrit knowledge, he was able to understand the Thai language perfectly well within a short period of time because 75% of Thai words are derived from Sanskrit or Pali. Within a mere six months, Swami Satyanandpuri was able to deliver a 30-minute speech on 'The Origin of Buddhist Thought' in the presence of the King and a Siamese audience at Chulalongkorn University. Everybody in the hall was stunned at his outstanding Thai.[27] After studying Thai for only two years, he started to write and publish books. During his ten years in Siam, he produced more than ten books in Thai, in addition to many lectures and articles. In the meantime, Swami Satyanandpuri's *ashram* became a center of the Indian community in Bangkok. These Indians would come to the *ashram* and discuss issues concerning Indian and Siamese relations. The topic of most interest was the necessity of establishing a permanent organization to promote Indo-Siamese cultural relation. Therefore, the Thai-Bharat Cultural Lodge was founded in 1940. Swami Satyanandpuri was declared the Founding President of the Lodge of which members included both Thai and Indians. Soon afterwards, when the official diplomatic relation was established in 1947, some staff of the Thai-Bharat Cultural Lodge also joined the Indian Embassy as local staff. The Thai-Bharat Cultural Lodge remains in operation to this day as well as shared cultural heritage and activities remain one among the main pillars of the relationship between Thailand and India up to the present.

Notes

1 An earlier version of this paper was presented in the conference to commemorate Rabindranath Tagore's 150th birth anniversary in 2011 at Chulalongkorn University, Bangkok, Thailand under the title 'Rabindranath Tagore in Thailand : His Visit and Its Impact' and will be published in the conference volume to be brought out shortly by ICCR (Delhi). However, in the analytical part, this version focuses on Rabindranath Tagore's diplomatic roles whereas the earlier version focused on the cultural and political impact of his visit.

2 During the time of Tagore's visit in 1927, Thailand was called *Siam*. This was changed to *Thailand* in 1939. However, as the time period covered by this article extends to the present, *Thailand* is used in the

title, while both *Siam* and *Thailand* are used in the main text in accordance with the practice in each time period.

3 'Rabindranath Tagore A Chronicle of Eighty Years, 1861–1941' in *Rabindranath Tagore: A Centenary Volume 1861–1961* (New Delhi: Sahitya Akademi, 1987), p. 487.

4 National Archive, Document of Ministry of Foreign Affairs.33.6.7/4: Rabindranath Tagore and his party will visit Siam.

5 *Lak Muang*, Monday, 17 October 1927, cited in Shakti Das Gupta, *Tagore's Asian Outlook* (Calcutta: Nava Bharati, 1961), p.121.

6 National Archive, Document of Ministry of Foreign Affairs.33.6.7/4: Rabindranath Tagore and his party will visit Siam.

7 Sonakul Dhani, 'Tagore's Visit to Siam' in *Rabindranath Tagore: A Centenary Volume 1861–1961*, p. 306.

8 Telegraph of 2 August 1927 from the Minister of Foreign Affairs of Siam to the Siamese Consul General in Singapore, National Archive, Document of Ministry of Foreign Affairs. 33.6.7 / 4: Rabindranath Tagore and his party will visit Siam.

9 *Siam Observer*, Monday, 10 October 1927, cited in Shakti Das Gupta, *Tagore's Asian Outlook*, p.104–5.

10 *Siam Observer*, Saturday, 15 October 1927, cited in Shakti Das Gupta, *Tagore's Asian Outlook*, p. 114.

11 Letter of 19 September 1927 from Mr. J. F. Johns to Luang Srivisalvaja, Deputy Permanent Secretary of Foreign Affairs, National Archive, Document of Ministry of Foreign Affairs. 33.6.7/4: Rabindranath Tagore and his party will visit Siam.

12 Prince Dhani Niwat also went by the name 'Sonakul Dhani' under which he wrote the article 'Tagore's Visit to Siam' in *Rabindranath Tagore: A Centenary Volume 1861–1961.*

13 Sonakul Dhani, 'Tagore's Visit to Siam,' p. 306.

14 *Siam Observer*, Monday, 10 October 1927, cited in Shakti Das Gupta, *Tagore's Asian Outlook*, p. 103.

15 Letter of 8 October 1927 from Prince Dhani Niwat, Minister of Education, to Prince Traidos Prabandh, Minister of Foreign Affairs, National Archive, Document of Ministry of Foreign Affairs. 33.6.7 / 4: Rabindranath Tagore and his party will visit Siam, Document of Ministry of Education. 22.4/5 : DR. Rabindranath Tagore and his party will visit Siam (30 September-13 October 1927), Shakti Das Gupta, *Tagore's Asian Outlook*, p.91-92, 94-96, 109, 114-115.

16 Chris Baker and Pasuk Phongpaichit, *A History of Thailand* (New York: Cambridge University Press, 2005), p. 112.

17 Shakti Das Gupta, *Tagore's Asian Outlook*, p. 90.

18 Shakti Das Gupta, *Tagore's Asian Outlook*, p. 128.

19 Shakti Das Gupta, *Tagore's Asian Outlook*, p. 129.

20 *Bangkok Times*, Monday, 17 October 1927, cited in Shakti Das Gupta, *Tagore's Asian Outlook*, p. 103.

21 King Prajadhipok's Museum, Photographs and Documents during the Reign of King Rama VII, 09

Lottery and Postcards.

22 Prem Chaya, 'Rabindranath Tagore' in Rabindranath Tagore, *Gitanjali,* trans. Karuna and Ruang-Urai Kusalasaya, 6th edition (Bangkok: Mae Kum Pang, 1991), p. 306.

23 *Siam Observer*, Friday, 14 October 1927, cited in Shakti Das Gupta, *Tagore's Asian Outlook*, p. 109.

24 One of Swami Satyanandpuri's colleagues. He was one of the guests invited to the welcoming function in honour of Tagore at the Phyathai Hotel. Later, he was one of the founders of the Thai-Bharat Cultural Lodge in 1940 and the President of Indian National Council after Swami Satyanandpuri's death in 1942.

25 Pandit Raghunath Sharma, 'Satyanandpuriji's Early Life,' *Thai-Bharat Journal* 18, 28 (1990): 40-41.

26 A Siamese scholar who later became Minister of Education.

27 Pandit Raghunath Sharma, 'Satyanandpuriji's Early Life', p. 42.

References

Baker, Chris, and Pasuk Phongpaichit. *A History of Thailand.* (New York: Cambridge University Press, 2005).

Gupta, Shakti Das. *Tagore's Asian Outlook*. (Calcutta: Nava Bharati, 1961).

King Prajadhipok's Museum, Photographs, and Documents during the Reign of King Rama VII, 09 Lottery and Postcards.

National Archive, Document of Ministry of_Education. 22.4/5: Dr. Rabindranath Tagore and his party will visit Siam.

National Archive, Document of Ministry of Foreign Affairs. 33.6.7/4: Rabindranath Tagore and his party will visit Siam.

Rabindranath Tagore: A Centenary Volume 1861-1961. (New Delhi: Sahitya Akademi, 1987).

Sharma, Pandit Raghunath. 'Satyanandpuriji's Early Life.' *Thai-Bharat Journal* 18.28, End Notes (1990): 38-64.

Tagore, Rabindranath. *Gitanjali.* Translated by Karuna and Ruang-Urai Kusalasaya. 6th edition. (Bangkok: Mae Kum Pang, 1991).

Vision Comes True: Rabindranath Tagore and Balinese Contemporary Religious Identity

Gde Dwitya Arief Metera

On April 1, 2012 in Jogjakarta, Central Java, there was a public lecture by Dr. Partha Chatterjee, the Bengali author, on Rabindranath Tagore and his vision of Asia. The talk was organized by Kunci Cultural Studies Centre, Sangam House, and Etnohistori, a collective of young History and Anthropology students of which I am a part, and was sponsored by the Embassy of India in Jakarta. In the talk, Dr. Chatterjee told the audience about Tagore's visit to several countries including Indonesia. Set against the historical background of the rise of Southeast Asia's nationalism, Tagore's visit had considerable importance for envisioning the future of Asia, as Dr. Chatterjee managed to convey. A specific part of the talk that aroused questions from the audience was the fact that Tagore was in Jogjakarta and Bali among the several places he visited in Indonesia. Stories began to unfold, and it turned out that very few in the audience realized this historical visit. A poet that they adored actually visited their hometown and even wrote poems inspired by his visit. Tagore was indeed in Jogjakarta and Bali in 1927 and met several of the elite as also important figures in the Indonesian nationalist movement. The fact that not many Indonesians are aware about this became an object of my curiosity. From Dr. Chatterjee's lecture it was evident that Tagore's visit was not one of a casual nature but more a visionary one. I was, hence, inclined to look at this visit more closely.

This article is about Tagore's visit to Bali and the vision he had during his visit. His vision is one of modernization, especially as it related to Balinese religion. More importantly, there were two Balinese figures that studied at Santiniketan who became important actors in the modernization of Balinese religion as we see it today in its contemporary version. As a student of Balinese religion, my contribution through this article is to call for in-depth research to investigate the influence of Santiniketan's education and Tagore's ideas on these two Balinese intellectuals.

The Historic Visit

Many Indonesians are not aware, or at best, do not remember Tagore's visit to the country. The historian, Arun Das Gupta, who wrote an excellent account of Tagore's visit to Indonesia, made the same observation. Cornell-educated and a Fellow of the Asiatic Society, Professor Das Gupta is an expert on the history of Southeast Asia and the Indian Ocean, and taught at the University of Indonesia in Jakarta from 1979-1981. As he wrote in his article, '*Rabindranath Tagore in Indonesia: an Experiment in Bridge-Building*[1],' this visit was 'well-nigh forgotten.' (Das Gupta, 2002: 455).

The poor effort to preserve documentation on Indonesia's side might be the reason for this lack of awareness among the younger generation of Indonesians. As Das Gupta also mentioned, a thorough search for resources in Indonesia and the Netherlands for colonial archives is needed to produce more information. However, from perusing sources mainly in India, Das Gupta has written an excellent account illustrating the motive and nature of Tagore's visit. My understanding of Tagore's visit, therefore, relies confidently on Das Gupta's account that I will retell here. I will specifically underline Tagore's stay in Bali.

Southeast Asia Tour and the Invitation

Tagore was in Indonesia for a little over one month from August 21 to September 30 1927. Half of that time, for around two weeks from August 26 to December 8, he spent in Bali, and the rest he spent in several towns in Java and Sumatra. He departed from Calcutta on July 12, 1927 with several friends. They were the philologist, Suniti Kumar Chatterji, the painter and musician, Dhirendra Krishna Deva Warman, and the artist Surendranath Kar. They all sailed from Madras on the French ship *Ambois*[2]. Tagore's travel intention was a Southeast Asia tour and not to exclusively visit Indonesia. He and his companions stayed in Singapore for around a month before arriving in Batavia in August. Their route of their travel in Indonesia was Sumatra-Batavia-Bali-Surabaya-Surakarta-Jogjakarta.

There were several reasons as to why Tagore was in Indonesia. We know for sure that he was invited by several institutions that took the initiative and worked together to make the visit possible. These institutions were the Cultural Society for the Promotion of Literature and the Arts, or the *Kunstkring*, in Batavia, and the *Java Instituut* of which the king, Mangkunegoro VII of Surakarta, was the Chairman. Another institution that was involved was the Dutch Indonesian Archaeological Service, the *Oudheidkundige Dienst.* There was also a notable individual that helped with the visit. Arnold Bake, a Dutch musicologist who studied music in Santiniketan, helped prepare Tagore's tour and was in Indonesia with his wife during Tagore's visit (Das Gupta, 2002:459).

There are at least two reasons why these institutions were eager to invite Tagore to come to Indonesia. First, Tagore's works had been read in Java because of a translation of his works to Dutch and Javanese. Das Gupta noted that Kalidas Nag, the historian, was surprised to find during his visit to Java in 1924 that Tagore was known on the island. Tagore was read in Java largely because of the work of Raden Mas Noto Soeroto who translated Tagore's works into Dutch. Raden Mas Noto Soeroto was a Javanese prince from the Pakoe Alam family. His grandfather was Pakoe Alam V who was the head of Pakualaman, one of Jogjakarta's autonomous principalities. Raden Mas Noto Soeroto considered Tagore his spiritual guru and even named his eldest son after Tagore: Rawindra Noto Soeroto[3]. Besides translating Tagore's work, he also wrote a biography of Tagore and a book on Tagore's educational ideals. Later on, his Dutch version of Tagore's work was translated into Javanese by his close friend Soeriosoeparto. Soeriosoeparto was also a Javanese nobleman who later became King Mangkoenegoro VII of Surakarta and played host to Tagore when he visited Surakarta (Das Gupta, 2002: 457). Moreover, after he was awarded the Nobel Prize for Literature in 1913, Tagore's reputation was even more elevated among his Javanese readers.

The second reason also related to his reputation. Tagore was a source of inspiration to the Indonesian nationalist movement. His school in Santiniketan that tried to find an alternative to Western education by integrating modern elements and local culture had inspired the same movement in Indonesia. Ki Hajar Dewantoro, an Indonesian nationalist leader, who struggled with the country's education and founded the Taman Siswa School, was especially eager to meet Tagore himself. The Javanese knew something else about Tagore – that Tagore had returned the knighthood he had received from King George V in 1915 as a protest against the killing of Sikh people in Amritsar by British soldiers. (Das Gupta, 2002:454).

These two reasons led to an invitation to Tagore to visit Indonesia in 1927. However, as the Dutch East Indies government was nervous that Tagore's visit might potentially strengthen Indonesia's nationalist movement, it did not let the nationalists arrange the visit. Instead, the Dutch government let the *Kunstkring* arrange it. This is understandable especially if we note that, in 1926, a year prior to the invitation, the Dutch government had just been challenged by a communist revolt in Batavia and several other places in Indonesia (Das Gupta, 2002:459).

If from Indonesia's side the reasons to invite Tagore were strongly related to his work and his reputation among Indonesian nationalist leaders, Tagore's motive to visit Southeast Asia and Indonesia was quite different. As Das Gupta mentioned in his paper, Tagore was curious about the Hindu religion and culture in Bali. There was something about the remains of ancient Indian culture in Southeast Asia that caught his interest, and as Tagore put it, he was 'going on a pilgrimage to India beyond its modern political boundaries' (Das Gupta, 2002: 456). Moreover, Tagore was especially inclined to discover an Asian cultural identity. This was a cultural mission of Tagore's with two goals. As Das Gupta wrote, the goals were: first, to study Indian civilization in the antiquarian relics in Indonesia as well to study its arts and the life of its people and, second, to forge cultural cooperation between India and Southeast Asia (Das Gupta, 2002: 456).

Tagore in Bali: His Vision of Religious Modernization

Tagore departed for Bali from Batavia after a welcome reception from the *Kunskring* and a dinner hosted by the British Consul in Batavia. He and his friends arrived in Buleleng on August 26, after a brief stop in Surabaya. From Buleleng, he traveled to several places in Bali: first, to Bangli via Singaraja and Kintamani, and then to Karangasem, Badung and Denpasar. He, however, stayed mainly at Tampak Siring and was invited to the palace of the King of Gianyar and to visit Ubud and Mundu (Das Gupta, 2002: 461).

What Tagore and his friends saw in Bali gave them such pleasure that it was reflected in the notes written by one of his accompanying friends, the philologist, Chatterji. The rituals, the people, the art performances, virtually every sceneries were memorable to them. Since the Hindu nature of Balinese culture was obvious, Tagore was intrigued by it. Das Gupta mentioned in his paper the lines that Tagore wrote in his letter about his impression:

> 'Along every village road processions of men and women were arriving with multifarious offerings. Some *puranic* age seemed to have come back to life before our very eyes, some picture from the Ajanta caves come out from the realm of art to the realm of life to revel in the sunshine.' (Tagore via Das Gupta, 2002: 462).

Tagore's meeting with the King of Karangasem also left him with a great impression on the strong link between Balinese culture and India. It was when the King of Karangasem, Ida Anak Agung Bagus Djelantik, took Tagore on a motor ride that the King uttered a Sanskrit word to refer to the sea that they saw along the way. The King said '*samudra*' which took Tagore by surprise since he did not know that the King understood Sanskrit. He and the King had not spoken much owing to the lack of a common language (Das Gupta, 2002:462). This Sanskrit word, however, was an indication of the cultural link between them. Das Gupta wrote about this moment when Tagore was both pleased and surprised:

> 'The *raja*, seeing that this was surprising and pleasing to his eminent guest, went on to recite the other Sanskrit names for the sea: *sagar, abdhi, jaladhya*. Then he said: '*Saptasamudra, saptaparvat, saptaban, sapta-akash*, (seven seas, seven hills, seven forests, seven skies). A little later, he recited the names of the Indian rivers Ganga, Jamuna, Narmada, Godavari, Kaveri and Sarasvati. India's geography seemed to be imprinted on the Balinese mind.' (Tagore and Chatterji via Das Gupta 462).

There was also another moment, as is noted by Das Gupta, in which Tagore and the King discussed a Balinese religious text and Tagore was impressed by the King's question. The King asked the poet to interpret some passages from the text and he surprised Tagore by asking a question about man's ultimate quest in life. Chatterji, who was also present and helping with the interpretation of the text, asked the King to answer the question himself. The King then gave an answer that startled both Chatterji and Tagore. Das Gupta recounted the King's answer as follows: '*Dewa-Dewa tida apa, Nirvana satu*.' It meant that *Nirvana* is the supreme goal and not merely worshipping the gods (Chatterji via Das Gupta, 2002: 462). This moment specifically left such a lasting impression on Tagore that he referred to it in his poem on Bali. Das Gupta translated the lines in Tagore's poem, *Bali*, about this encounter: 'We said the same mantras together, pondered upon the same question of Nirvana.' (Das Gupta, 2002:462).

Despite enjoying much of what he saw in Bali, Tagore reserved a critical view about the island that linked with his home, India. As Das Gupta wrote, Tagore sensed that Bali was trapped in the past in an 'archaic manifestation'. All the dances, rituals, and life of the Balinese people, to him, represented the past trapped in the present. Tagore, nevertheless, appreciated the remains of the Indian civilization he found in Bali. However, he was not a supporter of revivalism. Das Gupta wrote that he is a 'modern man with forward looking mind.' Tagore, according to Das Gupta, sensed there was something wrong with all this.

> The past was a great one [...]. But, still it is after all the Past and its duty was to fall behind the Present, not to come in front of it and stop the way of its self manifestation. (Tagore via Das Gupta, 2002: 463).

This vision of Tagore is particularly significant. His vision was that Bali should not linger in the past and instead move to the present. I take this as a vision of modernization. I believe this moment was at the heart of his visit besides the poems he wrote based on his experience in Bali. His discovery of the remains of the ancient Indian civilization in Bali yielded this vision. What he found in his Southeast Asia tour was India's past, trapped in the Balinese present.

Vision Comes True: Religious Transformation in Bali

The first time I read the story of Tagore's visit to Bali from Das Gupta's account, my initial response was one of surprise. As a student of Balinese religion, I had read the history of religious modernization in Bali and its contemporary form. I did not know that Tagore had his vision about this modernization. Moreover, I thought that the vision was a mere coincidence. However, further reading about the role of Balinese intellectuals in this process of modernization indicated to me that it might be more than a coincidence. Tagore's influence was present in the process of Balinese religious modernization since two Balinese intellectuals that played a prominent role in this modernization had been educated in Visva Bharati, Tagore's institution.

Becoming Modern: Balinese Religion in the 1950s

Thirty years after Tagore's visit, Balinese religion was indeed undergoing a transformation. Its transformation was in the nature of modernization, as it changed from a traditional and practice-based (orthopraxy) to a modern, rationalized orthodoxy. The anthropologist, Clifford Geertz, who was doing fieldwork in Bali in 1957-1958, first observed the symptoms of this modernization and this led to his essay *'Internal Conversion' in Contemporary Bali,* published in 1964. Having worked for several years in Java writing a manuscript that later became his dissertation on Javanese religion, he found a surprising contrast in the Balinese way of conducting their religious life compared to their Javanese counterparts. Java, in the context of mid twentieth century, had developed, in Geertz's words, 'relatively well-rationalized systems of belief and worships... giving a conscious sense of religious diversity, conflict and perplexity still quite foreign in Bali.'[4] Geertz was particularly struck by how indifferent the Balinese were to their own religious doctrines, rendering them to a state of 'total absence of either doubt or dogmatism, the metaphysical nonchalance.'[5] Advanced observation led Geertz to conclude that at the core of Balinese religion is its orthopraxy, meaning what matters to the Balinese is whether ritual details are carried out correctly, and not orthodoxy in which the conceptual side remains less important:

> 'beyond a minimal level, there is almost no interest in doctrine, or generalized interpretation of what is going on, at all. The stress is on orthopraxy, not orthodoxy—what is crucial is that each ritual detail should be correct and in place... But the conceptual side is of much less moment: the worshippers

usually don't even know who the gods in the temples are, are uninterested in the meaning of rich symbolism, and indifferent to what others may or may not believe.' (Geertz, 1964)

However, by the time Geertz wrote this essay he sensed that there was already a transformation well underway at the ideological level in Balinese religion. Geertz indicated that Balinese religion was experiencing a swing of the pendulum from orthopraxy to orthodoxy. Using Weber's theoretical framework of 'rationalized' and 'traditional' religion, Geertz managed to show that the Balinese changed from being concerned by the concrete elements of religion (i.e. the traditional way of practicing rituals) to the abstract questions of values and dogma (the rationalized).

Geertz observed the symptom of this change at a funeral gathering in a village. He was caught up in a discussion among the Balinese youth who asked questions such as which of their customs were secular and which were sacred. The discussion was apparently animated by enthusiastic responses from several educated and semi-educated young men in that funeral gathering. They took turns making philosophical speculations, sharing academic views or personal beliefs[6].

This *obvious* development at the personal level was also supported by an endeavor in terms of rationalizing the creed. As Geertz wrote, it was done through the agency of the printing business. There were several firms that published translations of classical palm-leaf literature available with the priests, books with interpretations of moral-symbolic values, and made them economically available to the keen masses.

Several decades later, anthropologists conducting research on contemporary Balinese religion see the result of this transformation. Shinji Yamashita (2003), in his research on the impact of the tourism industry to the development of religion in Bali, mentions several changes that took place after Geertz left. The most obvious was the establishment of the new orthodoxy mainly as a result of a campaign by the Hindu Council (*Parisadh*a) to reform traditional Balinese religion. This new orthodoxy was made manifest in the 'textbook religion', to borrow Yamashita's term and was different from the Balinese traditional 'ritualism'. Yamashita was, nonetheless, referring to what is now commonly understood as modern Hinduism or *Agama Hindu* promoted by *Parisadha*, and the Balinese 'ritualism' he mentions is the traditional Balinese religion that was more concerned with rituals.

Another anthropologist, Leo Howe, writes in his work (2001, 2005) that, besides modern Hinduism promoted by *Parisadha*, there were new religious movements such as Sai Baba and Hare Krishna existing in contemporary Hindu Balinese society. He argues that the proliferation of new religious movements in Bali was a development subsequent to the introduction of modern Hinduism in Bali in the late 1950's. This modern Hinduism itself went through its formative years in the 1960s in parallel with an increasing interest in Indian Hinduism within a small circle of Balinese intellectuals.

Struggling for Recognition from Indonesian State

What was the reason behind this religious transformation? In order to address this question, we have to look at the political situation in Indonesia at the time. Minority religions in Indonesia were struggling for recognition by the state, including the Balinese religion. The Indonesian state was in its formative era in which many issues were being negotiated. The question concerning the fundamentals of the state was at the heart of the debate. Religion was part of that question. Indonesia was torn between those who supported a non-confessional state that was secular in nature and those who supported an Islamic state. As it turned out later, the supporters of the non-confessional state won the debate and Indonesia became anything but an Islamic state. However, as part of the trade-off, a Ministry of Religious Affairs was established in 1946 that was heavily dominated by the Muslim faction.

In 1952, the Indonesian Ministry of Religious Affairs stipulated several requirements for a religion to be recognized by it. First, it had to be monotheistic; second, it had to have a codified law for the followers; third, it had to possess a Holy Book and a prophet; fourth, it had to enjoy international recognition; and finally it was not to be limited to a single ethnic group (Picard, 2004:56). With these requirements as parameters, the traditional Balinese religious practices and belief did not qualify as a 'religion' ('*agama*', as it is in Indonesian language), but only as a belief with a lesser status than a religion (in Indonesian language this belief with a lesser status was termed '*kepercayaan*'), a derogatory categorization that dismissed their religion and made them vulnerable to proselytizing from Muslim and Christians.

As part of the effort to gain recognition, the *Parisadha Hindu Dharma Bali* (the then Hindu Council of Bali) had been promoting reform to traditional Balinese religion by infusing concepts and theological understanding to realign the traditional Balinese religion with Indian Hinduism, the source of Balinese religion. Making religious texts available to the keen masses was part of the effort. Thus Geertz actually saw the effect of the struggle of the Balinese intellectuals to consciously transform their religion owing to pressure from an external factor: the Indonesian state. The importance of this transformation from non-rationalized religion to a rationalized one has been highlighted by Geertz as an 'internal conversion' (Geertz, 1964: 294).

The main task of the transformation was to introduce a new and modern theological framework to the indigenous Balinese religion (Picard, 2011). This new theological framework was imported from India by Balinese intellectuals, some of whom were studying in India. An interesting question pertaining to this religious transformation of the Balinese religion related to why that theological framework was imported from India and not derived from local texts. A possible answer might be that several important Balinese intellectuals at that time were educated in India and they were assigned a central task in the ongoing transformation. A study by Bakker (1993) argues that the Balinese intellectuals played a critical role in this transformation. Interestingly, among those Balinese intellectuals, two were educated in Visva Bharati, in Santiniketan. They were Ida Bagoes Mantra and Nyoman S. Pendit.

Ida Bagoes Mantra and Nyoman Pendit: Two Balinese Graduates of Santiniketan

Among several Balinese intellectuals that played a pivotal role in the transformation of Balinese religion were Ida Bagoes Mantra and Nyoman S. Pendit. They both were among the first Balinese intellectuals that studied in India, and specifically in Santiniketan, and then came back to become prominent leaders. Mantra went to study Indology in Santiniketan in 1950 and received his doctorate there in 1957, the first Balinese to receive a doctorate degree from a university. Pendit went to Santiniketan in 1958, a year after Mantra received his doctorate, to study History of Indian Philosophy. Pendit was largely a journalist, while Mantra was a towering figure as both an intellectual and a bureaucrat. In the following section I will focus on Ida Bagoes Mantra but with only a short account of Pendit given the limited sources available about him.

Ida Bagoes Mantra: Intellectual and Activist

Pendit was a prominent youth figure and was involved in the struggle for Indonesian independence. He was a lieutenant for the youth army of Sunda Kecil (a region comprised of Bali and West Nusa Tenggara) in 1945-1949 and become its head in 1949-1954[7]. Pendit went to Santiniketan in 1958 to study the History of Indian Philosophy. When he came back to Indonesia he became a prolific writer and translator. His works were important in the availability of religious knowledge that fueled the transformation of Balinese religion. Among his important books were the translations of the Bhagavad Gita, Ramayana and Mahabharata. Most Balinese read his translations and this helped the dissemination of new religious orthodoxy among the general Balinese population. His books also, in a way, shaped the understanding of the new generation of Balinese who read works such as the Bhagavad Gita.

Unlike the scarcity of sources on Pendit, there are several sources from which we can learn about Ida Bagoes Mantra. His biography that was written by his colleagues at the Udayana University is an excellent source. From this biography we can examine Mantra's role in the transformation of Balinese religion and learn about his study in Santiniketan.

Ida Bagoes Mantra's first encounter with Tagore's work was when he studied in Macassar, Eastern Indonesia. He was studying at the Dutch middle school AMS *(Algemene Middlebare School)* from 1947-1949 where he read world literatures and encountered Tagore's work for the first time. He was so taken with Tagore's work that he was determined to study at Visva Bharati in Santiniketan (Rama et. al, 1998:5).

Mantra eventually managed to study at Santiniketan in 1950. At Santiniketan, Mantra studied Indology. He was interested especially in Indian culture, an intellectual curiosity that was mirrored in his thesis. After four years of study, on December 24, 1954, he graduated with an MA thesis titled *'Ancient Indian History and Culture.'* He received his doctorate in 1957 by writing a dissertation titled *'Hindu Literature and Religion in Indonesia.'* (Rama et. al, 1998: 6).

From both his master's thesis and his dissertation, we can infer that Mantra dedicated much of his intellectual energy to Hindu religion and Indian culture. He also wrote several books on the religion and culture of Bali. Among his books we can find titles such as: The Foundation of Balinese Culture (*Landasan Kebudayaan Bali*) and Bali: Socio-cultural Problems and Modernization (*Bali: Masalah Sosial Budaya dan Modernisasi*). This remarkable background will help us understand his later activism and role in the transformation of Balinese religion.

Mantra's role was central in the establishment of the Hindu Council in Bali. As has been mentioned above, the history of the establishment of the Hindu Council cannot be separated from the demand of the new state of Indonesia for a certain religion to be recognized by the central government. The Balinese religion that did not yet meet these requirements then needed to receive recognition from the state. A delegation led by the Bali governor at that time, Ida Bagus Sutedja, went to Jakarta to discuss the recognition of the Balinese religion with the central government. The outcome of this meeting was then shared with important religious figures of Bali in 1958. One step taken after this meeting was the setting up of a commission that would deal with securing this recognition. The head of this commission was none else but Ida Bagus Mantra. The very reason he was selected to head this commission, *Hindu Bali Sabha*, was that he was the first Balinese to have a doctorate and his expertise was on Hindu religion. Moreover, he had studied in India, in Visva Bharati, Santiniketan (Rama et.al, 1998:169).

The first meeting of the new commission was in November 1958 to prepare for the setting up a new institution, a Hindu Council, which would concern itself with the religious matters of the Balinese. The second meeting, on February 23, 1959, was at the Faculty of Letters of Udayana University, where Mantra was the Dean. This meeting turned out to be an historical one as it managed to establish the first Hindu Council in Bali. At that meeting, there were eight delegations from all over Bali, the heads of religious offices in Bali, and several representatives of religious organizations on the island. The task of the newly-established council was to sustain, strengthen and develop Hinduism in Bali. Mantra became the Council's first secretary. The Board of the Council itself consisted of priests (*sulinggih*) and laymen religious leaders (*walaka*) (Rama et.al, 1998:168).

The role of the newly-established Hindu Council was important in the transformation of Balinese religion. It introduced new practices, it built a new temple, and it produced religious texts as guidance for the Balinese. The new practice that was introduced to the general Balinese population was the three-times-a-day prayer by chanting *Gayatri* mantram. This was followed by the establishment of a new kind of temple, *Jagadnatha*, in Denpasar, the capital city of Bali, before being replicated in each town all across Bali. The new temple, *Jagadnatha,* was built primarily to accommodate Balinese that stayed on in the city and could not go back to their own villages to participate in rituals and worship. This establishment of the *Jagadnatha* temple with its goal of serving the Balinese urbanites is central to the modernization of Balinese religion. The Balinese urbanites that could not always participate in the rituals held in their villages of origin needed a new form of religious articulation of which the new temple became an essential part (Metera, 2012). The idea of building *Jagadnatha* temple came from Ida Bagoes Mantra, among others. He was the one who insisted on the name *Jagadnatha* and he led the team that built the temple in Denpasar, the capital city of Bali. The building process started in 1964 and ended in 1975 (Rama et.al, 1998:173-175)

Mantra also took part in the establishment of an institute that taught Hinduism and produced young Balinese Hindu intellectuals. His initial idea was to set up an educational institution that would preserve the teaching of Hinduism. This was urgent because Mantra observed that there was not much effort to preserve Hinduism, which was a minority religion in Muslim-dominated Indonesia. He set up another meeting, in November 1961, with representatives of Hindu youth from all over Indonesia as well as some priests sponsored by the Hindu Council. The meeting took place for about a week from November 17 to November 23 and resulted in some policy recommendations. Among the recommendations was the setting up a Hindu institute that would teach Hinduism to the general public. Two years after the recommendation, on October 3, 1963, the *Maha Widya Bhawana Institut Hindu Dharma* was established, largely based on Mantra's idea (Rama et. al, 1998: 9).

At the beginning of the Institute's operation, people were skeptical about its graduates, largely about their ability to find jobs. However, as the Institute keeps operating, its graduates find jobs as teachers of Hinduism in schools all over Bali and Indonesia. The Institute became a university in 1993 and has expanded to include departments like Law and Literature (Rama et. al, 1998:181).

Mantra's role as secretary of the Hindu Council was pivotal as he managed to advance important ideas and policies. We can see that he was an activist actively promoting transformation and consolidation of Hinduism in Bali. However, Mantra's role was not limited to the Hindu Council. He went on to assume important roles in Bali and in Indonesia. He became the Rector of Udayana University in Bali from 1964-1968; assumed a position as the Director General of the Cultural Affairs Section under the Ministry of Culture and Education of Indonesia from 1968-1978; became the Governor of Bali from 1978-1988; and finally became the Indonesian Ambassador to India from 1989-1992.

During his role as the head of the Cultural Affairs Section in the Indonesian Ministry of Education and Culture and as the Governor of Bali, Mantra was very active in promoting Balinese art and supporting art events in Bali. The Art Centre in Denpasar and the Bali Art Festival (*Pesta Kesenian Bali/ PKB*) were established with his support. His passion for art was clearly cultivated during his study in Santiniketan. His biography mentions that the curriculum Mantra encountered in Santiniketan emphasized art and culture. This experience in Santiniketan seems to have had a lasting impact on Mantra, and he implemented his passion for art when he returned to his hometown, Bali.

Mantra passed away in 1995 but his influence and legacy continue to the present. His family is an influential one in Bali and one of his sons is now the mayor of the city of Denpasar. His name is immortalized in Bali with a street named after him, and his Bali Art Festival continues to be the most festive event in the year.

Conclusion: Tagore's Vision Comes True Through Two Graduates of Santiniketan

When Tagore had his reservations about Bali in 1927, Bali was still under colonial rule that deliberately preserved its archaic nature. This policy was called *Baliseering*, a policy to keep Bali in its seemingly pristine and untouchable form and secure from the influence of the outside world. The colonial government imposed this policy with a goal to develop tourism and attract more visitors to the beautiful island. However, as the nationalist movement defeated the colonial administration and a shift to a new Indonesian state took place, Bali also underwent radical changes. The new Indonesian state imposed new regulations on religious life that forced changes in the Balinese religion. A modernization of Balinese religion had to take place. This modernization, in a manner, is the realization of the vision that Tagore had decades earlier.

Interestingly, this modernization was carried out by Balinese intellectuals that had been trained and educated in Visva Bharati, Santiniketan. Ida Bagoes Mantra and Nyoman S. Pendit both studied in Santiniketan in the 1950s and came back to become prominent figures that shaped policymaking and influenced the development of the Balinese religion to its present state. They were the first Balinese intellectuals who studied in India and thus paved the way for later generations to come to India and strengthen the ties between the two countries. Tagore's mission of bridge building (Das Gupta, 2002) was successful, and Tagore's influence was felt in as far a place as Bali in Indonesia.

A future project is needed to shed more light on this link between Tagore's vision and its realization carried out by the two Balinese graduates of Santiniketan. The project should scrutinize what specific aspects of knowledge from Visva Bharati did Ida Bagoes Mantra and Nyoman S. Pendit implement in their writing and policies. What did they get from their study in Santiniketan that shaped their worldview? This short article, besides pointing to this link between Tagore's vision and its realization, also calls for a in-depth research on these two Balinese graduates of Santiniketan and what they received from Visva Bharati.

Notes

1 Das Gupta, Arun. 'Rabindranath Tagore in Indonesia: An Experiment in Bridge-Building'. *Bijdragen tot de Taal, Land en Volkenkunde*, Vol.158, No. 3, 2002, pp. 451-477.

2 Das Gupta, *idem*, p. 45

3 Das Gupta, *idem*, p.457

4 Geertz, Clifford. 'Internal Conversion in Contemporary Bali' *Malayan and Indonesian Studies Presented to Sir Richard Winsted*. Ed. Roolvink, J. Bastin & R. (Oxford: Oxford University Press, 1964), p. 287.

5 Geertz, 1964. *idem*, p.287

6 Geertz, idem pp. 294-5.

7 Short biography of Pendit by Gramedia books, the largest publisher in Indonesia. http://www.gramediapustakautama.com/penulis-detail/35045/Nyoman-S-Pendit

References

Bakker, F.L. *The Struggle of the Hindu Balinese Intellectual.* (Amsterdam, VU University Press, 1993).

Das Gupta, Arun. 'Rabindranath Tagore in Indonesia: An Experiment in Bridge-Building'. *Bijdragen tot de Taal, Land en Volkenkunde*, Vol.158, No. 3, 2002, pp. 451-477.

Geertz, C. 'Internal Conversion in Contemporary Bali.' *Malayan and Indonesian Studies presented to Sir Richard Winsted.* Edited by J. B. R. Roolvink. (Oxford, Oxford University Press, 1964).

Howe, Leo. *Hinduism and Hierarchy in Bali.* (Oxford: James Currey, 2001).

The Changing World of Bali: Religion, Society and Tourism. (New York: Routledge, 2005).

Metera, GDA. *Religious Commuters: Contemporary Balinese Religious Identification in a Northern Balinese Town.* Unpublished thesis. Center for Religious and Cross-Cultural Studies, Gadjah Mada University, 2012.

Picard, Michel. 'What's in a Name? Agama Hindu Bali in the Making.' In *Hinduism in Modern Indonesia: A Minority Religion between Local, National and Global Interests* . Edited by Martin Ramstedt. London: Routledge Curzon—IIAS Asian Studies Series, 2004.

'From Agama Hindu Bali to Agama Hindu and Back.' *The Politics of Religion in Indonesia*. Edited by Michel Picard and Remi Madinier. London, Routledge Contemporary Southeast Asia series, 2011.

Rama et.al, *Prof. Dr. I. B. Mantra: Biografi Seorang Budayawan, 1928-1925.* Denpasar. Upada Sastra, 1998.

http://www.gramediapustakautama.com/penulis-detail/35045/Nyoman-S-Pendit

Tagore in Sri Lanka

Reba Som

A light breeze ruffled the leafy branches of the trees atop the hillock overlooking the sleepy campus of Sri Palee. Located in Horana, a few hours away on a busy road from Colombo, Sri Palee seemed an abode of peace. Today taken over by the government, the institute had once been part of an exceptional experiment in rural reconstruction and education inspired by Rabindranath Tagore. A marble plaque marks the spot where Tagore had laid the foundation on May 20, 1934 naming the institute Sri Palee – 'Where Beauty Reigns'. A simple house surrounded by tall palm trees is where Tagore had rested.

Sri Palee was the brainchild of Wilmot Perera, born into a wealthy family of planters, who gave up his studies in archaeology in Cambridge University following his father's death to embark on a quest to rediscover his cultural roots. It was his dream to build a unique educational institution where students from rural and urban backgrounds alike could be in touch with the environment and realise their own potential. Perera's inspiration was Tagore, whose speech he had heard as a youth of seventeen when Tagore visited Ceylon for the first time in 1922. Tagore's direct exhortation to his Ceylonese audience to free themselves of a colonial mindset and acknowledge their deep cultural connection with India left an abiding impression on Wilmot's impressionable mind. Later in 1932, Perera visited Santiniketan with his wife Esme which proved to be the defining moment of his life. During his stay over several months, in close proximity to Tagore, he could implicitly understand the poet's lesser known work as a rural reformer and an educator. Tagore had sought to bathe his university in a cultural atmosphere of music and the arts so that students, even from other disciplines, could imbibe the holistic philosophy. This was to appeal particularly to Perera who was to write later, 'Every child cannot become a fully fledged musician or a dancer. But every child can be provided with the opportunity of imbibing the beauty of dance and music and learning to appreciate good music and good dancing. This too should become a part and parcel of a child's education.' His wife Esme benefited from her association with Kala Bhavan in Santiniketan to learn several handcrafts which she would then go on to introduce in Sri Palee.

At Perera's invitation, Tagore arrived in Ceylon in 1934. He had made two earlier visits - in 1922 when he gave lectures in Colombo and Galle and stayed for a week at the health resort of Nuara Eliya; and again in 1928, when he visited Colombo as guest of Dr W. de Silva after his visit to Madras. This visit in 1934 was possibly the one which left the deepest impact on Ceylon. It was also the last time that Tagore was to travel overseas. On this visit, which he intended to be a cultural tour,

Tagore was accompanied by a troupe of artistes and several teachers. From 11 to 18 May, Colombo was treated to a cultural festival from Santiniketan. An exhibition of Tagore's paintings along with those from Kala Bhavana, the art school of Santiniketan, was held while several performances of his dance drama *Shap Mochan* were staged. Bandaranaike, later Prime Minister of Ceylon, reviewed the performance *Shap Mochan* for 'Ceylon Daily News'. He was to write that the aesthetic simplicity of the stage setting with a long-robed, bearded poet seated on one side had the appeal of a Greek drama. The treatment of human emotions in its rich diversity through music and dancing was 'a revelation of art at its highest level.'

On May 19, Tagore accompanied Wilmot Perera to Horana to lay the foundation stone of his school modelled on Santiniketan's style of teaching. Tagore was very impressed with the institution which reminded him of his own work in Bengal. He was to write to Perera how it had made him very happy that a channel of communication of hearts had opened between their respective institutions. He particularly appreciated the presentation of Kandyan dance at the inaugural celebrations and was to incorporate many of the dance postures into his later choreographic compositions. Tagore's encouragement to this particular dance form was of considerable significance since it enabled it to emerge from the peripheral position it occupied and enter the mainstream of Ceylonese culture. Tagore's visit inspired many artistes, dancers and singers to develop their genres to classical forms. Prominent artistes such as Anand Sarnarakoon, Chitrasena and Sunil Shantha attended Visva Bharati University and were enriched by the experience. Music, dancing and crafts were to become integral parts of Sri Palee's curriculum. Santidev Ghosh, who headed Visva Bharati University's school of music, personally gave lessons in dance and music to students at Sri Palee. His repeated visits to Ceylon gave him the opportunity to personally supervise the dance of each member of the troupe, including the ladies. Even in art, the influence of Santiniketan was to leave an imprint as Nandalal Bose, who headed Kala Bhaban in Visva Bharati, had accompanied Tagore on the visit with his students.

Wilmot Perera was to face considerable criticism for following the Indian example but he made a spirited defence of his decision through an article in *Times of Ceylon*, 5th July 1938 titled 'To portray what is Indian is to reveal what is inherent in our Soul'. He wrote: 'Everything that forms the culture of Ceylon has been derived from India. Didn't Vijaya himself come from India and more particularly from Bengal? And what of Mahinda, and the gift of Buddhism? And in the more recent past, can it be denied that the awakening in our country to the necessity of developing the Fine Arts is due to Rabindranath Tagore and his students from Santiniketan? It was left to him to re-discover for us the beauty of the Kandyan Dance, which incidentally, is considered to have been brought to us from Malabar. We are proud to claim that our institute, christened by him on his visit four years ago, has contributed in no small measure to promote that cultural awakening.' [1]

Sri Palee was perhaps the most vital experiment in education initiated in Ceylon since the efforts by the Buddhist Theosophical Society of Olcott. For several decades it continued the seminal work begun by Wilmot Perera until it was taken over by the government and lost its exclusive character. Distinguished artists, intellectuals and speakers from India invited to Sri Palee over the years included Kamaladevi Chattopadhyay, Vijaylakshmi Pandit, Ravi Shankar, Alla Rakha, Radhakrishnan, Humayun Kabir, among many others.

Tagore's visit to Ceylon was of considerable importance not only because of the significant cultural exchange that came with it but also, perhaps equally importantly, in his lectures he tried to analyse Ceylon's experiences in the context of his larger vision of the Nation and the World.

As in Japan during his Nationalism lectures in 1916, in Ceylon too Tagore was very blunt and forthright in his discourses. However, the Ceylonese audiences responded positively and regarded Tagore's frank critique as an important wake up call to realise themselves.

The thrust of Tagore's lectures was on the detrimental consequences of imposing an artificial nation structure on a people and cutting off its civilisational and cultural roots. Ceylon he felt had been structured by the colonisers to become a closed community, cut off from its cultural links with mainland India and fed on the borrowed culture of the colonisers. He lamented the lack of indigenous literature in Ceylon and a disproportionate dependence on the language of the colonisers. He exclaimed: 'What saddens my heart is to find that though you are born to a beautiful land, there has risen no poet in your midst... to sing her serenade... The greatest of your problems is that of finding your own true voice – not that of your master – in your own language.' He felt it was his duty to give the Sinhalese the following message – 'Do not waste time and intellect in carefully imitating other people, however great they may be. You cannot belong to yourself if you do not produce your own literature as the truest document of the mastery of your mind. You must have a continental background of your culture, which is the Indian background that will vitalise your thoughts and enrich your imagination.'[2]

To Tagore, the civilisational continuity of contiguous countries was significant and stood in danger of being forgotten in the artificial creation of nation states. As he put it, 'a community, cut off from its parent stem by disastrous oblivion, is sure to forget the great meaning of its own personality.'[3] After all, he said, 'political division is merely a division of property through which we can change our residence, but not our brotherhood.'[4] During his visit to Ceylon it struck him with dismay that the people of the country may have 'lost the consciousness of their unity with their Indian kinsmen.' This unity was not political. It was entirely cultural, developed in the subconscious mind of a people with the accumulation of ages of creative memory producing 'in secret the racial genius which creates.'[5] Tagore explained that perhaps the best example of this would be the influence of Buddhism which spread from India, the land of its birth, to Ceylon.

The influence of Buddhism in the outer world fascinated Tagore because it stemmed not from narrow self interest but the highest ideals of mankind. As he put it 'Buddhism was the first spiritual force, known to us in history, which drew close together such a large number of races separated by most difficult barriers of distance, by differences of language and custom, by various degrees and divergent types of civilisation. It had its motive power, neither in international commerce, not in empire-building, nor in scientific curiosity, nor in a migratory impulse to occupy fresh territory. It was a purely disinterested effort to help mankind move forward to its final goal.'

It was not that Tagore was enamoured of all aspects of Buddhism. He was a self-proclaimed lover of the world and sought to live life to the full. He was no believer in ascetic withdrawal. Therefore the doctrine of control of passion and renunciation of self did not attract him. Instead what fascinated him was to see how Buddhism had grown and spread its branches far away from its mother soil because its message carried a universal appeal. Its basic philosophy was rooted in Indian tradition and in Buddhist Dhamma can be found resonances of Upanishadic Brahma since both had the highest ideals of spiritual enlightenment. Tagore felt that Buddhism, 'one of the greatest religious achievements of man', should be studied keeping in mind not only its wide outreach but equally its inspirational source which was India. Only by remaining rooted in one's own cultural foundations could one reach out to the world. By finding this interrelatedness of world cultures and civilisations one could develop a universal outlook. Like the two movements of the earth, diurnal and annual in rotation and revolution, man too moved on the axis of his own cultural roots while exploring the cultures of the world. It is through these two movements that man discovered himself. This was in essence the philosophy behind Tagore's university, Visva Bharati. Inaugurated in 1921, it sought to welcome all the cultures of the world in one nest offering to them the hospitality of India's own culture while accepting from them their best.

Tagore explained in his lecture to the Ceylonese audience, 'One of the objects of Visva Bharati is concerned with the personality of the people she represents – we must be fully conscious of what we are.' He made an impassioned plea for acquiring knowledge about one's own cultural pluralism. 'The internationalism of Visva Bharati must be the internationalism of India, with its own distinct character,' Tagore said, adding, 'I do not believe in an internationalism which is amorphous, whose features are broadened into flatness.'[6] This quest to discover himself made Tagore the eternal wayfarer, as he called himself. He travelled extensively through India learning about its vast cultural, artistic and linguistic diversity and across the world at a time when travel was arduous and difficult. It was this relentless travel that led him to realise the most important fact of his life – namely that he had been born to the great world of man.

The views of Tagore on the Nation and the World assume considerable significance today when moral and ethical values seem to be under threat in the face of aggressive consumerism. It is time to go beyond constricting boundaries of narrow geographical divisions and reach out to the universal man. Tagore had written that, from his infancy, he had a keen sensitiveness which kept his mind tingling with consciousness of the world around him, natural and human. He retained this child-like sense of wonder to the last day of his life.

Notes

1 Quoted in *The Splendour of Sri Palee*, 75th anniversary publication by Sri Palee College Past Pupils Association, 2010

2 Nityapriya Ghosh. (ed.) To Sinhalese – in *The English Writings of Tagore* : *Volume 4: A Miscellany*. (New Delhi: Sahitya Akademi, 2009), p.756.

3 Sisir Kumar Das. (ed.) 'To My Ceylon Audience', in *The English Writings of Tagore, Volume 3*. (New Delhi: Sahitya Akademi, 2008), p. 760.

4 Ibid

5 Ibid

6 Ibid, p 764

Rabindranath Tagore: His Worldview

Alim ur-Rahman Khan

The ICWA conference on 'Rabindranath Tagore - Envoy of India: His Vision of India and World' aimed to explore Tagore's global legacy as an envoy of India, highlighting his vision of independent India in a peaceful and unified world'. The idea was to 'discuss his philosophy of internationalism, universalism, humanism and spiritual unity and religion of man'.

The ICWA hopes to evaluate his ideas taking into account the current changed situation of the world while drawing upon Tagore's perennial philosophy and wisdom and then relate those ideas and concepts to the foreign policy of India.

In consideration of this noble idea 'The Religion of Man' by Tagore seemed to be the most appropriate work for the purpose. This is a very dense book and needs plenty of commentaries. These commentaries, in turn, may also need revision from time to time, in view of the fast changing conditions of the world.

This book contains the key to his philosophy of internationalism, universalism, humanism and spiritual unity from which springs his Religion of Man. It is not a straight-forward logical presentation but more of a metaphysical work and thus needs to be approached from a somewhat mystical angle. The message of this book transcends ordinary religious boundaries and delivers a message of wisdom that could unite, rather than divide, the various religious groups in the world; not necessarily to merge them as one religion but enabling the religions to keep the space for the 'other' and work towards a common goal in a united manner. There are many closed doors to be unlocked. Some of them cannot be opened today with words that carried a different shade of meaning eighty years back i.e. when the lectures were delivered. In a way, the book has to be read with the help of the peripheral vision of the intellect.

It is a rather esoteric book. There are layers of meaning, open to varying interpretations. The problem would be as to which interpretation is to be applied at what time to which problem.

The book dates back to 1930 i.e. 83 years back from the present. The book contains perennial truths. Whilst the truth remains perennial, the meanings of words, technology, society, knowledge of semantics and semiotics etc. change. In other words, the paradigm of a 'word' changes. This compels re-reading and re-interpreting of the same truths from time to time.

The world had never changed as rapidly as it has during the last fifty years and specially the last twenty years and the rate of change is accelerating. As the paradigms of words are changing so is the understanding of the physical world and, to some extent understanding of the worlds beyond the physical one. This fact has to be kept in mind always. The major clues to the understanding of the book, lie in the first chapter although the entire book is strewn with thoughts which glitter like jewels.

In Gunnar Myrdal's book, 'A Challenge to World Poverty' lies a very significant sentence. It says (quoting from memory) '...truth, like falsehood, takes the most opportunistic direction'. Myrdal is quoted just to clarify that there is no one truth and there is not just one direction for any of these truths to move. At any moment of time in the past, a truth could have taken another direction to move, producing another set of results.

In the late nineteenth century, when Tagore was growing up to be a young-man, both Malthus and Darwin were popular and so was their combined conception of the preservation of stronger groups in the struggle for life. T.H. Huxley added his bit in 1888, stating that the animal world is at about the same level as a gladiator's show where the creatures are fairly well treated, and set to fight; whereby the strongest, the swiftest and the most cunning live to fight another day. This was accepted and justified as the norm of life and a base for all kinds of exploitations. It suited the time. Technological inventions were proliferating and raw capitalism was making full use of it, justifying it in all possible manners. These were times when ruthless 'progress' was taking place. Colonialism, capitalism, exploitation of the poorer people, even of one's own people, was rampant.

However, by 1893, Huxley changed his tone by recognizing that in the cosmic process there is a place for the ethical process and that it could be the best and here he included ideas like goodness, virtue etc. There were other thinkers also who were already seeing the drawbacks of the principle of a wild and free struggle for existence. New argument began to appear to prove that this mode of thinking as wrong or limited.

Around this time and just after, the raw face of capitalism was exposed and in flowed ideas from people like Karl Marx and other thinkers about another kind of a society where, as they saw it, justice and fair play would prevail. The entire approach was somewhat based on a kind of dialectic materialism that came out more as a 'revenge' against the direction that the society had taken. This was followed by the First World War and the revolution in Russia. However, the western world was still strongly materialistic and based its beliefs and progress on materialism or on quantitative goals. Again arguments were created in favour of these ideas. These were not 'wrong' ideas but truth was taking opportunistic directions.

These were also the times when Mme. Blavatsky and her Theosophical Societies and their likes were also looking for a spiritual dimension to life and were trying to explore the esoteric world.

The Hibbert Lectures of Tagore which eventually took the shape of the book 'The Religion of Man' came about in 1930. It was like a fresh breeze carrying the fragrance of unseen wild flowers in this world struggling with absolute materialistic ideals whilst at the same time not accepting it fully and thus simultaneously struggling with the esoteric. In these lectures Tagore brought out beautifully and in quite a rational manner the existence of the spiritual dimension without any form of excess. It is not a logical exposition. It is an intuitive one. In Tagore's words '...logic is a machinery created by the mechanic man.'[1] Both logic and rationality have their limits; nevertheless, it is a good starting point.

The contents of the book is best put in Tagore's own words, 'The idea of the humanity of our God, or the divinity of Man the Eternal, is the main subject of this book. This thought of God has not grown in my mind through any process of philosophical reasoning. On the contrary, it has followed the current of my temperament from early days until it suddenly flashed into my consciousness with a direct vision.'[2] It is important to understand and to appreciate this.

It is an intuitional method of understanding/comprehending/apprehending or perceiving, the Truth. It is an acceptable method in almost all mystical or mystico-religious systems. It was accepted by ancient philosophers of Greece and Alexandria; Indic religions and philosophies; and Christian and Muslim mystics and philosophers. Only the modern Western philosophers may not accept this method but let us leave that issue for the time being. To quote Tagore again '...logic is a machinery of thinking created by the mechanic man'. Towards the end of the book Tagore acknowledges the fact that he has been influenced by Hindu (Indic which includes Buddhism and other Indian religions), Muslim, Christian, Zoroastrian and even Taoist thoughts. This is what gives the broad-base on which one can present and suggest the internationalism, universalism, humanism and a philosophy along with a philosophy of religion of Tagore, which can be accepted and acknowledged by the followers of other religions.

This idea needs a full and long commentary, but due to constraint of space only three important focal points are chosen to understand and comprehend Tagore's ideas of the Religion of Man. These, in turn, can be used by India and, in fact, by the entire world, to build a better world in future.

1. Evolution: Divinity or Nature first experimented with the creation and evolution of

    (i) Inert matter and thus created the entire cosmos full of 'dead' matter. This was found to be inadequate even along an evolutionary line and thus

    (ii) Life was introduced; initially as mono-cell units, which then multiplied as per the rules of life itself and then combined and co-operated with each other to evolve higher and higher patterns of life till it produced finally, the physical man but then this was also found inadequate and in the final moment it introduced

    (iii) Spirit, in the man. This then allowed another form of growth or evolution.

    The base has now changed from the quantitative to the qualitative.

2. Inter-relationship: There is an inter-relation of everything from the microcosmic to the macrocosmic. However separate things may appear to be, they are actually all working together in the same direction. However, antagonistically they may act at a superficial level; deep inside they are working in unison along the evolutionary line; the goal is to help one another to move towards a 'wholeness'. 'The divine principle of unity has ever been that of an inner inter-relationship.'[3]

3. Union: The entire creation is on a journey. The inert matter is moving forward in time. Living beings also are moving along in time. The Spirit of Man is also moving forward. It is the Spirit of Man that we are mainly concerned with. They i.e. inert matter, other forms of life and the human beings, all move in the same direction for the same purpose in a quantitative manner but, qualitatively, the spirit is a bit different. It exists neither in space nor in time, yet the purpose of the movement is identical. It is towards a grand union. The spirit leads this race towards this grand union. It is as if the purpose of the entire creation is to return to the Creator, however He or It may be imagined. Life is a journey, where the spirit has to be allowed to make a free movement towards this Unity. Merging back into this Unity is its fulfillment.

To quote Tagore once again 'The development of intelligence and physical power is equally necessary in animals and men for their purposes of living; but what is unique in man is the development of his consciousness which gradually deepens and widens the realization of his immortal being, the perfect, the eternal. It inspires those creations of his that reveal the divinity in him—which is his humanity—in the varied manifestations of truth, goodness and beauty, in the freedom of activity which is not for his use but for his ultimate expression. The individual man must exist for Man the great, and must express him in disinterested works, in science and philosophy, in literature and arts, in service and worship. This is his religion, which is working in the heart of all his religions in various names and forms.'[4] At the end he will find bliss but along the way he/she will find beauty, peace, harmony, love and goodness. There will be a sense of equilibrium and end of all opposites.

There are two more concepts used by Tagore. These two concepts also need a bit of explanation. The first one is 'creative imagination'[5] and second one is 'Man the Eternal.'[6] He also uses other words like 'Universal Man', 'Man the great'; to describe the same idea, the term/concept of 'Perfect Man' is also used in other traditions.

According to Plato, ideas are ultimately 'real'. They exist in a different realm. In this realm of ideas, the 'good' has a very high and unique place. It is not impossible that Tagore's concept of 'creative imagination' (this is a word he has used towards the end of the first chapter of his book and is of much significance) starts with a root from here. Much later, Henri Corbin, the French writer of mystical philosophy, had coined and used the word 'imaginal world' especially when dealing with the mystical-philosophy of ibn-Arabi. This (idea of ibn-Arabi) could have, again consciously or unconsciously, influenced Tagore. The second possibility exists from the fact that Tagore's father was a scholar in Persian and was very fond of Hafiz. He, therefore, could not have been free from the influences of the writings of ibn-Arabi, Rumi and perhaps of Jami, who was born sometime after Hafiz's death, since the Persian world of Hafiz and post-Hafiz, was based partially on this world of ibn-Arabi, and Rumi, to name two.

Tagore's Eternal Man or the Universal Man resides in this realm. It is the imaginal realm or the realm of ideas. Embodied in this Universal Man, who is also the Perfect Man, are all the perfections conceived by mankind for the 'good' and including all that could be needed for the benefit of mankind and through him, of the universe of mankind, ...all ideas, rules and visions are imprinted on this Eternal Man. This is also very close to ibn-Arabi's idea of the 'al-insan al-kamil' or the perfect man. It is also the idea of the Perfect Man of Taoism. More about this below.

The evolutionary process envisaged by Tagore states that the earlier efforts of nature to create, was with, what we would call the 'dead' or inanimate objects where physical size and bulk were the initial criteria. On completion of this (at least to a certain extent) nature found the fruitlessness of size only and hence '....life was brought into the arena in the tiniest little monocycle of a cell'.[7]

Tagore continued: 'With its gift of growth and its power of adaptation, it faced the ponderous enormity of things and contradicted the un-meaningness of their bulk. It was made conscious not of the volume but of the value of existence, which it ever tried to enhance and maintain in many branched paths of creation, overcoming the obstructive inertia of Nature by obeying Nature's law'.[8]

Here the next stage unfolds. Tagore writes as he 'sees' it: 'But the miracle of creation did not stop here in this isolated speck of life launched on a lonely voyage to the Unknown. A multitude of cells were bound together into a larger unit, not through aggregation, but through a marvelous quality of complex inter-relationship maintaining a perfect co-ordination of functions. This is the creative principle of unity, the divine mystery of existence that baffles all analysis'.[9]

'But this evolution which continues on the physical plane has its limited range. All exaggeration in that direction becomes a burden....'[10] Tagore states that sheer strength and size becomes a useless burden after a certain stage. Man appeared on the stage around this time, physically weak and bodily vulnerable but with a potential capacity for something more. Again in the words of Tagore: '...Man appeared and turned the course of this evolution from an indefinite march of physical aggrandizement to a freedom of a more subtle perfection. This had made possible his progress to become unlimited, and has enabled him to realize the boundless in his power.'[11] Here again is the qualitative factor as against the quantitative.

This (' ...to realize the boundless') is a process that Man achieved and is still achieving not just through physical intelligence, which he shares with other animals, specially the higher animals, but through the introduction of what Tagore calls the Spirit into this Man. It is the presence of this Spirit that moves the base for measurement from the 'quantitative' to the 'qualitative'. There is the man in his physical body, which has its own demands that require fulfillment; and there is the man in his spirit, which has its own urge and its own method of fulfillment. The introduction of the body and the spirit, though not a new idea, is a significant matter. The spirit takes over, of course after allowing the body the space that is needed to exist. The development of this spirit becomes the main purpose of the life of man.

Once this very vital point is understood and accepted, one can then move to the next important idea. It is that everything that exists and has what can be called 'life' is working in unison. This happens from the micro-cosmic to the macro-cosmic. Everything is working and moving towards one direction. Life appeared in the mono-cells and started to act and react. It multiplied and these cells organized themselves into larger units. The little individuals then co-operated to keep up the 'large' individual. Life starts its function but unlike a generally accepted Darwinian vision of the survival of the fittest, Tagore moves in the direction of thoughts that prompted (later) persons like Ashley Montague and some others to propose that co-operation and not competition is the fundamental basis of life.

Tagore goes further and, like a mystic that he was, he also sees 'unity' in the total appearance of the 'multiplicity'; as if multiplicity exists only for the purpose of the unity. In Tagore's vision, all the elements we see are actually multiplicity at the simple earthly/tangible level but in the higher/intangible level they are functions within and for the great Unity.

Once again in Tagore's words: 'The unity becomes not a mere subjective idea, but an energizing truth. Whatever name may be given to it, and whatever form it symbolizes, the consciousness of this unity is spiritual, and our effort to be true to it is our religion.'[12]

This can be further clarified by looking at it from other angles. It can be said that the unit cells living inside a greater body are all functioning towards a Unity, which in this case is maintaining the greater body in all its efficiency and fitness. By extension, this unity of working is exists/prevails in the entire cosmos. Both at the micro-cosmic and at the macro-cosmic levels everything is working towards this Unity both in the physical and in the spiritual senses. One has to develop a special kind of insight to 'see' this move towards Unity.

Diversity, or rather multiplicity, is an illusion or a maya. A good example perhaps would be the ocean. When one looks at the sea or the ocean, one sees… a single beautiful and huge mass of water; a single unit. But then when one wishes to see the waves then one has to concentrate and 'see' the waves, each a single separate unit. One can go further and breakdown the waves to see the water as separate from the froth on top and, each, almost each wave will have a crest of froth but then again one can wipe it all out and look at the beauty of the one ocean. This is just a simile and like all similes, it should not be carried beyond its limits.

In human beings, the meaning and purpose of life itself is a movement towards the fulfillment of this Unity. This can only be sensed through 'consciousness'. This is valid for all human beings and this is the commonality of human beings. It is the base of the equality and the similarity of each and every human being. This overcomes the difference of person to person, the difference of nations and is again the base of Tagore's philosophy of internationalism, universalism, humanism and spiritual unity and religion of man.

Tagore's family, though of Hindu Brahmin origin, had adopted the new eclectic religious thinking of the Brahmo Samaj, which wanted to re-establish the original Hindu monotheistic religion of the early Upanishad period. They were also influenced by the Semitic religions specially Christianity and Islam. Personally, Tagore had knowledge of Taoism and Zoroastrianism also. His final expression on philosophy, mysticism and religion thus had flowered by taking the best from all these systems.

To begin with, there is the Upanishadic monism. Tagore's father was one of the pillars of the newly-found Brahmo Samaj and his effort was to draw the basic thoughts from the Indic monism as expressed in the Advaitic principles of the Upanishad. Tagore quotes the Isha Upanishad and affirms that '... the Super Soul which permeates all moving things, is the God of this human universe whose mind we share in all our true knowledge, love and service, and whom to reveal in ourselves through renunciation of ourselves is the highest end of life.'[13] Whatever Indic philosophy, religion, mysticism and culture that penetrated into Tagore was probably well filtered by and through this Advaitic line of thinking.

Keeping in mind the purpose of this article, it is necessary to have a little look into the other religious cultures that influenced Tagore.

This may also be the right place to clarify further the two concepts mentioned by Tagore. One is 'creative imagination' and the other is 'Man the Eternal' or the Perfect Man.

Buddhism, being an Indic religion had also its profound influence on Tagore; but he was also aware of the Taoism of Lao Tze and perhaps Chuang Tzu. One can see how these thoughts also had been incorporated in Tagorian thinking. For example, quoting from Chuang Tzu ' ....the Perfect Man...is a man who is completely unified and united with the Way.'[14] The Way of course means Tao. Further on is Lao Tze's vision of the Perfect Man. 'In fact, he is not a 'man', if one understands by this word an ordinary human being. He is, in reality, an infinitely large cosmic human being.'[15] Even in conventional Hinduism, there is the concept of the 'Jivan Mukta Purush', a man who is free of life whilst being in life. He is one who, likes the Perfect Man of Tao, is completely one with the way, both unified with it and united. The concept of this Perfect Man or the Universal Man is a foundation of Tagore's mystical thinking.

With Christianity and Islam, Tagore shared the principle of monotheism. If we take Christianity first, then we find that many of the moral attitudes have come directly from Christianity, as it was popularized in Koltata; these were and are universally acceptable. Whilst his attitude of 'disinterested service' could come from the Bhagvad Geeta, it was also present in both Christianity and Islam where they insist that one does these 'for God's sake'. Even the conception of the Universal Man or Perfect Man could come out from the Bible: 'And God said, Let us make man in our image, after our likeness' (1 Cor. 11.7) or 'So God created man in his own image, in the image of God created he him; male and female created he them' (Mt. 19.4 · Mk. 10.6). The 'image of God' referred to cannot be just the six feet tall man weighing around a 100 kilos. What then is in man beyond the physical? There has to be more, when he is created in 'God's image'. Even the importance of 'Love' can be linked directly to the Christian teachings and indirectly to the Muslim sufistic influence as we shall find below.

In addition to this element of commonality with Islam, Tagore's father, could have come across ibn-Arabi's concept of the 'Oneness of Being', (Wahdat-al-Wujud) and of the 'Perfect Man' (al-insan al-kamil). These could have crept into Tagore's thoughts from these sources, consciously or unconsciously. Besides these, there was also the strong presence of sufistic teachings of Islam in India. Its influence on people like Kabir, Nanak and others was also appreciated, acknowledged and accepted by Tagore.

The Religion of Man that Tagore is referring to is the unified spirit of all religions that is embodied and encoded in this macro-cosmic man. In practice, there are different religions in existence in this world but the differences are more at the level of rites and rituals, each one having its own complete system but it is not these separate religions that Tagore is referring to.

The above approach of Tagore applies to all, men and women, irrespective of caste, creed or colour. Once this concept is understood and accepted then it naturally leads to internationalism, universalism and humanism. It must be noted that humanism covers an attitude both towards other fellow human beings and to the entire creation. Here man is the vicegerent of God in the Quranic (2:30) sense.

Now the challenge is how to dovetail this mystical-philosophical concept to the Kautilyian/ Macheavillian world. This is possible with patience and understanding; working with the belief that finally, it will be for the betterment of the world.

However idealistic Tagore's thoughts might have been, they carry universal truths. People of all religions should find some common meeting points in his philosophy of religion. Those countries or people who are firm upholders of secularism; even denying or refusing to acknowledge in policy-matters the existence of the non-material world, would still not deny the truth of the conclusions of Tagore. The conclusions drawn through intuitive visions nevertheless will coincide, in the final analysis, with the teachings of all religions and all philosophies worth considering. The secular humanitarians, who draw conclusions through logical, rational or scientific thinking will not be able contradict the truth of the conclusions, however much they may be in disagreement with the methods that led to the conclusions.

From the wide range of very valuable ideas that Tagore presents in his book, only three points have been considered here. These three seem to be adequate, comprehensive and relevant. It is also the minimum number required to elucidate the objective of the ICWA. There are many other points which are also relevant specially if an overall policy has to be determined.

This task of relating the foreign policy of India (or any country) to an ideal standard and then to apply it on the conditions existing, is a Herculean task. It is uphill and success will be very slow but it is worth doing it and to keep at it constantly, however low-keyed the effort may be. 'Truth does not win but its enemies die out.'

The first and important problem would be the implementation of the ideas selected. It will not be easy to implement altruistic utopian ideas in a Kautylian/Machiavallian world. Nevertheless, this has to be done.

There are some basic principles, which are to be considered. There is no immediate necessity to adopt all ideas right from the beginning but, if accepted, they should be implemented in the long run, gradually; otherwise there is a waste of energy through working in self-contradicting principles.

The first idea is the acceptance of the spirit of man as being more important than the physical man. The presence of this physical man is obvious and has to be satisfied but only up to a certain extent. Altruism has to be accepted and valued as the door to liberation; to reiterate in Tagore's words '...manifestations of truth, goodness and beauty, in the freedom of activity which is not for his use but for his ultimate expression. The individual man must exist for Man the great, and must express him in disinterested works, in science and philosophy, in literature and arts, in service and worship. This is his religion, which is working in the heart of all his religions in various names and forms.'[16]

The next problem would be that many countries would not like to consider a philosophy which is associated with 'spirituality', (in whatever manner this term is taken and explained). This will need a re-formulation of these Tagorian principles. They then have to be re-cast in a way that is acceptable to almost all countries of the world. This is possible, perhaps at the cost of sacrificing or limiting some ideas, which in the sub-continent are accepted as natural. In this connection, one point should be remembered. It is that the conclusions that Tagore has reached are important, and not how he reached it. The conclusions are such that they have to be accepted, in some form or other, by all groups; be they religious, spiritual or totally materialistic.

The influence of the thoughts of Tagore should be used like an umbrella to cast its shade in the sun and to keep away the rain when it pours, and also to act and as a beacon of light to look up to, when in difficulties. Actions will take place on the physical plane, where most participants may not even be aware of the 'spiritual' plane on which the main thrust of Tagore's philosophy operates. Thus India will have to act as the bridge from the intangible world to the tangible. Given India's history of philosophy and religion from time immemorial, it looks like a possible task.

The spirit of the book under consideration gives a direction to the spiritual man, in whom perhaps (mystically speaking)) lies encoded all the rules and guidance needed for the physical man and for his symbiotic relationship with the Universal or Spiritual Man, in whom in turn is encoded all required messages for mankind.

Tagore's ideology, which is very Indian at the core, can be elaborated and prepared as a set of desirable goals to be reached at some point in the future. One can easily build up this programme by considering the ideals suggested by Tagore. It is not necessary to find out how he arrived at these conclusions. As written above, the conclusions will be affirmed both by logical and scientific thinking. This perhaps goes to show that intuitive thinking, on which these lectures in the book are based, is also a valid method; those who do not accept it, may leave it aside but the conclusions, all the same, are firm and correct for all; whether they be countries where religion is strongly recognized or they be absolutely secular countries.

It may be worthwhile to reiterate and sum up the three messages from Tagore that will go to affirm his philosophy of internationalism, universalism, humanism and spiritual unity and religion of man.

1. The evolution of the spirit in man. This is what makes the human being different from all other animals. Man has a consciousness that allows him to think, ponder and work out a way forward, which is above the mundane. He can express this spirit in art, philosophy and selfless service and his growth is actually along these lines. This pushes the thrust of living away from being the consumer-oriented society and create a society based on other deeper values.

2. The next point is that everybody in this world should work in co-operation and not in competition. The object is to work for each other so that the quality of living is raised. It is unimportant to talk about rich and poor. Let there be the rich and the poor. What is needed is that no one is left behind in the race for existence and in the race for self-expression within the framework given above. All should have equal access to opportunities. This may turn up into framing a new philosophy for the future but that is exactly what we need.

3. Finally, the journey of life. Everything is moving towards an end, at least end in the form that it exists now. Those who believe in some kind of God and answerability to Him/It, have the direction set clear, but those, who do not believe in God and the after-life or answerability, still show signs of great selflessness. In simple terms, there are many such who contribute their eyes, liver or kidneys after death for others. There are those who make trusts and other such institutions for public good. These people are not always believers. What is it that encourages and pushes them to such selfless acts? There has to be a core in all human beings, which pushes them towards these selfless services. This spirit has to be harnessed and used for the betterment of mankind and the world in general for the future.

This article can be ended here by quoting from the Geeta (2:47)

Karmanne badhikarastu, ma phalushu kadachan

Ma karmaphala hetur bhurma, te sangwashte karmani

(You are responsible for your actions only, not for its results

Never act with the desire of the result, but never remain inactive.)

Notes

All references to Tagore are from: Rabindranath Tagore. *The Religion of Man,* (London: George Allen & Unwin Ltd, 1958).

1. The Religion of Man p. 19
2. Ibid p. 17
3. Ibid p. 15
4. Ibid p. 16 & 17
5. Ibid p. 24
6. Ibid p. 17
7. Ibid p. 13
8. Ibid p. 13
9. Ibid p. 13 & 14
10. Ibid p. 14
11. Ibid p. 14
12. Ibid p. 16
13. Ibid p.24
14. Sufism and Taoism by Izutsu Part II, Chp. XI
15. Sufism and Taoism by Izutsu Part II, Chp. XI
16. The Religion of Man p. 17

Travelling on the Edge of Sorrow - Tagore's Unseen Prison

*Ghada A. N. Mostafa**

The traveller has to knock at every
alien door to come to his own, and one
has to wander through all the outer
worlds to reach the innermost shrine at
the end.

- (Gitanjali)

When I first read 'Gitanjali' in an old Arabic translation in my grandfather's library at the age of fourteen, I was astounded both by the lyricism and the drawing on the book cover. My grandfather had lived and worked in India in the 1920's. We still have some of the golden-threaded tapestries he brought back home, plus numerous photographs of a place, which was to become, for his granddaughter, the model of a dream or fairy land. More than thirty years after his death, I would search out my dream and visit the same cities and temples that he wrote about in his memoir and notebooks.

This is a study about the themes in some of Tagore's works that seem to substantiate the hypothesis that the great artist and poet was more of a monumentally sad individual than he was a traveler.

Its focus is precisely this: Have Tagore's travels determined the 'content' or the nature of his ideas, themes and moods as a poet, essayist, short story writer, playwright, novelist, musician and painter?

To reverse the question – quite legitimately – would Tagore have formulated the same ideas/ideals and had the same inclinations towards some issues of his time and most importantly, the same feelings and stance regarding life and his own existence even with the omission of travel as an obsession for the unknown?

I shall not discuss the ideas of Tagore the educationist or Tagore as a man with his own economic views. They may tell us that he was a universal man and he obviously was. But even in those realms, his 'universalism' was Indian, since his influence by the Baul sect of Bengal meant that the ideas he lived by and was a firm believer of, were those essentially of 'encompassing' a humane approach where there is no distinction between men on the basis of colour, caste or creed.

I have not been able to have access to Tagore's writings before he started traveling. Without claiming to have read everything Tagore has written or to have seen everything he has painted, or composed, I propose and hope to show that the 'ideas ' and spirit which dominated works such as ' Gitanjali ', ' Stray Birds '; Thought Relics ' and ' The Religion of Man ' are likewise manifested in his last poems 'Shesh Lekha'. It is a circle of thought and feelings spanning a life where travel seems – with all its influences – to stand on the periphery of his psyche and mood.

The focus of this paper therefore is to prove, by drawing parallels and comparisons between Tagore's books cited here and by references to known personal incidents in his life, that the prolific and miraculously energetic Tagore would have formulated the same ideas of the religion of man, the ideas of love and oneness with the Supreme or Eternal Person, the themes of the quest for truth/ Truth and embrace of death even if he had not travelled as much as he did. I leave outside the realm and the scope of this thesis, Tagore the musician to experts in the field as I leave Tagore the painter, playwright and educationist to specialists and will allude only to certain observations by critics about his painting, that lend proof to the basic principle of this paper.

We know that the giant rise of painting later on in Tagore's mid-sixties may well have come about as a result of boredom and a fatigue with other forms of expression. Some of my poet friends in Egypt initially started from painting while others reversed from poetry to painting. The movement from something in art 'to' something else also in art is governed by reasons which the artist him/ herself may not be aware of. From the examples I have known around me, one art form has always been at the expense of another. I have no explanations for this.

The process of life intersects with the process of art, accumulating audio-visual pictures, obliquely burying memories by each other, drowning scenarios, dreams, dialogues probable and real, intensifying some sounds and omitting others. But while for Tagore, who was unusually blessed with the discovery of another god in his fingertips, in later years it was no problem to continue in various art forms, mastering them all, for many of us the effort of mastering one art form as much as can be done, may devour an entire life.

In the paper titled 'The Painter' by Sanjoy Kumar Mallik in 'Rabindranath Tagore: A Commemorative Volume', Mallik refers to the faces in Tagore's paintings which have 'expressions ranging from the sullen and somber to the calm and contained, and rare instances of the joyful or the merry'.[1] What is interesting is that Mallik thereby suggests the independence of Tagore's artistic inclinations and his moods such that in a time of nationalistic revivalism swaying the Indian subcontinent, he ' had the strength of will to propose a larger vision beyond the restrictive criteria of national/geographical boundaries in matters of creative expression '. [2]

But since painting emerged later on in Tagore's life this seems to suggest that this universalism was essentially the by-product of travel. It is here that Mallik insists that Tagore's images for the whole reflect a visual language that is so 'utterly individual'. [3]

An individuality, which continues to manifest itself as if rarely touched by the surroundings while on occasions projecting the opposite can be problematic for critics.

Ina Puri in 'Influence On Other Painters' in the same volume, observes that Santiniketan reflected a deliberate choice to live in isolated places, and that such a choice was reflected by the inner mindscape of Tagore which 'seldom used imagery from the world around, no matter how beautiful that outer world was.[4]

In a chapter titled 'England' in 'Rabindranath Tagore: The Myriad Minded Man', there is curious support for this critique. Krishna Dutta and Andrew Robinson tell us: 'The Houses of Parliament were almost the only great institution described by Tagore in his letters. He was never a sightseer, but his lack of curiosity about London's other great sights – he wrote not a word about Westminster Abbey, the British Museum, or Trafalgar Square, for instance – was surely significant. In later life, he would write not a single detailed description of any building, whether sacred or secular, seen during his worldwide travels, and only rarely would he write about those in India.[5]

The world of Tagore – being a poet, artist and writer – was the inner one where the architects are those close to his heart, the moments of his love, his art and his vision rather than sights, scenes or other matter coming from the external. I hope to investigate and prove this in the course of this paper.

Once Nietzsche wrote: 'Every great philosophy "is itself" a species of involuntary and unconscious autobiography'. [6]

I have always suspected so. The greatness of Nietzsche's mind is that he was able to think beyond norms and limits, and able to realize and say something like that. And to say it like a long forgotten fact.

Rabindranath Tagore was a versatile man with a god's weight of talent in him. Born in a family whose doyen detested idolatry, being himself an ardent follower of Raja Ram Mohan Roy and the revivalist movement of Bramhoism, Tagore's poetry, fiction and many articles reflect much of that stance.

The existential pain of Tagore's loneliness, and its lyrical expression in much of his poetry was clearly the result of personal mishaps too colossal to endure and overcome by either travel or art alone. The death of his mother when he was a child and the subsequent suicide of his sister-in-law Kadambari (a figure with interest in his writing) months after his marriage would mean a constant search for a possible mother figure/beloved sweetheart mingled with a possibly continued sense of guilt. The futile quest of his novel *Two Sisters* is this search for that mother figure/sweetheart.

Kadambari's ghost was to haunt Tagore. She would become for a long part of his life his demon/ muse. A curse that entails – by way of writing about the torment of her loss – a blessing, but a threat to his own peace of mind possibly for the rest of his life. It may even have involved a stifling of the experimental tendencies in his poetic approach or acted as a thematic block sometimes.

Topobrata Ghosh sees that his sister-in-law was, on the contrary, his 'life's deity', 'impelling him from a microcosm towards a macrocosm, but it was a human and humane macrocosm, not a cosmic or superhuman one '.[7]

In Ghosh's introduction in 'Rabindranath Tagore: Selected Short Stories', one footnote offers great insight into the Tagore-Kadambari relationship. Ghosh cites the poem titled 'After Death', written before the tenth anniversary of Kadambari's death, where the poet writes:

'In the boundless, silent land/ She has found eternal night/And endless consolation'.[8]

Later on in 1935, and at the age of seventy four, he would write poems recalling his sister-in-law and Ghosh refers to one of them which is translated as 'Drama's End' where he says : 'She suddenly went away in the night/ That never sees dawn'.[9]

These and other numerous examples testify to the continuing impact of personal tragedy on the body of his writing. Even the names and relationships of some of his short story heroines will be influenced by the name and place that Kadambari held in his life.

The Bengali folk life, and places like Shahi Bag Palace in Hyderabad, the Ghats, etc. are all evoked in his short stories confirming that what essentially inspired and moved him was that which was characteristically Indian.

Kadambari was considered as having staged a 'social revolution of sorts by going riding with her husband'.[10]

Tagore would even make the woman with the narrator of the poem 'Beside the Sea' ride together on horseback. In his volume of poetry, 'Wild Geese', there is a whole poem about Kadambari Devi. It was believed to have been inspired by the fact that he had come across one of her pictures. In one of his lines in Dutta and Robinson's book, Tagore had analyzed in simple terms his condition: 'I remain behind/burdened/by the past'.[11]

In the same book, we are told: 'Three, probably four of Rabindranath's books were dedicated to Kadambari while she lived. After her death, he dedicated at least two more. In 1901, he portrayed her almost directly as Charu in Nashtanirh (the basis of Charulata).

For the rest of his life, he wrote about her in his poems. In his paintings a mysterious woman's face appeared again and again: he admitted to an artist friend that she might well be Kadambari-whose 'glowing eyes present themselves before my eyes not unoften' when painting, he said. Her presence can also be felt in his songs.'[12]

Reading about Tagore's life, we learn that his deceased sister-in-law was unhappy with her husband Jyotirindranath (his brother) who was the center of female attention, being himself an actor, musician, writer of historical plays and translator. Kadambari was an interested admirer of Tagore's poetry and writing and some scholars believe that she did not really want to see Tagore married, even though she took part in the socially-accepted role of an Indian sister-in-law at the time who engaged herself with the other sisters-in-law in searching for a wife for Rabindranath. In Dutta and Robinson's book the fact that Kadambari committed suicide months after Tagore's marriage to his child wife is more than mere coincidence, even though she herself was childless.

It is sad to know that the rumored letter Kadambari was supposed to have left was destroyed and biographers have noted that the family's aim was to avoid 'scandal'. What supports this contention is the fact that the police were informed of her having poisoned herself, yet her body was not sent to the morgue for an autopsy. Here was a situation of continued emotional suppression, since the fact that both Rabindranath and Kadambari cared greatly for each other while she was living, (perhaps both were suffering a silent mutual doomed love) he alone would have to deal with her absence and probably with the knowledge of some degree of his 'responsibility' for her suicide. Such forced silence would find compensation in his art.

'The Ghat's Story' is seen by Topabrata Ghosh as drawing explicitly on personal disaster, specifically Kadambari's suicide which, we are told, occurred months before the story was written. Ghosh believes that Tagore started his career as a teller of tales, a short-story writer from such a sparking incident of sorrow. He explains that in Tagore's *The Poet's Tale* the figure of a 'guardian deity' … 'merging the pagan demon with the Christian guardian angel in the character of Nalini'[13] was a poetic construct in the first volume of poetry but that construct was later given life following the suicide of his sister-in-law.

The weight of the horror of a loved one's violent death may well have transformed the 'human' that Ghosh refers to, first into the godly and then slowly over time into something approaching the nightmarish. In Tagore's words, his 'life's deity' linked his life 'to the Vast, the Immense through separation and pain'. [14]

Separation offers the artist the best opportunity for the essentially tragic attitude to existence which he/she may assume. The artist becomes the hero of his own project, the living expression of the anger, the lingering anguish he mourns. His energy in rejecting the cruelty and meaninglessness of life is thus not wasted. In fact, it is supported by necessary 'proof' for the transformation of the artist into a tragic hero and it is no secret that critics have agreed that the character of 'Nikhil' in Tagore's novel *The Home And The World* is evidently Tagore himself.

For Tagore, it was separation – as he expressed in letters to friends – which ennobled the life and work of a man, since he was reaching for a 'union of the homely and the alien, the inward and the external'. [15]

Perhaps this partly explains the call of travel.

'*Gitanjali*' is a kind of spiritual exodus mounting to a confessional rite:

'I have roamed

from country to country

keeping her in the core of my heart

and around her have

risen and fallen

the growth and decay

of my life'. [16]

According to Ghosh it was really with Kadambari in mind that Tagore wrote a poem where he expresses:

'The two of us

have flowed upon the

stream

of a conjured love

from the heart springs

of primordial time.'[17]

Personal bereavement then had a monumental effect. And though in India, the word 'mother' can be said to Mother India, mother cow and one's natural mother, it seemed Tagore was mourning the latter when he wrote in *Gitanjali*:

'Mother, I shall weave a

chain of pearls for thy neck

with my tears of sorrow.' [18]

Many of Tagore's children's poems were written by him for his own children who became orphaned, after the untimely death of their mother at the age of twenty-nine.. In that respect, the fate of his own children was repeating his own fate, since he had lost his mother, as some sources put it, at the age of thirteen whereas other sources claim that it must have been between the age of eight to ten.

After his wife's death, Tagore's second daughter became ill and died a year later. Two years later, his father died and following that his youngest, favorite son died of cholera at the age of thirteen. By marrying, he had lost Kadambari who committed suicide. He was struggling with grief and anger in the writing of two great poems: 'The Beginning' and ' The End' that would depict separation and mourning. In these poems, Tagore wishes to reverse the deprived situation and punish the mother figure with abandonment. Later on, Tagore will lose his only grandson to tuberculosis in Germany. No amount of travel could undo such losses or heal such deep wounds.

We do not forget in the sense of shedding memories unless we are aided with the passage of time and advent of new incidents, or emergence of new individuals. The initial drive for travel in Tagore's life was thus strengthened by the new force of immediate calamity, of renewable loss. In his 'Thought Relics', he is explicit in his explanation that our sufferings seek us out in our protections thereby taking away our artificial props and ultimately leaving us faced with our naked loneliness. It is the same 'loneliness' he described as 'infinite' in 'Stray Birds'. By 'our protections' he may have meant love and art. By 'our artificial props' he may have meant travel.

The fact that Tagore chose to live beyond his pain is the brave man's choice but it can also be the mystic's and the artist's choice. Between the two cross-currents of living and not living, he, as an artist, made his handicapped freedom. Travel falls into that sphere. It carries the hope of forgetfulness and rediscovery, a shoveling of the past and the opening of a vista for some new experiences, some hope even of love! Later on in life he will meet the literary figure Victoria Ocampo and we allow ourselves to believe that perhaps Tagore was in need of feeling that he was more to her than a great mentor and artist. In 'Shesh Lekha' there are many lines that can apply as easily to the chair Ocampo is believed to have shipped for him at one time as to a chair which may have been used by Kadambari Devi.

Reading Tagore and reading about Tagore leaves us with his lot of unanswered questions. His search for love and his search for truth seem to have moved in parallel lines reaching the same end somehow. He realized that truth is not afraid to come to an end, or to be small, and that a poem – as a project or specimen of beauty -- is likewise not really dead once finished owing to the nature of its goal: the ideal or model of perfection. He too was not afraid to 'come to an end' if that meant 'perfection' as a unity with truth and liberation from torment. The mystic (Tagore translated 'The Songs of Kabir') sees the end as a hope for another happier existence perhaps, or an embodiment of all that had remained unanswered and unknown in his/her life. Even for the nihilist, death can mean a release. The choice of this translation by Tagore or any artist is not random. It may attest to a feeling of affinity with the ideas of the translated corpus. Indeed many of Tagore's ideas about Divinity and man's relation to it seem to be the ideas of leading Moslem Sufis in older times.

In *Gitanjali*, Tagore expresses that he dives into 'the ocean of forms hoping to gain the perfect pearl of the formless', [19] and in 'Shesh Lekha', nearing his life's end he submits:

> 'He who can effortlessly bear your guiles
>
> receives at your hand
>
> his imperishable right to peace.' [20]

As he seeks perfection in a 'finale' or involuntarily reaches the end of a poem or painting, so his life must seek its end. Truth, we are reminded is not afraid to come to an end and he must face his death as he faced life - with courage. Right, for him, unlike Wrong, can withstand defeat. Such an ethical stance, such formidable idealism can love the world and not merely tolerate it. In *Gitanjali* he tells us: 'And because I love this life, I know I shall love death as well'. [21] However, it is a mystic's or hermit's type of love. It is a love vanquished by what it could not have, what remained unrequited, unrealized or suppressed, leaving a void and creating the constant thirst where 'the perfect ending is in the endless'. [22]

It is an aesthetic begotten of successive loss and experimentation with life's denials and pain, written almost as if it were an artistic manifesto that lives on, where the subtraction or suspension of meaning (as opposed to either its heavy accumulation to the point of explosion or its nullification once drowned in mystery) adds to it in all artistic forms of expression. It helps us understand what he perhaps meant when he told us that his was a poet's religion of sorts.

Such a soul does not 'need' to travel, not because it claims to be a traveler in the heart, but rather because its truth is attainable elsewhere. As early as when Tagore is able to write in 'Stray Birds', 'the world puts off its mask of vastness to its lover. It becomes small as one song, as one kiss of the eternal'. [23]

In other words, travelling within India as he did extensively, coming in contact with Indian villagers and the songs of the Bengali Baul sect, listening to stories about Indian medieval poet saints and watching folk operatic performances would seem to equate if not surpass in size and effect, the travels abroad. His early trip with his father in the Himalayas, and later independent local travelling had the stronger influence on the psyche and caliber of the young Tagore if only because we are all immensely influenced by what experiences, people and impressions we are subjected to early in life.

For the artistic mind or temperament, thinking about Creation poses an almost existential dilemma which enriches – as it aggravates – the artistic moment or the poetic condition. This is only because it creates a series of questions formulated by our deepest psyche for which there may be no answers, or they may be answers which we reject and try not to reach! This is the high moment of art when the artist is either unable or unwilling to continue to subdue or ignore the probable answers concerning Truth, Love, Reality, Freedom, The Other, Death etc.

The artist is an initiator of a vision, an expounder of a different dream. He or she may spend their entire life trying either to live or express that indescribable dream. I may venture to claim – boldly – that 'all' art is an attempt against futility, and for meaning – even when it disguises its 'quest' by claiming that it has no 'mission' and is not in need for meaning but rather for its violent abolition. The artistic or poetic product is itself testimony to this urge, to the artist's struggle for something beyond our surrounding reality. It reflects our anger and bewilderment in its voracious search for the beautiful, anger at the possible 'loss or absence' of meaning in our existence which is in dire need of reparations by way of 'creating' a substitute reality 'with a meaning' in the world of art.

Tagore tried to tell us:

'God's silence ripens
man's thoughts into speech'. [24]

In many of his works, he beautifully gives vent to this dilemma. In 'Stray Birds' he writes:

'What language is thine, O sea?
The language of eternal question .
What language is thy answer, O sky?
The language of eternal silence'. [25]

He repeats in 'Shesh Lekha':

'The first day's sun
asked
the world's first emergence:
Who are you?
There was no answer.

Years Passed.
The last day's sun
asked a final question
near the shores of the western sea
amidst the silence of dusk:
Who are you?
There was no answer.' [26]

From the early works to the last works, I believe Tagore remained possessed by the same ideals, dreams, unquenched thirst, the same longing for a 'salvation', the same knowledge that love and art, both of which demanded perfection as well as completeness – paradoxically for him – proved the most evasive bearing their 'completeness' in a perfect ending of incompletion. This is the eternal journey which strangely seems like incarnation theory. It prompted the poet, dramatist and artist that Tagore was to profess in 'Stray Birds': 'I shall die again and again to know that life is inexhaustible'. [27] And also in the same book:' Truth seems to come with its final word, and the final word gives birth to its next'. [28] As such, Tagore himself was what he called the Eternal Traveler. In other words, the eternal seeker for whom love, perhaps more than art, remained more beyond his control.

To 'die into the deathless'[29] resonates with a mystical philosophical wish. It is also a poet's wish that sees death as a solution to pain. It is life's misfortunes which make it, for the unlucky few, desirable.

The lines of 'Shesh Lekha' are characterized by two features: an apparent serenity and a remorse-filled pathos where the head and the heart are turned away from the present moment toward the past. Tagore assumes a Kierkgaardian tone when he writes: 'If I can once again…'[30], 'escaped dreams of the past',[31] and 'Recalling pleasant memories.'[32] In 'Shesh Lekha' he writes of 'the loving message of a distant land'[33], and adds : 'with the love of a distant land/she who has waited for long/will sustain her nearness forever/her whispers resound in my ears'.[34]

As if all time were not a continuum but one moment! These lines are tantamount to a proclamation of the death of time, the negation of place. We invent time and live with the invention and its outcomes all our lives. The approach of death opens our eyes and souls before an eternal silence premeditated in much of Tagore's early poetry. In both *Gitanjali* and *Stray Birds* there are allusions to this state. In the latter he writes: 'The small truth has words that are clear, the great truth has great silence.'[35]

He wrestles with this fact -- the victory of silence, which has the final word – throwing himself on a life's quest for love, 'other loves', for other mediums of expression, hoping to quench the unquenched thirst of his soul through travel but he knew his deliverance was elsewhere. In 'Stray Birds' he was clear in his indecision: 'That which oppresses me, is it my soul trying to come out in the open, or the soul of the world knocking at my heart for its entrance?'[36]

Perhaps he knew the answer, inwardly. He wrote in 'Stray Birds', 'My heart has spread its idle sails for the shadowy island of Anywhere'.[37] The starting point is his heart. This is how he formulated the problem and it seems like a statement of priorities.

Death is passionately called upon in the early works, but serenely accepted and welcomed in a primordial way in the last poems, in 'Shesh Lekha'. It is no longer a mere poetic theme for the young romantic whose idealism suffers psychic-quakes, but an impending reality standing at the door. Now he summons all his fortitude and forbearance waiting for the release from the loneliness so interwoven with his soul.

'No alarm' is the most accurate way of describing Tagore's attitude toward death in 'Shesh Lekha'. Maybe we can go further and venture into saying that by now the weary soul of the traveler had had enough of life as a ' tapasya of pain till death'(38) and was beyond verbal flirting with it, committing a sin against life by 'recommending' death as if it were an eleventh commandment.

'to gain the terrible value of truth

settle all debts by death'.[39]

And also:

'When I set sail in my raft

towards that silent festival of the end'.[40]

It is thus a solution, a means to a higher state where 'Truth' resides and it is a 'festival'. It is even addressed as 'king' in his poem 'Borderland'. I do not see any difference between the line in *Gitanjali* for example which stated: 'In this play house of infinite forms I have had my play and here have I caught sight of him that is formless', [41] and the lines in 'Shesh Lekha', near the end of Tagore's life: 'The primeval dust you belong to with supreme indifference claimed you in its soundless chariot racing into the unknown'. [42]

Not surprisingly too, death is the hero and solution in various of Tagore's short stories such as 'The Ghat's Story', 'The Wedding Garland', 'The Living and the Dead', and 'Punishment', whereas 'The Devotee' contains ideas resonating with layers of meanings from the writer's 'The Religion of Man'.

'The Religion of Man' and the books cited in this paper represent a continuum of ideas in Tagore's weltanschauung.

He refers to his family's development of a monotheistic religion based upon the philosophy of the rishi authors of The Upanishads, in a chapter titled 'The Vision' in the aforementioned book. In such a philosophy, the divinity of man and the humanity of God are part and parcel of an understanding of the Supreme as synonymous with truth and love. In that realm both 'the subject and object are perfectly reconciled'.[43]

He expressed this in his poetry in *Gitanjali* when he wrote: 'Thou art the sky and thou art the nest as well'.[44]

This is very close to the poetry of the martyred Persian Moslem Sufi saint, Al Hussein Ibn Mansour Al Hallaj, who lived during Abbasid rule and was crucified and beheaded in 921 CE.

Al Hallaj was a poet and searcher for truth who travelled to many parts of the world including - it is said - India and China. He is reputed to have written a description of the Divinity that led to his crucifixion after a thousand flogs – but in reality the reasons were political since he wanted to bring about reform. It was maintained that he believed an individual can engage himself with goodness and celibacy to a point of purification where his soul becomes one with The Holy Ghost, in which state he can perform all that the Divine is able to perform.

Tagore's idea of the fusion of subject and object bears a strong resemblance to many ideas in Al Hallaj's books and poetry in which the latter wrote about the Divinity: 'I am who I love, and who

I love is myself/we are two souls incarnated in one body/if you see me you see Him/and if you see Him you see us'.[45]

Likewise, for Kabir, Nanak and others by whom Tagore was influenced, man's body was the temple of God.

Tagore described his moments of sudden vision, as a result of coming in contact with nature. It was like an experience of sudden light, as if some fog had suddenly lifted, an exaltation. These would seem to have been moments associated with poetic outbursts and artistic expression whatever the form.

A vision is like love. It is considered the result of an agitated state which leads to a certain 'discovery' that combines fervor before/and or during its experience, ending in a sense of relief and joy. To 'see' as if for the first time is a revelation that seems to bestow new meaning onto an entire existence or relationship. Spiritually and artistically, it involves a sudden surge of the beautiful and sublime. It is the delight of a momentary 'absolute' freedom where all the senses, the spirit, the imagination and mind are invoked in the tension, the energy of the liberating state. It is a moment of light or 'enlightenment'. An artistic moment, in short, is necessarily a 'spiritual' one which may come as near as can be to a trance since, in that rare and fleeting moment of creation, the experience is that everything halts, and the fragments and parts seem to revel in a unity of vision – a peace – which Tagore simply described as 'a luminous unity of truth'.[46] It is a moment where we 'see' and 'accept' the truth of things, detaching ourselves from pretension. It is a strangely sorrowful-joyous moment, a moment of compassion as well towards the world and the self.

Tragedy for Tagore was the consequence of truth being a fragment. Remarkably, while his poet's religion was the result of vision more than knowledge or dogma, his spirituality cultivated reason and meditation.

What Tagore's soul became intensely conscious of through 'the illumination of joy', [47] he recognized as contact with the infinite .The creative moment can be a medium of that infinite. The pulse of that infinite is measured by the truth of love and goodness within us. It is to be heard in nature as well since the song (reverie or the music he composed) feels that infinite in the earth, feels the picture (painting?) in the earth and feels the poem in both the earth and the air to paraphrase Tagore. This is his formula in 'Stray Birds' as if preaching a religion of nature. The infinite is with us in all stages of our life – as we are reminded in both 'The Religion of Man' and 'Thought Relics'– since just as our soul aims to realize our individuality, so likewise does it have as an inherent desire to surrender itself in love, obviously to the Great Soul.

In both 'Stray Birds' and 'Gitanjali' we find notions of the ideas dominating 'The Religion of Man'. Tagore allows himself anger in 'Stray Birds' when he writes:

'Your idol is shattered in the dust to prove that God's dust is greater than your idol'. [48]

In 'Gitanjali' he writes something that appears as if it were in the same book.

> 'Leave this chanting and singing and telling of beads! Whom dost thou worship in this lonely dark corner of a temple with doors all shut? Open thine eyes and see thy God is not before thee.
>
> He is there where the tiller is tilling the hard ground and where the path maker is breaking stones....'[49]

In 'The Religion of Man' he refers to the song of a beggar belonging to the Baul sect by which he was influenced:

> 'It spoke of an intense yearning of the heart for the divine which is in Man and not in the temple, or scriptures, in images or symbols.........
>
> He does not follow any tradition of ceremony, but only believes in love'.[50]

This is very close to the ideas of the Arab Sufi poet, Mohei'i El Din Ibn Arabi, whose poetry defining love as opposed to all religious houses, ceremonials and temples also reminds us of some of the famous Baul songs which impart the ideas of the sect about pilgrimage.

> 'I would not go, my heart, to Mecca or Medina,
>
> For behold, I ever abide by the side of my Friend.
>
> Mad would I become, had I dwelt afar, not knowing Him.
>
> There's no worship in Mosque or Temple or special holy day.
>
> At every step I have my Mecca and Kashi; sacred is every moment'.[51]

And in 'Shesh Lekha' Tagore writes:

> 'But he finds truth
>
> bathed in the inner light within himself.'[52]

And again in 'The Religion of Man':

> 'The truth that is infinite.................
>
> This truth of realization is not in space, it can only be realized in one's own inner spirit'.[53]

And more precisely, in 'Shesh Lekha' we read:

> 'He who knows this world exists
>
> realizes himself
>
> as a witness of all that exists
>
> his truth is in the truth of that ultimate self :
>
> this I know for sure'. [54]

This bears very close resemblance to Al Hallaj's outcry: 'I am the Creative Truth/ Right', Truth or Right being one of the names of God in Islam. When writing about truth, Tagore's tone oscillates between sadness and reasoning, or it is the artist's inevitably sad reasoning, sadness being a consequence of a sense of incapability or impotence in a continued existence which harnesses our impatience, subjugating it as well as our quests and needs to delay, to postponement and disappointment.

Tagore's universalism was mystical and idealistic .The more I read him, the more I saw such universalism had its roots in the ideas he formed, the 'solutions' and theories he conceptualized prior to his travel, the sensations and tastes he had acquired long before he went abroad. It seems to me to have been more the product of the ideas circulating in the home of his father, his own readings and contemplation and also the result of his cultural background that was immensely rich with Hindu mythology (despite those aspects of Hinduism he rejected), the Upanishadic and Vedantic literature and eastern wisdom.

The Bauls welcomed in their ranks people, devotees from the lowest strata of all Moslem and Hindu communities. Theirs was a sect which did not believe in aloofness, contempt or renunciation of any person, and they were dependent on Yoga which Tagore makes references to in 'The Religion of Man'. In his educational project, Visva-Bharati, he aimed at the fostering of an alternative, 'Indian' education which hoped to develop what he called a 'sensitiveness of soul' [55] in his students as regards their relationship with their human surroundings. It was to be an early encouragement (and utilization) of the freshness of the child's sentiments and imagination (maybe close to what he experienced in his endless days of solitude as an orphan forced into contemplation of the wild scenery around him in his father's house). What would help would be literature, festive ceremonies etc.? This emphasis on a schooling which enables us to 'come to the nearer presence of the world through the soul'[56] was 'Indian' enough, 'Buddhist' enough, 'Sufi ' enough not to be branded as an exclusive following of any of these realms alone. It was an amalgamation of principles drawing on all that fed his mind and senses thereby forming the tenets of his ideas and his spirit – a spirit which had been embracing the universal long before he was to start formulating the idea of internationalism. Indeed, critics have written that he wrote his most innovative poetry as a result of coming in contact with the folk music of Bengal. Tagore himself told H.G. Wells, when he met him in Geneva in 1930, that he had composed (by that time) more than 300 pieces of music. His imagination and his inspiration had already been pre-shaped by what was profoundly Indian and subsequently, English. Nearly all the examples Tagore draws in 'The Religion of Man' are from his culture, Sufi, Baul, Vedas, Upanishads, while also drawing on the ideas of Zarathustra and Lao-tze. It was a universalism that would not be against the necessity of maintaining one's cultural identity. He wanted his country, as it were, to assimilate the west, and a great deal of western civilization but not at the expense of national dignity, not to the point of becoming an object of western mockery.

As early back as 1892, Tagore's universalism was revealing itself as a balanced cultural attitude when he had advocated education 'in the mother tongue, rather than in English'. [57] Such pride was also manifest when later in 1908, while presiding over the Bengal Provincial Congress Session in East Bengal, he delivered his speech in Bengali, thus breaking the tradition of delivering speeches in English (in those sessions).

This is not to say that the effects of travel were negligible. Travel contributed to a broadening of outlook as it does with all of us, but the basic ideas were there, the basic 'feelings', attitudes had already been formed. Travel may have contributed to a refining of some Tagorean ideas. But I firmly believe that its effect on the realm of Tagore, the human being, was limited and marginal. Travel perhaps helped with some experimental works such as 'Africa and Camalia', but Tagore's poetics were certainly rising to what we may assuredly claim as their 'finesse' by the time 'Gitanjali' (the book that was to earn him the Nobel Prize in Literature) was published in 1913. If anything, travel sometimes would reaffirm to Tagore some of 'his' own ideas and positions. He was dubious of the rising heated nationalism both in the world and in his own country and was opposed to imperialism. The rise of fierce nationalism in Japan (which he visited just as he visited the United States of America and other countries) would increase his fears for a world racing headlong toward disaster, a world which was in urgent need of his 'Message of the Forest'- the title of his famous lecture. But it was already too late for the world to listen, with the characteristic of such a world (perhaps to this day) being the exercise of pressure on the individual to take sides.

It has been said of Tagore that in the colonized India of the caste system, the anti- women society, the society of untouchables of his time, he sided with the weaker groups. In other words, he sided and sympathized with women (even though from a patriarchal standpoint), sided with the colonized (despite his famed skepticism of nationalism and his rejection of economic non-cooperation against the British) as when he asked to be relieved of his knighthood following the massacre of Jallianwala Bagh (13 April 1919) and sided with untouchables. He both lectured extensively and wrote drama on India's problem of 'untouchability'. Critics have noted how many of the Tagorean heroes in his works act neither for their own class nor country but for others.This is the artist's unavoidable universalism, partly an ethical choice, partly a philosophical and aesthetic one.

In his poem, 'The Earth' (Basundhara), composed in November 1893, we see the germinating seeds of Tagore's ideal dream that would later express itself philosophically in 'The Religion of Man' (The Hibbert Lectures in 1930). In 'The Earth' he writes:

> 'I long in my heart
>
> To be kin to all human minds
>
> In every land.' [58]

As a poet myself, I fail to understand how any 'genuine' artist can be parochial, narrowly nationalistic or even fascist.

In his universalism, Tagore saw that irrespective of man's religion, we are essentially tied together (according to his ideal of human perfection) by a bond of unity between all individuals that would culminate in the Supreme or Eternal Person representing the eternal in man. If anything, this is idealism at its peak.

The arts and all literature, for Tagore, are the consequence of our impulse to express such a Universal Man, since they are inspired by what he saw as the Universal Mind. The individual striving for excellence must therefore - according to him - seek an ideal with some value of truth, a truth which ideally belongs to all men, as he put it. For him, the history of the growth of freedom was no other than the history of the perfection of our human contact. Is love not another name for that?

To Tagore, as an idealist and artist, the idea of the inevitability of human strife was anathema. At the age of thirty, he would comment on a verse play he had written when he was eighteen years old. Its title was 'The Broken Heart'. He stated, in what appears to have been self-criticism, that the condition of his immature mind was fostered by the precept and example of the time, noting that he was not sure that the effects of these were not continuing on to 'the present day'. The emphasis, for me in reading this, would seem to be on time, not place. He confesses in 'The Religion of Man', that he felt he had found his religion at last, 'in which the infinite became defined in humanity'.[59] This is a prophet's and ascetic's contention. It promises something deeply needed: that we immediately touch the infinite around us and within us only when we perceive the pure truth of love and goodness, not through the explanations of theologians'.[60]

The reciprocity of the relationship between the finite and infinite, elaborated in 'The Religion of Man', where the infinite is described as coming close to the poet/writer in the 'need' of his love and co-operation suggests a degree of mutuality, which comes close to the Hindu belief that everyone Can become a God. I say this despite his belief in monotheism. Tagore was even expressing a degree of responsibility toward the Creator, 'For the creation in me is his as well as mine'.[61]

In 'Gitanjali', passionate supplications are not wanting.

> 'O thou Lord of all heavens, where would be thy love if I were not?'[62], and in 'The Religion of Man' he could not be more explicit: 'The infinite One is infinite love'.[63]

Again, this is a Christian and Sufi Moslem conviction. What's more is the very curious line by Tagore in his 'Stray Birds' that 'God finds himself by creating.'[64] It is a Quranic statement. In the Quran, God expresses that He created humans and Jin to worship Him.

Tagore's universalism is thus a reflection of his cultural mélange, the prototype of a calling or message, even an artist's 'Appeal' to humanity, a humanity in which he saw men as cruel and Man as kind.

Another of Tagore's ideas worthy of examination in the books cited here is his idea of freedom.

The legendary cry of 'Gitanjali' that 'Freedom is all I want!' [65] may well be the lamentation of the tragic heroes of all time, since Greek mythology to the present, before and after Tagore, whatever the culture.

The absolute for the artist and philosopher is an attractive end in itself. Yet, Tagore was also specific. Freedom within the self must shine first before any outside light can touch the dimness of one's soul. He explains in the chapter titled 'The Artist' in 'The Religion of Man', that it is man's dreams and creations which comprise his freedom, according to the nature of such dreams and the quality he gives to them, to his creations and his choice of existence. There is a world of difference between dreams of accumulation of wealth or power and dreams of love. It is exactly in the realm of man's freedom (depending on the quality of the dream) in which man realizes his divine dignity and human truth, and it is there that he has his opportunity to rise and make his civilization. Of course, Tagore is much simpler in 'Thought Relics' where he repeats what many of us know – that it is only love which is true freedom. Accordingly, we may paraphrase his cry: 'Freedom is all I want' to become: 'Love is all I want'. The freedom that love grants us is far from the negation of restraint. Indeed it may be the exact opposite, where the acceptance of 'bondage' may be testimony of the measure of the 'truth' of love. He has written in *Gitanjali*: 'I feel the embrace of freedom/ in a thousand bonds of delight'.[66]

I share this belief with Tagore.

One definition of freedom may be the victory of our will in the face of all things that lie outside our control. Such a dream of freedom is quintessentially the artist's. The freedom of the freedom fighter, the philosopher and the stoical mystic may be different even though all of them may want to demolish the world as it is and create a new one in its place.

In 'The Religion of Man' there is a very direct simple question: 'What is Art?' to which Tagore presents an answer no less simple and direct. He says: 'It is the response of man's creative soul to the call of the Real'. [67]

The experience of the Real does not need travel. When we try to transcend our finitude and our incapacities with love or art, we are aiming for freedom, indeed for the 'Real', so that we may die but not perish – as Tagore would seem to draw on the ideas of Lao-tze in 'The Religion of Man'.

Art and love represent the highest forms of the resistance of our being against the 'thrown-ness into being' with which we are faced from birth until death. They reflect the struggle of the human will against cruelty, loneliness, death, separation, and ugliness of reality. Art is an attempt at non-perishing, a dream of reaching the Real, reaching some absolute assuring truth. – a dream of embodiment of that 'Real'. What is inherent in it is a hope for eternity, for happiness everlasting. Travel, however, is but one component in the equation of living with forbearance.

Tagore clarifies in 'The Religion of Man': 'In the world of art, our consciousness being freed from the tangle of self interest, we gain an unobstructed vision of unity, the incarnation of the real, which is a joy forever. As in the world of art, so in the spiritual world, our soul waits for its freedom from the ego to reach that disinterested joy which is the source and goal of creation.' [68]

Where art and travel meet is the conjunction of Tagore's vision of some paradise which he found in nature's sunlight, its repose of a winter morning, flowing streams, the beauty of the human face and human love. The spirit of such a paradise was awake for those whose inner ears were willing to listen. If travel was an answer, the sadness of his heart and poetry would not have persisted. In reality, his dream was poisoned by his memories which - at one time - were dreams themselves. This sadness that he carried in his heart was expressed in all his poetry and in many of his short stories, as we have shown. Indeed, it seeped through all his art, but particularly in his poetry we feel the anguish of unfulfilled love. In his poem, 'A Posy', from the book, 'Lover's Gift', the ardent lover cannot hide behind the poetic persona:

'Can it be true that I forget you? ....

You have moved from my world, to take seat at the root of my life, and therefore is this forgetting - remembrance lost in its own depth'.[69] And, again, in the same page: 'Dying, you have left behind you the great sadness of the Eternal in my life'. [70]

How can we allow ourselves to believe that it was not life's torrential experiences with death, particularly Tagore's sister-in-law who 'chose' to commit suicide, that imprinted his soul with such great sadness, forever? How can we not accept the poet's own confession, his exclamation and anguished lament, that someone very precious to his heart (in effect, the beloved herself) by moving from his world has consequentially become his idol, the possessor of his soul ?. She/the beloved has taken her seat at the root of his life. And, is anything deeper than the 'root'?

In reality, I feel that Tagore was touching the tangent of his own inextricable pain when he was reminiscing in his poetry, as all idealists and failed dreamers do. The façade of art offers freedom of expression. It would seem that the great Gurudev Tagore knew inwardly the cause of his ongoing ailment. His love was not only unfulfilled, it was 'martyred'. The 'iconizing' of a lost sweetheart is doubled by the ferocity of losing her to death. After all that had been written or all that remained unwritten and unsaid, the mystery is revealed in all simplicity, as a genuine Truth/fact overpowering and overshadowing all other truths, realities, achievements and failures. Since truth is not afraid to be small, it is also not afraid to be 'defeated', as was his love.

In 'Stray Birds' he writes: 'One word keep for me in thy silence. O world, when I am dead, I have loved'. [71]

It is a line that reminds us of his initial tragedy. It may have been summed up by his early submissive poetic explanation in *Gitanjali*: 'only there is the agony of wishing in my heart'. [72]

If this line explains – among other reasons – the motivation for travel, if it betrays the artist's fatigue with life's misfortunes, it is also a confession, a summation of all that has not been told or done and could only be partly unveiled by the poetic. It sheds light on the poet's assessment of what his life's contributions and glories amounted to, even suggesting a negation of the importance of anything beside the testimonial it provides: 'One word. ... I have loved'. It is an utterance, not only of un-fulfillment but of all human un-fulfillment, on our behalf.

As such, it is a line which assures us that the ethos of such an artist and his psyche could not have been drastically influenced by travel any more than his ever-growing pain could have been alleviated by it.

***Egyptian Published Poet, Writer and Columnist**

Notes

1. Sanjoy Kumar Mallik. 'The Painter', in Singh, Udaya Narayana and Suri, Navdeep (eds.). *Rabindranath Tagore: A Commemorative Volume,* First Edition 2011. Published by (New Delhi: Public Diplomacy Division, Ministry of External Affairs, 2011), pp. 189, 190.
2. Ibid, p.187
3. Ibid, same page.
4. Ibid, chapter titled: 'Influence on Other Painters'. p.200
5. Dutta, Krishna and Robinson, Andrews. *Rabindranath Tagore: The Myriad-Minded Man*, First published (Great Britain: Bloomsbury Publishers Ltd, 1995). First Edition, January 1996. p. 74.
6. Kukar Sudhir. 'On Rabindranath Tagore', in Singh, Udaya Narayana and Suri, Navdeep (eds.). *Rabindranath Tagore: A Commemorative Volume,* op.cit. p.223
7. Chaudhuri, Sukanta. *Rabindranath Tagore: Selected Short Stories*, Introduction and Notes by Topobrata Ghosh, (New Delhi: Oxford University Press, 2000), p. 29.
8. Ibid, p. 17
9. Ibid, same page
10. Ibid, introduction, p. 20
11. Op. cit. Dutta, Krishna and Robinson Andrews. Rabindranath Tagore, p. 192.
12. Ibid, p. 89
13. Op. cit. Chaudhuri, Sukanta (ed.). Rabindranath Tagore: Selected Short Stories, p. 7.
14. Ibid, p. 28
15. Ibid, p. 7
16. Tagore, Rabindranath. *Gitanjali - Song Offerings,* Translation by the author from the original Bengali with introduction by W. B. Yeats (Chennai: Macmillan Deluxe Edition 2000), p. 67.
17. Op. cit. p. 16.

18. Op. cit. p. 84
19. Op. cit. Tagore, Rabindranath. *Gitanjali - Song Offerings*, p. 100.
20. Tagore, Rabindranath. *Shesh Lekha: The Last Poems of Rabindranath Tagore*, trans. Pritish Nandy from the Bengali (New Delhi: Rupa & Co., 2002), p. 36.
21. Op. cit, p. 95.
22. Tagore, Rabindranath. *Stray Birds*, (Madras: I.S.S.D Press, 1917), p. 29, stanza 111.
23. Ibid, p. 1, stanza 3.
24. Ibid, p. 79, stanza 305.
25. Ibid, p. 4, stanza 12.
26. Op. cit. Tagore, Rabindranath. *Shesh Lekha: The Last Poems of Tagore*, trans. Pritish Nandy from the Bengali (New Delhi: Rupa & Co., 2002), p. 36.
27. Op. cit. Tagore, Rabindranath. *Stray Birds*, (Madras: I.S.S.D Press, 1917), p. 73, stanza 282.
28. Ibid, p. 292, stanza 295.
29. Op. cit. Tagore, Rabindranath. *Gitanjali - Song Offerings*, p. 100.
30. Op. cit. Tagore, Rabindranath. *Shesh Lekha: The Last Poems of Rabindranath Tagore,* p. 10.
31. Ibid, p. 10.
32. Ibid, p. 10.
33. Ibid, p. 11.
34. Ibid, same page.
35. Op. cit. Tagore, Rabindranath. *Stray Birds,* P. 46, stanza 176.
36. Ibid, p. 44, stanza 168.
37. Ibid, p. 57, stanza 218.
38. Op. cit. Tagore, Rabindranath. *Shesh Lekha: The Last Poems of Rabindranath Tagore,* p. 28.
39. Ibid, same page.
40. Ibid, p. 26.
41. Op. cit. Tagore, Rabindranath. *Gitanjali - Song Offerings*, p. 96.
42. Op. cit. p. 23.
43. Tagore, Rabindranath. *The Religion Of Man: Being The Hibbert Lectures For 1930,* (London: Bradford and Dickens, 1931), The chapter titled 'The Vision', p. 93.
44. Op. cit., Tagore, Rabindranath. *Gitanjali - Song Offerings*', p. 68.

45. Mitz, Adam. *Islamic Civilization in the Fourth Hijri Century or The Age Of The Renaissance Of Islam*, (Arabized by Mohammad Abdel Hadi Abu Reeda). Part two, chapter 19 titled 'Religion'. (Cairo: Al Khangi and Beirut: Dar Al-Kitab Al-Arabi, 1967), p. 63.

46. Op. cit. Tagore, Rabindranath. *The Religion of Man: Being the Hibbert Lectures For 1930,* (London: Bradford and Dickens, 1931), The chapter titled 'The Vision', p. 96.

47. Ibid, p. 107.

48. Op. cit. Tagore, Rabindranath. *Stray Birds*, p. 13, stanza 51.

49. Op. cit. Tagore, Rabindranath. *Gitanjali - Song Offerings*, p. 11.

50. Op. cit. Tagore, Rabindranath. *The Religion of Man: Being The Hibbert Lectures for 1930,* The chapter titled 'The Man Of My Heart', pp. 110-111.

51. Ibid, p. 215.

52. Op. cit. Tagore, Rabindranath. *Shesh Lekha: The Last Poems of Rabindranath Tagore*, p. 36.

53. Op. cit., the chapter titled: 'Spiritual Union', p. 66.

54. Op. cit., p. 5.

55. Op. cit. Tagore, Rabindranath. *The Religion of Man: Being The Hibbert Lectures for 1930,* The chapter titled 'The Teacher', p. 180.

56. Ibid, same chapter and page.

57. Op. cit. Singh, Udaya Narayana and Suri, Navdeep (eds.). Rabindranath Tagore: A Commemorative Volume, p. 239.

58. Op. cit. Chaudhuri, Sukanta (ed). Rabindranath Tagore: Selected Short Stories, p. 9.

59. Op. cit. Tagore, Rabindranath. The Religion of Man: Being The Hibbert Lectures for 1930, The Chapter titled 'The Vision', p. 96.

60. Ibid, same chapter p. 107.

61. Ibid, same chapter, p. 96.

62. Op. cit. Tagore, Rabindranath. Gitanjali - Song Offerings, p. 57.

63. Op. cit., p. 66.

64. Op. cit. Tagore, Rabindranath. Stray Birds, p. 12, stanza 46.

65. Op. cit. Tagore, Rabindranath. Gitanjali - Song Offerings, p. 26.

66. Ibid, p. 74.

67. Op. cit. Tagore, Rabindranath. The Religion of Man: Being The Hibbert Lectures for 1930, The chapter titled 'The Artist', p. 139.

68. Ibid, the chapter titled 'Spiritual Freedom'. pp. 183,184.

69. Tagore, Rabindranath. A Lover's Gift, (New Delhi: Rupa & Co, 2002), The part titled 'A Posy', p. 34.

70. Ibid, same page.

71. Op. cit. Stray Birds, p. 72, stanza 278.

72. Op. cit. Tagore, Rabindranath. Gitanjali - Song Offerings, p. 13.

References

Chaudhuri, Sukanta (ed). *Rabindranath Tagore: Selected Short Stories*, Introduction and Notes by Topobrata Ghosh, (New Delhi: Oxford University Press, 2000).

Dutta, Krishna and Robinson, Andrew. *Rabindranath Tagore: The Myriad-Minded Man*, (Great Britain: Bloomsbury Publishing Ltd. 1995).

Singh, Udaya Narayana and Suri, Navdeep. *Rabindranath Tagore: A Commemorative Volume*, (New Delhi: Public Diplomacy Division, Ministry of External Affairs, 2011),

Tagore, Rabindranath. *Shesh Lekha: The last Poems of Rabindranath Tagore*, trans. from the Bengali by Pritish Nandy, (New Delhi: Rupa and Co, 2002).

Tagore, Rabindranath. *Gitanjali- Song Offerings*, trans. by the author from the original Bengali with introduction by W. B. Yeats. (Chennai: Macmillan Deluxe Edition, 2000).

Tagore, Rabindranath. *The Religion of Man - Being the Hibbert Lectures for 1930*. (London: Bradford and Dickens, 1931).

Tagore Rabindranath. *Stray Birds* (Madras: The I.S.S.D Press, 1917).

Tagore, Rabindranath. *A Lover's gift*, (New Delhi: Rupa.co. 2002).

Mitz, Adam. *Islamic Civilization in the Fourth Hijri Century or The Age Of The Renaissance Of Islam*, (Arabized by Mohammad Abdel Hadi Abu Reeda). Part Two. (Cairo: Al Khangi library and Beirut: Dar Al- Kitab Al- Arabi).

The Echo of Sweet Songs of Bengal's Parrot in Iran Impact of Rabindranath Tagore's Presence in Iran

Ehsan Abbaslou

Introduction:

Rabindranath Tagore (7 May 1861 – 7 August 1941) was an Indian poet, philosopher, musician and playwright and the first Nobel Prize winner in Asia.

Rabindranath was the youngest son of Debendranath Tagore. He was educated at home and at the age of seventeen was sent to England for formal schooling, but he did not succeed to finish his studies there. His family was noble and religious and his father knew Hafiz's sonnets by heart and loved Persian poetry as did his own forefathers.

Tagore's success in world literature started with his poems which made him a universal figure who intended to carry the message of love, peace and friendship all over the world.

Tagore has enjoyed a dual position in Iran. Some respect him for his ideas and beliefs whereas some view him doubtfully because of his relationship with Reza Shah who has been known as a dictator both by intellectuals of his own era, especially the religious ones, and also by present regime of Iran. As a consequence of this latter view, the poet has not been welcomed as warmly as he should be and deserved. But, at the same time, owing to his global literary position and due to the soul and life of his verse that resembles and recalls old Iranian counterparts as Hafiz and Rumi, he and his place could not be ignored.

The Iranian people, on the other hand, are deeply fond of mysticism since it has been mixed with their flesh and bone. Nowadays, a mechanical life has indirectly led people towards mysticism to find peace and comfort in it. Mystic literature helps its reader to escape from the daily tension and stress of life. Tagore's poetry is of this kind to Iranian people. The soothing, consoling poetry of Tagore that talks of kindness and friendship, peace and beauty, love and emotion finds its way to the heart of the Iranian people, these days more than ever. This naturally will leave an impression, an unforgettable one. But it seems that in spite of his presence in Iran, Tagore has not been as leading a figure for high-ranking literary figures as he has been to the ordinary lover of literature.

Whether deep and grave or partial that his influence may be, it is necessary to state precisely the likely and the certain impact he has had in Iranian history of literature.

Tagore in Iran

Invited by the Iranian government, Tagore travelled to Iran twice. Once in the spring of 1932 when he arrived in Bushehr province through Jask Harbor accompanied by Jalaloddin Keyhan, who was the head of the Iranian consulate in Bombay in that time. Two other figures who accompanied him were his own daughter-in-law and also Dinshah Irani, one of the Zoroastrian leaders.

In his trip to Iran, Tagore defended the national identity and independence of the country while also talking of the equality of nations. Perhaps that is why the literary view towards him faded into some political considerations in some periods.

Abolhassan Forooghi , the then Head of the Higher Center for Teachers, and Malekol Shoaraye Bahar (Poet Laureate), the representative of Literary Society of Iran, went to welcome him.

Kalidas Nag, a close friend of the poet, writes: 'In 1932 the poet received a personal invitation from the builder of Modern Iran, Reza Shah Pahlavi. Tagore, then in his seventy first year, flew to Tehran and to Baghdad and amidst the glorious roses of Iran, his birthday was celebrated with banquets and poetic recitals, evoking truly Iranian grace and glamour.'

He was warmly welcomed by Iranians and so every where he arrived, a huge number of literary figures gathered around him to get familiar with his point of view regarding philosophy, mysticism and literature. The trip was a good propaganda for the Pahlavi regime that was to introduce itself as a pioneer in middle-eastern progress.

Tagore accepted the invitation because:

1. Farsi was a part of literary heritage of Bengal and the number of Persian words found in Bengali language was estimated to be more than three thousand. Mohammad Aleem in 'Tagore: His historic visit to Iran' writes: The language Persian and its literature were never alien to Indian people ever. People who came from the Urdu background especially understand it very well. I vividly remember my Maktab days when our education was first started with the Persian alphabet and after getting some knowledge of reading and writing of our mother tongue, Urdu; soon we were given a book of Persian grammar and then the great books of Sheikh Sadie, Gulistan and Bostan. We could read those books in beautiful Urdu translation along with its Persian texts.' Language is the gate to friendship and understanding. Language unearths similarities rather than differences. Tagore was familiar with Farsi and it could be a good and sufficient motivation to accept the invitation for coming to Iran.

2. His father was familiar with Persian literature and this could make him interested and flame his inner enthusiasm to come and see the land of Persia. Aleem's quotation from Tagore is interesting here: 'My father was a great scholar. He was intoxicated with Hazfiz's verses. When I was a boy I often used to listen to his recitation of those poems, and he translated them to me with fervor of

enjoyment that touched my heart.'

At the same time, Ram Mohun Roy (1772-1833) who was a source of inspiration for Tagore in philosophy and was considered as his spiritual father, knew Persia and this may have fueled his interest more than ever.

3. Hablolmatin, an Iranian newspaper, managed and edited by Seyyed Jallaloddin Kashani, was published in Calcutta. It began publication in the last years of Ghajari king, Naseroddin Shah and was mainly political in its approach. Fourteen years later, the newspaper was published and circulated in Tehran as well. In Tehran it was directed by Jallaloddin's brother, Seyyed Hassan. The newspaper could have laid the foundation for people getting to know more about Iran and Persian literature and language in India especially among those who had already had a previous acquaintance with this land. Everything about Iran could have made Tagore eager to set off and visit the land in person.

Tagore has noted somewhere that: 'I had my knowledge of Persia even before my coming. I had read something about your history and geography, and formed some idea of your people and your country. My imagination was aroused through your great poets whose call had come to me even though I had no direct access to them. I used to dream of a Persia where Bulbuls made love to the roses, where in dreamland gardens poets sat round their wine cups and invoked visions of ineffable meaning.'

4. His enthusiasm and intention in defending oppressed countries and supporting their struggle to achieve independence can justify his acceptance of Reza Shah's invitation. In 1933, delivering his welcome speech to Mr. Poordavood for his presence in India to teach Farsi, Tagore referred to the awakening of Asian countries directly and said,'... now that the awakening of Asia has started we succeed to discover old interests...'. It seems that he supported the revival of Ancient Persia away from Arabic Persia and it implies the fact that Tagore knew Iran from a historical point of view, not from a religious stand point.

Tagore also on his arrival to Iran tells us of a fact behind his motivation to come: 'The vision of Persia was invoked in my imagination by the voice of your own poets, who brought to my mind's sky breath of your spring breeze with the enchantment of its blossoming roses and nightingales' songs. My arrival in your land today is a continuation of the same enchantment and I am glad to mingle my voice with the rejoicing of life which has broken out in the air of your beautiful country fragrant with the perfume of orange blossoms.'

It was during the first trip that his 70th birthday party was held in Tehran. He wrote two poems within the occasion and the second tells us of this event and Iranian hospitality regarding his birthday:

IRAN, All the roses in thy garden
and all their lover birds
have acclaimed the birthday
of the poet of a far-away shore
and mingled their voices in pair of rejoicing.
Iran, thy brave sons have brought
their priceless gifts of friendship
on this birthday of the poet of a far-away shore
for they have known him in their hearts as their own.
Iran, crowned with a new glory
by the honour from thy hand
this birthday of the poet of a far-away shore
finds its fulfillment.
and in return I bind this wreath of my verse
on thy forehead, and I cry: Victory to Iran!

In Tehran, Tagore delivered two speeches, one in Masoodia Building and the other in Literary Society of Iran, in both of which he had noted 'I am Iranian and my ancestors emigrated to India from this land and I am happy for returning to my ancestors' homeland.'

Malakolshoara-ye Bahar composed a long ode for him in which he praised Tagore's position in poetry and called his pen 'God's gift' and addressed him as 'Bengal's Parrot'. The parrot is famous for being a sweet speaker both in Iranian verse and prose and has always been metaphorically taken as a confidante and companion to Man in Persian literature.

Tagore was in Iran for about forty days and visited some historical and ancient cities such as Isfahan, Shiraz. The poet asked Reza Shah to send some Professors to teach Farsi in India, as a result of which Mr. Ibrahim Poordavood was sent as the first formal Iranian professor to teach Persian language and literature.

Tagore was not as well-known at the time in Iran as he deserved to be, and he himself had reminded Mr. Poordavood of it. The reason was clear; because his works were not translated into Persian before and Mr. Poordavood writes: 'Since I heard this a few times from him I replied I will do the job in India by the help of one of the professors… Therefore, with Mr. Ziaoddin, who knew Farsi as well, translated one hundred stanzas of his poems from Bengali into Farsi.'

It was a beginning for the translation of most of Tagore's works into Persian.

Two years later, in 1913 (?), Tagore came to Iran for the second time to participate in the ceremony held in Mashhad for the inauguration of Toos Tomb or Firdousi's Tomb.

These two visits, though being short, formed an unforgettable event in the history of this land and made the Iranian people keep alive the memory of a man who was God-filled in his heart and a lover of mankind.

Impact of Tagore's Presence in Iran

Reviewing the great event of Tagore's presence in Iran, it will be easily seen that no direct impact has been recorded as the aftermath of his presence; since everyone here in Iran, due to his own cultural and national orientation, has made attempts to show the impact that Persian literary figures have left upon Tagore. No one has attempted, until now, to refer to the probable impact Tagore has had upon his Iranian readers.

On the other hand, the very close resemblances and similarities between Tagore's lines and those of Iranian poets have veiled the impact that he may have had in favor of the Iranian figures.

Paying no attention to the characteristics of his poetry, his words have been always, by mistake, compared to and measured with those of old Persian men of poetry. It seems that without considering old Persian poets, it would be impossible to approach Tagore solely. This has always placed him in the shadow of great Iranian poets.

What remains from his journeys to Iran has been mostly emotional. People got Tagore by heart and have kept him there eternally. He is read here not for the sake of an odd or novel style or for any given modern or post-modern attribute of his form. He is read for his spiritual themes, for his sincere tongue. It shows that Tagore's heart is emotionally attached to the hearts of people in Iran.

Mr Hossein Gholi Mostaan, in a magazine of the time titled Siyah-o-Sefid (Black and White)[1], retells his memory of first meeting with Tagore: 'One day we were let known that Tagore the famous poet has come to Iran. The news moved us specially me for I was fascinated by him who was told to be a philosopher and a mystic figure. I was fascinated because I was interested in mysticism and philosophy.'

As it is clear here, Iranians were interested in mystic side of Tagore's personality rather than in his literary aspect. That is actually owing to the Iranian conception of India. Iranians mostly know India as the land of various religions and creeds, its mysticism and Hinduism and Buddhism. Iranians are very interested in the mysteries of that land and naturally, even now, are not very familiar with Indian literature.

What they know began with getting familiar with some political, philosophical and literary figures including Tagore, of course. However, before the translation of Tagore's works, they knew little of his literary side.

One impact of his presence has been establishing him as a literary figure in addition to the previously engraved mystic conception of him.

Mostaan also adds that when he asked Tagore of mysticism and philosophy the poet answered that he knew nothing of them. Finally, when the former objected that Tagore does know then Tagore had replied 'I am a man of wine and the Beloved and know nothing of what you talk about.'

Actually Tagore had presented them with the best answer but it seems that they were ignorant of the purpose and meaning of this sentence. Tagore had tried to answer them on their own cultural and mystic ground noting that love should be felt, not talked or asked about. Mysticism means establishing a direct contact with the Beloved and not be engulfed in dialogic debates.

With regard to what transpired in this meeting, we find out why Tagore has not had the impact we expected him to have at that time. Most people, among them men of literature, did not understand him as he really wanted them.

Of Tagore, what is also said to have left a major impact on Iranians has been his physical appearance and gestures. He has been portrayed as being tall, with a long robe that made him look taller, silver hair, a wide and big forehead, piercing eyes, a long beard, a soft and pleasant voice and distinguished and dignified movements and gestures. He was also said to remind Iranians of their own old great mystic figures.

Tagore after the Islamic Revolution in Iran

Why has Tagore not been welcomed so warmly and widely by Iranian religious figures after the Islamic Revolution? This is a question we can ask after so many years of his presence in the Persian land. Well, considering the religious structure of the Islamic Republic of Iran and its ideals proposed by its leaders, some answers instantly come to mind:

1. His defense of Zoroastrian teachings and doctrines can be a major objection to Tagore on the ground that Islam is considered to be the first and only must-be dominant Faith in Iran. Tagore once declared Zoroaster as the greatest prophet coming at the beginning of the history of the people. He finds the unity of Iranians and Indians in Zoroaster's beliefs. Of course, the Zoroaster that Tagore pictures is very Nietzschean in concept depicting him as standing in front of his cave and waiting to start a down-fall to teach the people of Overman.

2. Further, as I have mentioned earlier, his presence at the time of Reza Shah and defending Royal Iran released from Arabic boundaries can make Islamic government doubtful about the poet and this means there is another impediment to welcoming him.

In spite of all this, Tagore has been praised by many literary and non- literary figures in Iran. The Secretary to Farsi Language and Literature Academy of Iran mentions him as someone whose ideas and thoughts should be a model for all thinking minds.

Abdoljabbar Kakai, a famous Iranian poet, ranks Tagore with old Iranian poets as Mowlavi (Rumi) recalling his themes of purification of heart and soul or being at the service of the people.

A problem with the true position of Tagore in Iran has been being always compared with Iranian counterparts. Researchers, in respect of Tagore, have based their work on comparing the divine and spiritual or formal aspects of his poetry with those of the Persian poets, mainly Rumi and Hafiz. Having a look at the papers, essays, theses, etc. approves the claim. Abolghasem Radfar in 'The Position of Mowlana Jallaloddin Mohammad in Subcontinent' compares Tagore's thoughts with those of Mowlana (Rumi). Radfar himself, in his research, mentions of the book 'Sufism in India' written by Amir Hossein Abedi who has magnified the same points.

Reza Mostafavi Sabzevari in *Tagore in Iran*[2] follows the same track and selecting lines that are similar with those of Hafiz, Rumi and Naser Khosrow compares Tagore with those poets.

Tagore in Today's Iran

In Iran nowadays, Tagore's name and verse lines can be found in teaching and educational materials. In some literary exercises , for example, a few lines of Tagore's verses are referred to and one is asked to unveil the true (deep) meaning of the line(s) or mention the literary devices used in the line(s). In Iran his teachings are also used as a lamp to shed light on the right path towards the Mystic Truth. The people who take steps along the mystical path use Tagore's quotes and sayings as the illuminating and guiding star to prevent them from going astray. Many researches and papers or even stories start with a quotation from Tagore on the top of the first page as their own prologue.

On clicking Tagore's name, the number of records that opened in the National Library of Iran ran to 12 pages. On each page we have 15 titles including books, papers, etc. The number of books on Tagore or translated from his works are 10 in the first page, 8 in the 2nd, 2 in 3rd, 5 in the 4th, 4 in the 5th, 8 in the 6th, 9 in the 7th and 8th, 10 in the 9th, 11 in the 10th, 12 in the 11th and 10 in the 12th ,which add up to 98 books with the name Tagore, of which some are by him and some are on him. Of the books by him, some are the same title(s) translated by different translators. In the 1960s, a good number of his works, from his poetry to his plays, had been translated.

In 2003-2004, some of his works have been published in the English language. In the recent decade, we have again witnessed a good deal of translations and written works from/on Tagore, as compared to preceding years.

The number of records, of course, does not appear satisfactory for a literary figure who has won the Nobel Prize and has many themes in common with the cultural and literary heritage of the nation. That is why Tagore has not obtained his true place in Iran though there may be many who deny this view.

Another problem has been with the translators. They have not been as professional as they were required to be; therefore, the reader is unable to read Tagore correctly.

Further, they have formed their own anthology of Tagore's poetry based on their own interest and using personal criteria in doing the job without perceiving the soul of the lines.

Lack of any governmental support from India is involved herein to make the situation worse.

Why Tagore is Welcomed by the Iranian People

The soft and simple language, lucid to all, coming out of a pure heart and blended with sincerity is the major feature of Tagore's poetry and the answer to why he has been generally acclaimed here in Iran. A kind of humility, for instance, common in Persian culture, can be traced in the Gitanjali. God's mercifulness, the infinity of His compassion and mercy, the sense of nostalgia between God and Man, Mystic drunkenness, a sense of lover and Beloved in the dialogues are all portrayed in Tagore's poems and are recurrent in mystic Persian poetry.

Apart from his form, perhaps, Tagore's themes are a good answer to his popularity in Iran. His themes are those of popular Iranian poets like Hafiz and Rumi. He might be classified as a mere romantic poet for a western eye but is divine and spiritual for eastern men such as the Iranians.

In the heart of Tagore's poems there lies a juxtaposition of contradictions. The beautiful elements of the nature, for instance, are not found to be the usual in his verse lines. In his poem 'Home', the sunset is likened to a 'miser'. A notable thing in these lines is that Tagore takes the surrounding as something that is alive and this is a feature of mystic poets such as Hafiz. Personification has been a constant device in the mystic poetry of the East and Tagore makes the best of this in his divine lines.

The poems of Iranian poets like Hafiz are popular for many reasons of which one is their closeness to romanticism. The themes current in romantic poetry are:

- **Intermingling of Joy and Sorrow**: In the heart of death you can face life, in hard times you will find tranquility and ease, inside darkness you see light. These are binary oppositions common in romantic poetry. When sorrow has surrounded you fully, a saving happiness arrives to inform you of hope and bright future. Both conceptions are taken as transient, but Man, relying on his divine side, finds rest in God's Will, ignoring whatever it may be.

  Tagore never denies the hideous face of life or its difficulties. Indeed these are to him facts of life and being; and that is why they sound beautiful to him.

  His poem, 'The Home', tells of the narrator (Tagore himself, perhaps) walking home at the sunset when everything seems and sounds dull and dead. Just then, suddenly, a boy's shrill cry rises to the sky. Consequently, the narrator stops to watch the houses veiled by the darkness of the night that has arrived and then is drowned in deep thought leading him to finally find life in the heart of the darkness surrounding him. The picture is a romantic one conveying the concept of loss and regaining, death and life (having the child as the bringer of life).

  The simplicity and, at the same time, the meaningful picture that Tagore draws is unique.

  The color of the sun is gold and the word is used in a brilliantly ambiguous manner to mean both a color and a precious stone. And now sunset arrives to act as Scrooge to hide this gold.

  Juxtaposition of death and life is well demonstrated in these lines. The pictures Tagore depicts are all present and in front.

- **Leading role of Nature**: It is in nature that you can find peace and tranquility. Nature is noble and here the theme 'noble savage' steps forth. Purity lies in nature and poetry turns to be the poetry of purity because it is inspired by nature.

  M. H. Abrams, in *The Mirror and the Lamp,* draws the triple relationship between nature, the writer or poet and the reader. The relationship between nature and poet or writer is imitative. These people make a model from nature and imitate it in all its aspects, reviving the neoclassical idea of taking art as the mirror upheld to nature. Man is a part of this same nature and when one gets spoiled the other gets too.

  Nature in Tagore's poetry has again two different roles as both giver and receiver.

  It gives life and joy while at the same time it can turn into a disaster and take life. But it is the way of nature as it is the way of life. Everything coming from nature should be accepted satisfactorily, this is what Tagore teaches us.

- **Loneliness and Solitude:** Man is alone on the earth, as always has been, and in his solitary moments he finds his true self and then he finds himself close to his Creator. Mystic poets believe that finding God inside, Man should never feel alone. The loneliness referred to here is a spiritual one. It is loneliness full of God.

- **Children and their Innocence:** Child is the father of the Man. This is what Wordsworth told us. It is the spontaneously unselfish love of little children that forms the transition to adulthood possible. This is the love Tagore asked for, a love that is pure and asking nothing in return.

- **Carpe Diem:** Carpe diem denotes the fact that Man should grasp the time and make the best of present. Carpe diem meaning 'seize the day' is Latin in its root and borrowed from Horace but it was Ovid who used it in the sense of seize or enjoy. Horace once wrote: *Carpe diem quam minimum credula postero* – 'Seize the Day and put little trust in the next day'. Man should seize the time to love and to be loved, to make friendship, to bring peace. This is what Tagore believed in and did during his life time and with his lines.

- **Transcendentalism:** The concept emphasises 'the role and importance of individual conscience, and the value of intuition in matters of moral guidance and inspiration.'

The great Indian Poet also shares with Persian poets this source of inspiration for writing his verses.'My force lies in renunciation… like the bird who renounces its shell in order to come out in the light of sky to breathe more life.' Yang-Shik remarks that by renouncing, Tagore did not mean rejection or escapism. 'It is the renunciation of the lesser joy for the sake of the greater.' The greater joy the poet speaks of is getting close to the source of everything, i.e. God.

- **Unity in Plurality:** This is one of the themes found in Islamic mysticism and in Tagore's lines as well. Everything in Iranian mystic view is summarized in One. The holy mind, the spiritual pure heart and inner eye never find any discord or disunion in the world though things are diversified. All phenomena and the whole being are gathered in one name and that is God.

'True knowledge is that which perceives the unity of all things in God'. This is the same unity that the mystic poets of Iran believed in and have referred to through their lines.

Now, keeping all these characteristics of spiritual romantic poetry in mind and turning to the poetry of Hafiz, Saadi and Rumi as well as to the lines of Tagore, we find many similarities that make Tagore popular to the Iranian reader and reading Tagore's verses would be as pleasant as the poetry of those Iranian figures. These features are enough to dedicate to Tagore's form and theme an Iranian mood.

Man and the Universe

The place of Man in the universe has been considered high and divine while ignorantly he degrades his own position. In the Quran, Man is introduced as the successor of Allah on the Earth. He has got unlimited power both in mind and heart. In the Islamic chain of being, he has got the top rank.

'Thou Hast made me endless'. This is the first sentence and the opening line of Gitanjali. A close reading of Tagore's worship shows the idea behind Islamic doctrine and Tagore's teaching to be one. Coming to the Beloved, Man loses himself and whatever he has. The first lines of Tagore's masterpiece describe this fact and show that coming to Him, 'at the immortal touch' of His hand, Man's 'little heart loses its limits in joy and gives birth to utterance ineffable'.

Saadi Says: 'Man will go to the position where he sees nothing but God'. The ideal of Man, as Saadi believes and Tagore admits, is being with Him and losing the heart to Him and His presence.

Children in Tagore's Poems

The role children play in Tagore's verse is demonstrated beautifully in another poem of him 'On the Seashore'. Again, these are children who bring and give life to the surroundings while 'The infinite sky is motionless overhead and the restless water is boisterous'. They know nothing of whatever man-made or superficial in life:

> 'They know not how to swim, they know not how to cast nets.
>
> Pearl-fishers dive for pearls, merchants sail in their ships,
>
> While children gather pebbles and scatter them again.
>
> They seek not for hidden treasures, they know not how to cast nets.'

They know just the simplicity and purity of life. What matters to them is not to be merchandised and is not to be hidden in the depth of the oceans. It is ever-present and one should just feel it and touch it with the heart. Death has no meaning to them as it didn't to the little girl in 'We Are Seven' by Wordsworth. Nothing can make children scared, not even the roaming tempest. Even the waves yield themselves to the children.

In Iranian art, children enjoy the same role of being innocent and pure in heart. The best example of this is the film 'Where Is Friend's Home' which was directed by great Iranian director, Abbas Kia Rostami, who was inspired by a poem of Sohrab Sepehri of the same name. Sohrab is very close in meaning to Tagore. He has been the one whose poetry in some cases could be analyzed on the basis of Indian mysticism.

Sohrab Sepehri (October 7, 1928 - April 21, 1980) is one of the greatest and the most popular poets of Iran. Some believe that the contemporary poetry of Iran at its zenith has witnessed only five great poets among whom the name of Sepehri shines. Most of his poems are translated into different languages. He died in Tehran in 1980 as a result of leukemia. Of his works, Green Space (or Green Volume), East of Sorrow, The Death of Color, and Water's Footfall are but a few.

Some Iranian researchers have found common elements and themes in poetry of Tagore and Sohrab Sepehri; and since Sohrab was born and lived after Tagore, it seems reasonable to imagine that Tagore had somehow had an impact on this Iranian poet. Sepehri had been to India for a short time and mentioned his experience several times in his works.

Sohrab Sepehri was fascinated by eastern cultures and mysticism and this interest motivated him to visit Japan and India. In 1961, Sohrab travelled to India. It is said that after his trip, Sohrab made use of Buddhist philosophy and eastern beliefs, especially those of India, in his versification so that some call him ironically Sohrab the Indian.

Ali Moosavi Garmaroodi, one of the great contemporary literary figures of today's Iran and former Cultural Counselor of the Iranian Embassy in Tajikistan, in his speech, 'About Sohrab Sepehri', said: '… of view point of thought, Sepehri is charmed by his own special mysticism, something that can be called natural mysticism…. But Sepehri searches God through Nature. … The third feature of Sepehri's poetry is personification; he personifies all phenomena and things. Everything in his poems is alive, has got emotion, conscience, and takes breath.'

I personally think Sohrab had known Tagore though it has not been ever mentioned. The gateway to get familiar with Indian literature for the Iranian people has always been Tagore's writings. It seems farfetched that Sohrab had not read Tagore.

In Water's Footfall many eastern mystical features can be traced:

> I am a Moslem.
>
> My Mecca is a rose.
>
> My mosque is a spring, my prayer stone the light.
>
> Fields make my prayer rug.

I make ablution with the heartbeat of the windows.

Moonlight flows through my prayers, the spectrum too.

My Kaaba lies by the water,

My Kaaba lies under the acacias.

My Kaaba travels like the breeze,

From one garden to the next,

From one town to another.

Like Tagore, Sohrab considers children innocent and the only ones who know the right path, the path leading to Friend's home. To Sohrab, a wayfarer who wishes to get to God's threshold should face many difficulties and dangers and then arrive where children inhabit and they will tell him the address of that threshold.

Conclusion:

No precise research has been done about the impact Tagore may have had on Iranian literature. It is necessary to propose the idea to researchers and support them financially etc. to fulfill it.

Tagore has not found his true place here in Iran as a result of some misconceptions and misunderstanding. Most Iranians love him and his versification and use his lines in their own writings while a few are there who object to him for his close contact with Reza Shah.

Further, a poor translation of his works has degraded the quality of his works and it is necessary to let readers know about the best translations done so far.

His language and his themes have charmed Iranian readers and they find his verse very close, especially in meaning, to those of the old Persian poets.

The importance of Tagore's position in world literature and also as the link between Iranian and Indian cultures (as stated by the Deputy Chief of Mission at the Indian Embassy in Tehran, Mr. C. B. George, at the Tehran International Book Fair, 2011) necessitates both Iran and India to have a joint project on Tagore and support more research on this world figure.

Notes

1 The magazine is different from Sepid-o-Siyah published by Ali Homayun

2 *The Quarterly Journal of Persian Culture, Language, Literature*, No. 49-50, Autumn-Winter, 2010-2011, Special Issue on Tagore and Bengal published by Center of Persian Research, Office of the Cultural Counselor, IRI, New Delhi.

APPENDIX:

THE HOME

I paced alone on the road across the field while the sunset was
hiding its last gold like a miser.

The daylight sank deeper and deeper into the darkness, and the
widowed land, whose harvest had been reaped, lay silent.

Suddenly a boy's shrill voice rose into the sky. He traversed
the dark unseen, leaving the track of his song across the hush of the evening.

His village home lay there at the end of the waste land, beyond
the sugar-cane field, hidden among the shadows of the banana and
the slender areca palm, the cocoa-nut and the dark green
jack-fruit trees.

I stopped for a moment in my lonely way under the starlight, and
saw spread before me the darkened earth surrounding with her arms countless homes furnished with cradles and beds, mothers' hearts and evening lamps, and young lives glad with a gladness that knows nothing of its value for the world.

ON THE SEASHORE

On the seashore of endless worlds children meet.

The infinite sky is motionless overhead and the restless water is boisterous.
On the seashore of endless worlds the children meet with shouts and dances.

They build their houses with sand, and they play with empty
shells. With withered leaves they weave their boats and
smilingly float them on the vast deep. Children have their play
on the seashore of worlds.

They know not how to swim, they know not how to cast nets.
Pearl-fishers dive for pearls, merchants sail in their ships,
while children gather pebbles and scatter them again. They seek not for hidden treasures, they know not how to cast nets.

The sea surges up with laughter, and pale gleams the smile of the sea-beach. Death-dealing waves sing meaningless ballads to the children, even like a mother while rocking her baby's cradle.
The sea plays with children, and pale gleams the smile of the sea-beach.

On the seashore of endless worlds children meet. Tempest roams
in the pathless sky, ships are wrecked in the trackless water,
death is abroad and children play. On the seashore of endless
worlds is the great meeting of children.

References

- Tagore, Rabindranath. *Gitanjali*, New Delhi: Macmillan, 1974).
- Cuddon, J. A. Penguin, *A Dictionary of Literary Terms*, (London: Penguin Books, 1979).
- *The Quarterly Journal of Persian Culture, Language, Literature*, No. 49-50, Autumn-Winter, 2010-2011, Special Issue on Tagore and Bengal , Centre of Persian Research, New Delhi.
- Dehbashi, Ali. *Collected Articles on Life and Works of Rabindranath Tagore*, (Tehran: Nika Pub., 2010).
- Yang-Shik, Kim. *The Theme: The Idea of Rabindranath Tagore.*
- The English Writings of Rabindranath Tagore, Volume 3, page 19, (New Delhi: Sahitya Academi, 2008).
- Aleem, Mohammad. Tagore: His historic visit to Iran.
- Verma, Rajendra. Rabindranath Tagore, (Bombay: Asia Publishing House, 1964).

Tagore's Visit to Iran

Mohammad Ajam

Rabindranath Tagore wrote in 1932, while on a visit to Iran, that 'each country of Asia will solve its own historical problems according to its strength, nature and needs, but the lamp they will each carry on their path to progress will converge to illuminate the common ray of knowledge.

Rabindranath Tagore was very interested in Iran and visited the country twice, first in 1932 and then in 1933 when he celebrated his 70th birthday in Tehran. Tagore's visits to Iran and his meetings with the Iranian writers and poets in Boshehr, Shiraz, Tehran and Esfahan led to an extensive cultural and literary exchange between the Iranian and Indian writers, scholars and poets that continued for years.

The visit to Iran was in response to an invitation extended to Tagore by Reza Shah Pahlavi who captured power in 1925. For a number of reasons, Tagore felt that he had to accept this invitation. Persian language and literature, which were propagated in India first by Zoroastrians after the advent of Islam in Iran, and later by the Muslims and then by the Gurkanian Mogul kings, formed part of Bengali literature, and around 3000 Bengali words and phrases have their roots in Persian. Raja Ram Mohan Roy, the spiritual father of Tagore and many other great personalities of India at that time, used to speak fluently in Persian. Although Tagore himself did not know Persian, he was familiar with Persian stories and Iranian poets, particularly Hafez, through their translated texts. Tagore's grandfather knew Persian and had himself even composed a poem in Persian. Tagore's father also knew Persian and he used to read Hafez for the young Rabindranath.

No doubt, Tagore had some inclination for the lifestyle of the desert. His first poems on Kosot (nomadic life) and his later references to the subject, his eagerness towards Hezar-o-Yek Shab (1001 Nights) and a translation of the Waste Land by T.S. Eliot in 1932 all prove this claim. He was curious and wanted to know how Iranian traditions intermingled with the westernizing tendencies of Reza Shah.

But traveling to Iran during August, one of the hottest months of the year, was a major problem for Tagore. The train or ship was not a suitable mode of travel for the old poet. So, he took a Dutch mail plane. He flew from Kolkata to Allahabad, then to Jodhpur and Karachi, next to the Iranian Jask Port and finally landed in Bosheher Port. His trip took two days during which he complained of being tired. The Maharaja of Jodhpur, who himself was a pilot, accommodated Tagore in a hotel specially built for air travelers there. In Karachi, he was served with Bengali food specially prepared by a Bengali lady residing there. This was Tagore's second air travel. The first time was in 1921 when he had flown from London to Paris.

Tagore's Visit to Sa'dat School in Boshehr

In early 1932, Tagore arrived in Iran for the first time and he then visited a number of places, including the Sa'dat School in Boshehr. In a comprehensive reference book on Boshehr it has been said that the school was cleaned up and the Chairman, the principal and the teachers of the school held a meeting to discuss the visit of an eminent personality of India to the school who was to arrive in Boshehr across the sea.

This great religious, scientific, literary and cultural personality and well-known philosopher was Tagore. Accompanied by the Consul General and the Consul of Britain in Boshehr Port, Mirza

Ahmad Khan, Chief of the Navy, Daria-biagy, and the civilian and military chiefs, Tagore finally arrived in the National Sa'dat School and was received by the Chairman, the principal and teachers of the school. In his honor, students first sang songs, played music and marched in front of Tagore. With great pleasure and interest, Tagore visited the classrooms and spoke to the teachers and students through Mirza Ali Farjad Khan, the First Secretary of the Consulate General. Thereafter, Tagore delivered a speech at the big hall of the school which was used as a community center and a theatre. During the speech, Tagore thanked the founders and the office bearers of the cultural institution for setting up the Intermediate National Sa'dat School in Boshehr. Honouring Tagore's visit to Boshehr and the National Sa'dat School, the Consul General distributed gifts and awards among the students of the school. Tagore's visit to the National Sa'dat School was reported in the Tehran publications and the Persian daily 'Hablol Matin' of Kolkata carried detailed reports of the visit and this important event was discussed for long in various meetings and gatherings. Even today, the founding day of the Sa'dat School is observed as the Boshehr Day and a portrait of Tagore is put on display on the occasion.

Tagore's Visit to the Tomb of Hafez in Shiraz

Many Iranian scholars and officials left Tehran and went to Shiraz to welcome Tagore.

After Boshehr, Tagore traveled to Shiraz (the Persian culture capital of Iran) to pay his respects to the great Iranian poet and philosopher, Hafez Shirazi. Hāfez (1325–1389) and Tagore are very similar to each other in their ideas of tolerance and respect for other cultures.

In his speech in Shiraz, Tagore said that his grandfather had learnt the poems of Hafez by heart and that he had also composed a Persian poem. He added that his father also had a copy of the Divān of Hafez and frequently read the poetry of Hafez.

In one of his poems Hafez wrote: If that Shirazi Turk would take my heart in hand (becomes my beloved) - I would remit Samarkand and Bukhārā just for her bindi. (The red dot on the forehead)

On the arrival of Tagore to Shiraz, Mohammad-Taqí Bahār (1884-1951), a famous poet and Minister of Culture, presented a poem to Tagore and described him as a person who, like Indian rice and Egyptian sugar, belonged to the whole world.

Tagore's Visit to Mashad

Tagore traveled for the second time to Iran to pay tribute and his respect to Hakīm Abu'l-Qāsim Ferdowsī Tūsī, known as the great Firdausi (940–1020 CE), a highly-revered Persian poet. Tagore was invited to participate with forty other eminent poets and scholars from Europe, Soviet Union, Germany, USA, Egypt, Yugoslavia, UK, France and many other countries in the Ferdowsi Festival. In his travel to Mashad, Tagore was accompanied by another Indian scholar, Jamshid J. Onvalla. In the festival, Tagore met several famous scholars and Iranologists from all over the world.

Firdausi and Hafez are the two most popular poets and philosophers of Persian culture.

Tagore described them in his speech in a very beautiful and respectful manner, and praised Iran and Iranian culture on his arrival at the tomb of Firdausi.

Tagore, like Hafez, worked for one supreme cause, the union of all sections of humanity in sympathy and understanding, in truth and love. They both believed that the national identities inherent in us need not create contradictions and that promoting one's culture should not preclude acceptance and respect for the culture of others. Hence Tagore, like Hafez, heralded the cultural rapprochement between communities, societies and nations much before it became the liberal norm of conduct.

That is why all Iranians love Hafez and all songs and Sufi music based on the poems of Hafez. The famous ghazal of Hafez is: For peace and prosperity in this world and the next, understand what I advise: loyalty and magnanimity to friends, and with foes, toleration and respect.

This was the reason why Iranian scholars admired Tagore so much. Tagore's travels to Iran contributed to a great cultural exchange and there were several articles on him in the newspapers of that time.

Each of these visits reinforced the resolve of our scholars to further strengthen people-to-people links. Tagore's travels made Iranian scholars pay more attention to Indian culture and many Iranian scholars travelled to India to rediscover India. One of them was Bahār who composed many famous poems about the greatness of India. On Indian culture he wrote: 'Once again, the white horse of my imagination travelled to India. My life be sacrificed to the earth of India. I sang, O beloved Hindustan, be happy and ever prosperous. And may every month and every year be free from sadness.'

On April 11, 1933, Rabindranath Tagore and Dinshah Irani visited Iran. Tagore was in Tehran in the second month of spring and his daughter-in-law, Pratima Devi, accompanied him to look after her father-in-law. The 70th birthday of Tagore was celebrated in Tehran. On his way back to India via Baghdad, Tagore was the guest of King Faisal. Shortly after the visit, Tagore received the news of the death of his only grandson, Nitindranath. No doubt, it was very difficult for Tagore to bear such a grievous loss as his lineage came to an end with the death of the young Nitindranath.

Professor Pur-dawood says: 'During his stay in Tehran, Tagore asked Reza Shah Pahlavi to dispatch a teacher to India to teach in his college. At that time, I was staying in Germany. I received a message from the Iranian government that I had been chosen for the purpose. On December 21, 1932, I left Berlin for India and returned to Germany on March 27, 1934. Santiniketan College, which is called Visva Bharati today, is one of the government universities in India. When Tagore was alive, a number of scientists used to teach free of cost at Santiniketan. Many Western experts, such as Sylvain Levi of Paris in 1920, Winternitz and Lesny of Prague in 1921, Carlo Formici and Giuseppe Tucci of Rome in 1925, visited Santiniketan and delivered speeches there. Carlo Formici and Giuseppe Tucci, who were sent by Mussolini, presented hundreds of volumes of Italian books to Santiniketan. In response to an invitation of Mussolini, Tagore visited Italy on May 15, 1926. Tagore spent the cash prize of his Nobel Prize for the development of his college. All income from the sale of his books was also spent for the same purpose. A considerable amount of the expenses of Santiniketan were met by donations received from the Nizam of Hyderabad, the Zoroastrians, etc.

During the discussions, Tagore told me many times: I feel that nobody in Tehran knows me as none of my works have been translated into Persian so as to introduce me to the people. Since he repeated this two or three times, I told him that I will do the job with the help of one of the teachers there. With the help of one of the teachers, Ziauddin, who knew Persian to some extent, I translated some 100 lines of Tagore's poems from Bengali into Persian. Tagore translated most of his poems from Bengali to English in his own way. The English text of Tagore's poems has been the basis for their translation into many other languages. So, everybody has chosen his own method of translating Tagore's poems.

On the eve of the Iranian New Year (Now-Ruz), I was in Santiniketan. Tagore, who had come to know the coming of the Iranian New Year, ordered that the occasion be celebrated in a big way. One of the Indian artists, Nandalal Bose, decorated the place wonderfully. Tagore and all the teachers and students participated in the festivities. I was given a greeting card which was written by Tagore himself in red ink. Red colour, among Hindus, is a sign of prosperity.

There I read a few poems that I had written to celebrate the arrival of spring. In the poems, I referred to the freedom of India and Hindu Gods. One of my friends, Krishna Kriplani, translated them into English. Tagore was very impressed by hearing the poems as well as terms such as Guru from an Iranian.'

Tagore was one of the most creative exponents of an Asian sense in the early twentieth century. He wrote in 1932, while on a visit to Iran, that 'each country of Asia will solve its own historical problems according to its strength, nature and needs, but the lamp they will each carry on their path to progress will converge to illuminate the common ray of knowledge.

Rabindranath Tagore was a multifaceted genius whose creativity spanned every form of artistic expression. His concerns ranged from global and universal ones to practical and mundane issues. He established institutions to reflect his vision and, through his writing, travels and actions, demonstrated that the international brotherhood of man was not merely a theoretical construct. Tagore was a man ahead of his time. His ideas on culture, gender, poverty, education, freedom, and a resurgent Asia remain relevant today.

During the last two decades, Iranians have paid greater attention to what Tagore has written. Many books and articles have been published about Tagore's life and ideas in Persian. Last year, I wrote an article on the occasion of the 150th birth anniversary of Tagore and it was well received by the Iranian mass media.

The IRNA official news agency of Iran and the Iranian TV, IRIB carried many features on the occasion.

****

The Poem of Mohammad-Taqí Bahār, Minister of Culture, about India and Tagore

(روگـات) .نکـشرکـش یطوط نآ درذگـب   نخس نیریش ظفـاح نطو زا
لد هب دیارگـب ناریا بناج   دنه ز دیآرب ملاگـنب یطوط
متفـاتشب و هدرک رس ز یاپ   متفـای ربخ هدژم نیا زا نم نوچ
دید «روگـات» هک هدید زا میضار   دید رون وا زا نم مشچ و دمآ
دنرف یدنه و یرصم رکـش نوچ   دنه صوصخم هن ،تسا ناهج ِناز
میتخیر شمدق رد لگ و ملال   میتخیمآرد ملال و لگ هب ناج
وردنت دیفس بسا = گـنخ) .تفـرگ ناتسودنه دای مٌعبط لیف ---تفـرگ نالوج مترکـف گـنخ زاب
تسم سوواط مولج مقوذ تفـای ---تسب دنه یور شقن ملایخ ات
دنه ریگ نماد کـاخ یادف ناج ---دنه ریجنز مقلح ریسا لد
انشآ یادن دناد انشآ --- ام یاهیانشآ دوب یدزیا
دن مج و نودیرف لسن زا ود ره ---دنمه نایانشآ ناریا و دنه
دوب قوش و روش و قشع رسکـی دنه ---دوب قوذ دیرخ رازاب دنه
تاره) .تفـای بیترت یره و ناهافـص زا ---تفـای بیکـرت رنه و قوذ تعنص
نزم نماد مشتآ رب نیا زا شیب   نزم مد نم اب دنه زا رامب یا
ما هتسب دوخ رب هک نیا تسا دنه مان ---- ما هتسخلد سب دنه قارف زک
میظع دنه تقرف زا ناپت لد ---- میقم یر رد مکچوک رامب نم
شاب دازآ مغ و دنب زا هام و لاس --- شاب داش یمارگ دنه یا میوگ

Tagore's Dark Vision of Humanity

Alok Bhalla

'What shall I do with that which will not make me immortal?'

Maitreyi asks in the *Brahdranayaka Upanishad*

'What am I to do with these, which are not of immortal spirit?'

- Rabindranath Tagore[1]

Between 1903 and 1909, Tolstoy wrote, 'The Restoration of Hell,' 'After the Ball,' and 'The Law of Love and the Law of Violence,' amongst his last and bitterest denunciations of the orthodox Christian Church and Tsarist Russia.[2] The reasons for his rage and despondency were many: in 1891-92, an estimated forty thousand peasants had died of famine in the Russia and thousands more had been forced into labour camps of torture and stagnation; around a hundred and fifty thousand soldiers had been killed in a pointless war between Russia and Japan in 1904-05; and, Father Gapon's peaceful march to the Winter palace in 1905 asking for political reforms was met with violence by the Tsarist forces. Like the later prophets of the Old Testament, Tolstoy's three texts have the conviction of a visionary who knows how simple the good and the true can be, but also how difficult.[3] However, they also have the wrathfulness of a man who recognizes at the end of his life that society is now deaf to the call of human sympathy. Thus, in 'The Law of Love…,' he thinks that all the citizens and institutions of Russia – thieves, soldiers, universities, ballets, synods, prisons, taxes, judges, gallows and conservatories (*Confession*, 178) – are 'corrupted by wealth, power or civilization' (*Confession*, 172), and practice 'deceit through violence, and violence through deceit' (*Confession*, 162). And, in the brilliant short story, 'After the Ball,' the narrator is filled with nausea as he watches a naked soldier, who is being pushed through a line of army cadets and whipped on a winter morning, plead: 'Have mercy on me, lads. Have mercy on me lads.'[4]

It is as if late in life, Tolstoy has understood with a shocked sense of finality that the conversations about love, peace and justice which he had once thought were possible either in his great novels or with his friends like Chekhov and Gorky, have ceased to inspire Russia.[5] With the scathing wit of a defeated moralist, in his play, 'The Restoration of Hell,'[6] he imagines a people who finally stage a revolt against Christ's vows of poverty, chastity and *caritas*[7] and establish a Church where they can at last worship the Devil. With Beelzebub as its new presiding deity, the Church can display its power through the ornate splendour of rituals, teach with clear conscience the virtues of killing for national pride, sanction flogging of the peasants for discipline, support prisons for the safety of the state, advocate drinking for the lucidity of the soul, and encourage class snobbery for the maintenance of a predatory but cultured aristocracy. Since the people had finally accepted the legitimacy of the Satanic Church and found in it a justification for their 'habitual and favourite vices; vengeance, avarice, envy, ambition, pride, cowardice and spite' (*Confession*, 216), the State, Tolstoy declares darkly, can carry on with its rapacious functions with greater confidence. It comes as a surprise to realize how abrasively close Tolstoy is placing his two texts beside Dostoevsky's Grand Inquisitor in *The Brothers Karamazov*; how dangerously near he has come to erasing his own long abiding faith that because human beings are innately good, they would eventually display their natural innocence just as trees burst into leaves. Dostoevsky, we feel, at least framed the ferocious logic of the Grand Inquisitor's contempt for Christ's sermons of peace within the gentle faith of Alyosha on the one hand and Ivan's intellectual passion for the truth on the other, thereby leaving the possibility of conversations about love and goodness open.[8] Tolstoy seems to have concluded that 'the whole structure of the world [is] founded on violence,' (*Confession*, 176) and leaves people locked in their own special areas of hate or anguish. So deeply dyed he appears in his melancholic appraisal of 'human becoming' that one cannot imagine him even concede with Chekhov's sadly defeated characters in *The Three Sisters* that 'a time will come when everyone will know what all this was for, why there is misery...' Perhaps, that is why there is unredeemable pathos, both in the homelessness of his death at Astapovo on November 7, 1910, and his final whisper as if to his own soul: 'Search, always go on searching' (Wilson, 513). Tolstoy, of course, had no idea of the barbaric century in which Tagore would make his life-history.

Tagore's imaginative journey towards a darkening vision of humanity in the twentieth-century is analogous to Tolstoy's. Like Tolstoy, he too struggled against skepticism to maintain his faith in the idea that the earth was a place of hierophanies and that human life had a divine purpose.[9] He failed. He also called, again and again, but with ever deepening sense of gloom, for peace, equality and the restoration of the loveliness of the earth. As early as 1919, in a piece entitled, 'A Cry for Peace,' he confessed his despair: 'I know I am crying in the wilderness, when I raise my voice of warning; and while the West is busy in its organization of building its machine-made peace, it will still continue to nourish its iniquities...' (*EWRT*, vol. 3, 410). As he traveled through Bengal and, then through war haunted and hungry Europe and Asia, he inevitably found himself confronted by a world which was strange, threatened, 'virulently cruel' ('At the Cross Roads,' 1918, *EWRT*, vol. 3, 380), and so obstinately tribal that it could only leave behind a 'legacy of ruin.' ('At the Cross Roads,' 1918, *EWRT*, vol. 3, 382). Early in his imaginative journey, he wrote to C.F. Andrews on July 16, 1915, 'The world is wonderfully beautiful, but you cannot help feeling that there is a lurking pain in its heart...' ('Letters to a friend,' *EWRT*, vol. 3, 246) Indeed, even as he continued to assert till the very end of his life that emotionally he could never 'commit the grievous sin of losing faith in man,' he had to admit that his reasoning self would not let him turn away from 'the crumbling ruins of ... civilization strewn like a vast heap of futility' all around him ('Crisis of Civilisation,' 1941, *EWRT*, vol. 3, 726).

In 1909, Tagore published *Gora*, a sad and forlorn account of the dangers that threaten a society when it becomes arrogantly narcissistic about its religious and national identity, but fails to confront a contemporary reality scarred by rural hunger, religious contempt, and a decadent aristocracy. Years before the First World War, in a text like *The Post-Office* written in 1911, he had premonitions of a world where innocence is perpetually lost. In the same year, in a play entitled *Achalayatan* (*The Institute of Fixed Belief*), he showed how humiliating it is to be in the thrall of narrow religious certainties and sectarian politics.[10] Religious bigotry and ideological arrogance can only light 'unholy fires' fed by 'human sacrifice' (*EWRT*, vol. 3, 773). As the twentieth-century turned into a slaughter-house far surpassing the bleakest fantasies of evil human beings, Tagore's social and moral analysis became more and more desolate. Of course, he continued to assert, even in his last stories and essays, that there was an indestructible part of us which was made of enchantment, of 'forests' not reason.[11] But he also knew, in the deepest recesses of his soul, that human thought had failed once again, as always, to withstand 'the demon of barbarity' striding across the earth with its 'unconcealed fangs, ready to tear up humanity in an orgy of devastation' ('Crisis of Civilization,' *EWRT*, vol. 3, 726). Tagore, I would like to argue, was not merely a maker of melodies and a dreamer of supernal dreams, he was also a skeptical thinker who knew how easily and often human reason could go to sleep[12] and so breed monsters of colonial greed, racial hate, religious bigotry and xenophobic wars.

Of course, my assertion that Tagore was, throughout his creative life, a harsh and despairing critic of his times is contrary to received opinion about his life and works. Perhaps, I can make this assertion because I am an outsider who does not know Bengali. Further, I have only made a few fleeting trips to Calcutta. I have never been to Santiniketan – a place which is still reported to resonate with the idea of a university of creative freedom that has little to do with the teaching of instrumental rationality which structures contemporary politics, commerce and everyday practicality and, at the same time, makes our world noisy, obsessive, threatening and, perhaps, fascinatingly complex. In Bengal, for most intellectuals, Tagore is still a quintessential mystical poet whose music aspires to a state of transcendence which reason cannot reach. In Santiniketan, he is Gurudev, divine teacher; a prophet who knows that the primordial word waits patiently to be heard again just as a seed lies patiently for its turn in the cycle of seasons to sprout again; a hierophant whose presence inspires a sense of holiness. Indeed, that is how he is often painted. Elizabeth Brunner, for example, uses soft water colours in her painting of Tagore as a poet-sage at work in the last years of his life. In Brunner's painting Tagore is the archetype of an inspired Biblical thinker working in a ritual space sanctified by a garland of flowers. Nandalal Bose, in a mural, shows Tagore as a plowman showing ordinary humanity the way to a more creative life free from the desperation a farmer feels when there is famine and he does not know if his prayer would be answered.[13]

To an outsider, it seems as if every man, woman and child in Bengal thinks about Tagore as the earth thinks of rain; he is as necessary to their daily lives as ritual is to worship; as inseparable from their daily journey as dreams are from sleep. Indeed, Nirad C. Chaudhuri says as much in his

Autobiography of an Unknown Indian, the only time when he is in awe of someone worthy of his adulation in 'native' India:

> Even now I cannot read the words of these songs, far less whistle the tunes, without instantly bringing back to my ears and eyes all the sounds from the soft rumble of rain on our corrugate-iron roofs to the bamboo pipe of the cowherds, and all the sights from the sails of the boats on our great rivers to the spreading banian [sic] tree – the sounds and sights which embody for me the idea of Bengal (Dutta and Robinson, 144).

Not having grown up in a Bengal, I can be fascinated by Tagore's music and yet acknowledge that throughout his life and frequently, he slipped into despair because he could not endure the knowledge of so much inhumanity. My heretical impulse tells me that my position of a critic of Tagore as a non-Bengali outsider gives me a slight, though momentary, advantage. It helps me to resist becoming a mere acolyte of Tagore the singer of earth's sacred beauty. I can, instead, be a respectful, historically attentive and skeptical reader who knows from experience that a mere song will not push the world's brutes back into the abyss. Because Tagore is not for me primarily an inspired crafter of Bengali poetry and prose, I am perhaps more conscious of the tragic resonances which give to his writings and paintings during a ravaged century, through which he lived, their continuing importance and human worth. I can sense why Tagore's hymns to nature move Bengali speakers deeply, but I am alert to his own skepticism when he adds almost immediately that humankind has always been impatient to sweep away the cover of the green earth and the blue sky is lifted – the veil of nature – so that it can either 'lose itself in the sands of the self' ('The Message of the Forest,' *EWRT*, vol. 3, p.391), or in the 'fury of passion' devour the wealth of the earth through loot and slavery ('The Message of the Forest,' *EWRT*, vol. 3, 388).[14]

My quest for Tagore teaches me that, as a poet, Tagore self-consciously echoes India's religious and classical texts as well as invokes the history of the country's saint-poets in his songs. But it also never allows me to ignore the fact that his poetry, fiction, music and paintings, from their earliest days of fame with *Gitanjali* to the end of his life, is infected by each historical moment in which human beings have broken their faith with the earth and made a fool's bargain with murder, lust and greed. It is Tagore's skeptical reason which makes him subtly transform the Upanishadic statement by Maitreyi, with which I began. Maitreyi rejects all that impedes human progress towards immortality. Tagore's skepticism, however, makes him ask what a poet can do on 'the sin-laden dust of the earth' (*EWRT*, vol. 3, 43)[15] where human beings refuse to have anything to do with immortality?

When, as a critic at the margins of a culture, I hear such despondent sentences repeatedly, I begin to catch, beneath the cadences of grace, holiness, and innocence in Tagore's music, darker tones of loss of home, betrayal of trust, sectarian arrogance and the spirit of persecution which has made the earth into a charnel-house. Sharp-clawed, bloody-beaked monsters did not suddenly crash into his godly and enchanted landscape late in his life when he began to paint at the age of seventy as many of his Bengali critics claim.[16] As he traveled restlessly through what was 'without doubt the most murderous century of which we have a record,'[17] Tagore was always aware of the ghastly and grotesque figures of terror gathered around him everywhere in Europe, India, China or Japan. How could it have been otherwise? How could he not have smelt a European sky filled with poison gas; or, not have seen fertile fields crisscrossed with barbed wires and mud-filled trenches; or, not have consciously recorded the fact that during the battle of Verdun between 21 February and December 18, 1916, artillery fire had killed 306,000 German, French and British soldiers; or, ignored the famines of Bengal which left skeletons of animals and human beings scattered across the land; or, the imperial wars of Japan in China whose savagery could shock even the callous; or, the irrationality of ethnic, religious and linguistic nationalisms on the rise everywhere which were predictably paranoiac and genocidal. Unless, one prefers to remain an innocent lost in the mist of enchantment, it is not a surprise to hear Tagore interpret his 1915 poem, 'The Boatman' (Poem 41 in *Fruit Gathering, in EWRT*, vol. 1, 173-74) as an anti-war allegory. He tells C.F. Andrews that the seafarer, who carries only 'a white rose in his hand and a song on his lips,' as he travels through a night 'poisoned with black fear,' is seeking an innocent country ('Letters to a friend,' *EWRT*, vol. 3, 237). The woman, whom the war-weary traveler finds on the shore in the following line represents, he told Andrews, Belgium which was neutral in the First World War: 'I know not at what shore, at last, he lands to reach the silent courtyard where the lamp is burning and to find her who sits in the dust and waits' (*EWRT*, vol. 1, p. 173).

Of course, Tagore sang of the earth's extraordinary beauty, but he also continuously mocked 'the physical brutes' and 'the intellectual brutes' ('Letters to a Friend,' New York, December 20, 1920, *EWRT*, vol. 3, 273), whose 'political and commercial ambition is the ambition of cannibalism, and through its years of accumulation it must get ready for its carnival of suicide' ('At the Cross Roads,1918,' *EWRT*, vol. 3. 381). Asserting continuously that the aims of a truthful life must always be freedom and a sense of responsibility towards all things, he was in actuality tormented by nightmares of a deceased Europe and Asia where everything seemed to smolder with rottenness even as it wrapped itself in the most glowing forms. In the first version of his English translation of the poem, 'The Oarsman,' published in *The Times* on 28 January, 1916, he says that the cause for the present evil is, 'The cowardice of the weak, the arrogance of the strong, the greed of fated prosperity, the rancor of the deprived, pride of race, and insult to man…' (*EWRT*, vol. 1, 608). [18]

He is, however, not confident that many in war-crazed Europe or a viciously nationalist India, China and Japan would heed his call: 'Stop your bluster of abuse and self-praise, my friend…' ('The Oarsman,' *EWRT*, vol. 1, 608). In one of his bitterest attacks on modern European civilization and its imitators in Asia, delivered in Japan in 1915, he says that everything vital in Europe is in 'the exclusive possession of the devil' ('The Spirit of Japan,' *EWRT*, vol. 3, 371). He chides Japan for its militant nationalism. He tells his audience that nationalism's 'eruptive inflammation of aggressiveness' first led Europe into violent colonial adventures across the globe, and then so imbued it with an 'abnormal vanity of its own superiority,' that it began 'to take pride in its moral callousness and ill-begotten wealth…' ('The Spirit of Japan,' *EWRT*, vol. 3, 370). First, in the spirit of caution against Japan's new ambitions, Tagore renews his attack on the degrading violence of colonialism which sustained Europe's unappeasable 'earth-hunger' ('The Message of the Forest,' *EWRT*, vol. 3, 393) and the spirit of hatred, half-truths and narcissism inherent in nationalism which can only lead to 'moral death' ('The Spirit of Japan,' *EWRT*, vol. 3, 371). Then, with mounting anger and in growing notes of despair, he tells his audience: 'The furies of terror, which the West has let loose upon God's world, come back to threaten herself and goad her into preparations of more and more frightfulness… To the worship of this devil of politics she sacrifices other countries as victims. She feeds upon their dead flesh and grows fat upon it, so long as the carcasses remain fresh, -- but they are sure to rot at last, and the dead will take their revenge, by spreading pollution far and wide and poisoning the vitality of the feeder' ('The Spirit of Japan,' *EWRT*, vol. 3, 371). And, then he adds, much to the annoyance of his hosts: 'The bloodhounds of Satan are not only bred in the kennels of Europe, but can also be domesticated in Japan and fed on man's miseries' ('The Spirit of Japan,' *EWRT*, vol. 3, 371). Even as his anger at Japan's incursion into China mounts, he tries, both lyrically and hopelessly, to offer his alternative vision of a Japanese civilization which thinks of nature and other human beings, not as alien others who need to be subjugated as they are in a degraded Europe, but as sources of inspiration where the 'rhythm of a perfect grace can be heard' ('The Spirit of Japan,' *EWRT*, vol. 3, 367). He is suddenly overwhelmed by the beauty of Fujiyama at sunset 'growing faint against the golden horizon, like a god overcome with his own radiance, – the music of eternity welled through the evening silence, and I felt that the silence and the earth and the lyrics of the dawn and the dayfall [sic] are with the poets and the idealists…that after the forgetfulness of his own divinity, man will remember again that heaven is always in touch with the world which can never be abandoned for good…' ('The Spirit of Japan,' *EWRT*, vol. 3, 375). But immediately afterwards, as it happens throughout Tagore's poetical journey, forebodings of disaster wreck his vision for he knows the sources of pain and understands that they have stained the human self so deeply that they can neither be erased nor evaded. As he ends his talk, he hears 'the hounding wolves of the modern era, scenting human blood and howling to the skies' ('The Spirit of Japan,' *EWRT*, vol. 3, 375).

Even though I cannot sing, I am not deaf to Tagore's enticing music about the beauty of nature, the special benediction of reason bestowed on man, and the eternal gracefulness of the divine everywhere. But I am also alert to the fact that all the diverse range of forms he worked in are broken into by lamentation; his public lectures speak about the human soul as a tormented victim who 'walks through a bloodstained path of violence and crime' ('Philosophy of Fascism,' *EWRT*, vol. 3, 773); his painted manuscripts are filled with acrid smoke that eats through words and strange beastly forms that leave trails of blood and slime. Thus, for example, in a 1929 poem appropriately entitled 'A Weary Pilgrim,' he says that in his innocence he thought that he would find a place where 'the spirit of man' was 'great as the grass that blesses the lowly dust,/ And meek as the mountain under the stars' (*EWRT*, vol. 3, 953). Now, having traveled through part of the century and looking prophetically to the future, he can only say, with unredeemable melancholy in his voice:

A weary pilgrim, I traveled across the haunts
of iron-limbed monsters,
prolific of progeny,
shrieking and stinking,

befouling heaven and earth,
devouring life
to change into piles of deadly peril.

The path is intricate,
Unfriendly the night,
The barred gates guarded by snarling suspicion
That growls at the shadows of strangers seeking home.

(*EWRT*, vol. 3, 953).

To a non-devotee like me, Tagore seems like a man who is so helplessly entrapped by the endless barbarity of the world around him that, when two of his dearest friends, C.F. Andrews and W.W Pearson, set out for Fiji to document yet another history of human servitude in Fiji, all he can offer them are a few lines from the Upanishads as a runic charm to ward off despondency: 'From Joy all things have their origin; in Joy they subsist, and unto Joy they return…I meditate upon His glory, who created the earth, the sky and the stars, and sends into our minds the power of comprehension' ('Letters to a Friend,' *EWRT*, vol. 3. 244).

Sometimes, talismanic charms may be all one has when one is utterly vulnerable in times of the extreme civilizational crises which Tagore lived through most of his adult life. Indeed, the Upanishadic chant must also have helped Tagore to keep his own spirits up during all those times when he felt that he was 'suffering from…deep depression and weariness' and was on 'the brink of a breakdown' ('Letters to a Friend,' *EWRT*, vol. 3. 241).

Similarly, his own early poetry, so full of hope and beauty and prayerfulness, often worked like a 'charm' and gave a fleeting hope to many in India and elsewhere who found themselves standing at the brink of annihilation. We are familiar with Gandhi singing in times of personal distress and rejection by his followers, 'If no one answers to your call walk alone, walk alone,' and with Wilfred Owen, the young English poet, waiting for death in the rat-infested trenches of France on the last day of the First World War, reading Verse 96 of *Gitanjali*: 'When I go from hence, let this be my parting word, that what I have seen is unsurpassable' (*EWRT*, vol. 1, 76). What we did not know, till recently, was that soon after the anguish of the First World War, Tagore offered a vision of the earth as an incandescent place where insulted humanity could still dream of peace, to a young and amazingly gifted Slovenian poet, Srečko Kosovel (1904-1926) who died at the age of twenty-two. Kosovel read Tagore and Gandhi in a Europe still torn between fear and self-loathing on the one hand and tormented by ancient inner demons of racial contempt, religious bigotry and military arrogance on the other – a contradiction that would take another twenty-five years of war to resolve. Kosovel was stranded on a wasted land. Tagore's poetry helped him ask if modernity could be so renovated as to ensure that societies live with each other in 'perpetual peace' and, pledge never to violate the great Enlightenment promise of the autonomy of all human beings and their covenant with the earth. Like Tagore's writings during the war years, Kosovel's poetry shifts between a sense of hopelessness and a lyricism inflected with visions of a gentler and a more tolerant world. And much like Tagore in his skeptical moods which overwhelmed him almost always from 1910 onwards, Kosovel is not convinced that Europe and all the other nations damaged by imperial wars, class contempt, gendered violence and other passions which can only derange the human soul, will ever find a place of calm understanding:

> In green India among quiet
> trees that bend over blue water
> lives Tagore.
>
> Time there is spellbound, a cerulean circle,
> the clock tells neither month nor year
> but ripples in silence
> as if from invisible springs
> over ridges of temples and hills of trees.
> There nobody is dying, nobody's saying
> Goodbye – life is like eternity, caught in a tree…[19]

It is clear that Kosovel does not read Tagore to evade the miseries of a war-torn Europe through devotion and song. Given that Kosovel has the same bitterness of experience and bleakness of vision as Tagore, my suggestion is that we learn to read Tagore's lyrical works differently as part of a new visionary programme of hope in times of great despair. Tagore is, I think, *writing beauty back into the social, political and physical landscape.* In his prose he offers a harsh analysis of why the world has become a wasteland. He is, after all a contemporary of the great modernists like T.S. Eliot, W. B. Yeats, Ezra Pond, Sigmund Freud, George Grosz and Bertold Brecht for whom the world's miseries were a consequence of an evil so deeply dyed in the soul of human beings that it was no longer possible to erase it; for them the world could only end in a scream. Tagore is not blind to their reality nor dismissive of their analysis of life as meaningless which he echoes when he says, for example, in a letter to Andrews dated 19 December, 1920, that he has to try his best to 'prove that Man has not been the greatest mistake in Creation' ('Letters to a Friend,' *EWRT*, vol. 3, 273).'

However, unlike many thinkers of modern despair, Tagore's visionary politics also urges him to record the continuing existence of a supernal space and a unique time where the 'miracle of Being' (Vaclav Havel's phrase)[20] manifests itself even amongst the powerless and even in the present. In the physical world, rivers still glow in the sunset and the music of the Baul singers still mingles with the laughter of children and birdsong. In the social and political realm, a courageous human being can still assert: 'I shall not commit the grievous sin of losing faith in man;' and, can still affirm his faith in the moral dictum: 'By unrighteousness man prospers, gains what appears desirable, conquers enemies, but perishes at the root' ('Crisis in Civilisation,' *EWRT*, vol. 3, p. 726).

The poems in *Gitanjali* or *Fruit-Gatherer* are testimonies to those fleeting moments when the beauty of the earth manifests itself, and human beings step out of the structures of power and refuse their humiliations and predatoriness. When Tagore catches those lovely 'spots of time,' (the phrase is William Wordsworth's) he feels that the earth is open to the glory of the divine, and human beings are so transfigured that they can fulfill their imaginative and rational promise – they can be source of happiness and its guardians for the rest of the earthly habitat. The struggle of his life's work is to somehow find ways of making these 'visionary gleams' of internal calm and peaceful responsibility toward all things that are on the earth more permanent.

But, contrary to a kind of consensus among the acolytes, Tagore is not a mystic who has detached himself from the worldly. His commitment to analytic reason is so fierce that merely 'quoting scripture against conscience' (The Changing Age,' *EWRT*, vol. 4, 445) does not satisfy him. He wants to know, as the poet Anne Carson puts it in another context, 'What it is like to be a human being in a family, in a fantasy, in longing, in a mistake.' Such a complex human understanding is crucial because Tagore has experienced human sorrow and knows what it means to stumble upon the earth. That is why, after the light of revelation, when he looks at the world again, it darkens. The 'demon of barbarity,' he bemoans in his final note, 'has given up all pretence and has emerged with unconcealed fangs, ready to tear up humanity in an orgy of devastation. From one end of the world to the other the poisonous fumes of hatred darken the atmosphere. The spirit of violence... has at last roused itself and desecrates the spirit of Man' ('Crisis in Civilization,' *EWRT*, vol. 3, 726). It is not surprising, then, that his last series of self-portraits, made when the world had once again plunged into its relentless wars, are blotted by dark shadows. In these self-reflective images, red blood-marks suddenly appear from nowhere to disfigure the face; one portrait acquires the rigidity of a death-mask; in another the surface of the skin is cut deeply by crevices before it finally flakes and decays; and in the next, the eyes seem to know so much horror that they can no longer express pity. There is no light that softens these images, no brightness of colours which gives them the quality of hope. The faces appear like unfamiliar ghosts still hovering between hope and death. Tagore's skepticism is too relentless to let him find easy comfort in the assertions of peace and beauty; his faith is too persistent to let him vanish with quiet acceptance into the heart of silence.

One has to admit that Tagore, despite the consistent notes of despair that can be heard throughout his works, is a visionary who struggles, like a righteous man of wisdom and love, to shake off the nightmares of his times and place, and become the advocate for all created things.[21] 'Against the tranquil light of morning,' he affirms in one of his last poems, 'I can see myself as a conqueror of sorrows' (Dutta and Robinson, 363). There is in him, as the following poem-painting he composed in Baghdad in 1932 indicates, an incandescence of faith which refuses to be extinguished. It is as if, like the small flame in the painting still burning in some wayside shrine, his works shall continue to add to the enigma of our survival on a violent earth and be the mysterious sources of hope which urge us to carry on.

> The night has ended
> Put out the light of the lamp
> of thine own narrow corner
> smudged with smoke
> the great morning which is for all
> appears in the East.
> Let its light reveal us to each other
> who walk on the same path
> of pilgrimage.[22]

Notes

1 Tagore, Rabindranath. *The English Writings of Rabindranath Tagore*, edited by Sisir Kumar Das, vol. 1. (New Delhi: Sahitya Akademi, 2008). 32. Hereafter, all further references to Tagore's English works are from the four volumes published by the Sahitya Akademi and are referred to as *EWRT*.

2 See, Leo Tolstoy, 'The Law of Love and the Law of Violence,' in *A Confession and Other Religious Writings*, translated by Jane Kentish. (Harmondsworth: Penguin, 1987). pp. 155-230. and, 'After the Ball,' in *The Kreutzer Sonata and Other Stories*, translated by David McDuff. (Harmondsworth: Penguin, 1985). pp. 255-266.

3 In a diary entry dated May 19, 1905, Tolstoy says: 'What is invaluable, important and difficult, is a good life.' Quoted by Wilson, A. N. *Tolstoy*. 1988. (New York: Ballantine Books, 1988), p. 473.

4 In *The Kreutzer Sonata and Other Stories*. 264.

5 When Chekhov met Tolstoy with Gorky in 1901-02, Chekhov's usual sense of irony did not allow him to become a disciple, but did not also prevent him from recognizing that Tolstoy was 'a great moral force for good.' See, Wilson, *Tolstoy*. 466.

6 For a translated version see http://www.nonresistance.org/docs_pdf/Tolstoy/Restoration_of_Hell.pdf.

7 In his emphasis on *caritas* as the highest of virtues Tolstoy clearly bases himself, as Mahatma Gandhi and Tagore were to do later, on the resonant Chapter 13 of *1 Corinthians* in the New Testament.

8 On conversation as a way of evading skepticism see Cavell, Stanley. *Conditions Handsome and Unhandsome: The Constitution of Emersonian Perfectionism,* (Chicago: University of Chicago Press, 1990).

9 Skepticism can, perhaps, be defined as that feeling of profound melancholy which burrows itself so deeply within us when we feel that life's finitude is 'inadequate to the demands of the spirit.' Life, as it moves through its diurnal course, seems strange to skeptical reason; it seems isolating, stifling or false to the one who does not know how to deal with human complexity. Emerson calls this refusal to struggle towards a life of the imagination as the 'unawakening' of the soul; Thoreau speaks about it as a life of mere 'business'; Nietzsche contemptuously dismisses it as 'philistinism.' In such a mood we begin to accept a life of injustice or violence as inevitable as death. For a detailed discussion of skepticism see Cavell, Stanley. *Philosophy the Day After Tomorrow.* (Cambridge, Mass.: Harvard University Press, 2005)pp. 44-51.

10 Dutta, Krishna and Andrew Robinson. 1995. *Rabindranath Tagore: The Myriad-Minded Man.* 1995. (London: Bloomsbury, 1995), p. 157.

11 See Tagore's story, 'Last Word,' (1939), in *Three Companions*. Translated by Sujit Mukherjee. 1992. (Hyderabad: Disha Books, 1992), p. 29.

12 I am thinking here of Francisco Goya's inspired etching, 'The Sleep of Reason Produces Monsters,' (1799) *Capricio*, No. 5.

13 Max Beerbhom draws a lovely parody of Tagore's portraits by his painterly disciples.

14 Tagore speaks of 'the mailed fist of earth-hunger' which has laid 'waste heaven's own kingdom ('The Message of the Forest,' *EWRT*, vol. 3, 393).

15 It is surprising to note that this collection, 'Thoughts from Rabindranath Tagore,' edited by C. F. Andrews. 1929 (*EWRT*, vol. 3), put together from Tagore's religious discourses at Santiniketan, is completely cleansed of his darker thoughts on colonialism, exploitation, and war which are so conspicuous in his lectures in Europe, China, Japan, and America throughout the twenties.

16 See, for example, many of the articles in *Rabindranath Tagore: Collection of Essays*. Edited by Rattan Parimoo. 1989; (New Delhi: Lalit Kala Akademi, 2008).

17 Hobsbawm, Eric. *Age of Extremes: The Short History of the Twentieth Century, 1914-1991.* New Delhi: Viking, 1994), p. 13.

18 The translated version published in *Fruit Gathering* is slightly different. See, poem no. 84, *EWRT*, vol. 1, 190.

19 'In Green India.' Translated by Ana Jelnikar and Barbara Siegal Carlson, in *Look Back, Look Ahead: The Selected Poems of Srečko Kosovel.* (New York: Ugly Ducking Press, 2010), p. 97.

20 Quoted from a speech delivered by Vaclav Havel on July 4, 1994 by Bruce Rich in *To Uphold the World: A Call for a New Global Ethic from Ancient India*. New York: Beacon Press, 2010. 169,

21 See, Benjamin, Walter. 'The Storyteller: Reflections on the Work of Nikolai Leskov,' *Illuminations: Essays and Reflections*. Translated by Harry Zohn. (New York: Schocken Books, 1969), p. 104.

22 *EWRT,* vol.4, p. 108. This printed version of the translation is slightly different from the one written in Tagore's own handwriting in the painting.

'A World-wide Commerce of Heart and Mind': Tagore as Cultural Envoy

Radha Chakravarty

'I love India,' declares Tagore, in *Letters to a Friend*, 'but my India is an Idea and not a geographical expression. Therefore, I am *not* a patriot – I shall ever seek my compatriots all over the world' (Das 1996: 294). In this statement, Tagore makes a distinction between two ways of looking at India: as a geopolitical entity, and as an idea. In the *idea* of India, he sees a potential link between the India he loves and the rest of the world. At this interface between the domestic and the international, Tagore locates his self-perceived role as a cultural ambassador. To think of Tagore as envoy in terms of current theories of international relations would of course amount to a distortion, for he was neither a political scientist, nor a diplomat. He did not formulate a systematic theory of national governance or international relations; in fact, after his initial enthusiasm for the Swadeshi movement, he shied away from active engagement with politics, and expressed his views through his creative writings instead. In place of administrative structures, he advocated people-to-people contact between nations. 'The human world is made one, all the countries are losing their distance every day,' he says, during a visit to Paris in 1930. '. . . Politicians struggle to exploit this great fact and wrangle about establishing trade relationships. But my mission is to urge for a world-wide commerce of heart and mind, sympathy and understanding.'[1] Tagore's evolving attitudes towards nationalism, internationalism and cosmopolitanism were not always consistent, and there were gaps between his theory and practice. All the same, his role as cultural mediator between nations remains worth considering, in terms of its possible value for international relations today. For Tagore, culture is not an alternative to politics, but an alternative approach to politics; not an answer to all the problems of the world, but a different way of posing the questions.

A compulsive traveller, Tagore visited numerous countries in his lifetime. He describes travel as his dharma: 'The heart wants to go on; that is its dharma, for unless it moves, it dies' (Tagore 2011: 762). He declares: 'Remaining in the same place, the same system of practices renders life immobile and senseless . . . without any awareness of this vast world. Now it is time to seek out the golden wand, to voyage out to faraway places. Now there is a need for a consciousness that will constantly knock on the closed doors of our eyes, ears and hearts with sundry new versions of the new, shredding our threadbare veil to reveal the ever-new' (ibid.). Tagore journeyed to other countries for a variety of reasons: for education, to cope with his loneliness, out of curiosity about other cultures, to raise funds for Visva-Bharati, or to present his works to audiences abroad; but often, he also self-consciously adopted the role of a cultural mediator. For instance, during his visit to Colombo in 1934, Tagore projects himself as a representative of Indian culture, declaring: 'I also come as a messenger from the land of Ram Chandra, to build to the best of my ability a bridge, not with stones and rocks but with expressions of beauty in poems, songs and dancing which are the best materials for the path of communication that may span the differences between two peoples'.[2] During his sojourn in South East Asia, in the poem 'To Siam, he salutes Thailand's ancient civilizational links with India:

> I come, a pilgrim, at thy gate, O Siam,
> to offer my verse to the endless glory of India
> sheltered in thy home, away from her own deserted shrine,
> to bathe in the living stream that flows in thy heart . . .
>
> (Das Gupta 128)

In Teheran, he was greeted as 'the greatest star shining in the eastern sky' (Dutta and Robinson, 1995: 317). The nations of the West perceived Tagore in a similar light, as a sage from the East with a mystical message for mankind. In his Nobel Prize Acceptance Speech (1921), Tagore said: 'I know that I must not accept that praise as my individual sphere. It is the East in me which gave to the West' (Tagore 2011: 186).

Yet Rabindranath's self-representation as cultural envoy raises several questions. How, for instance, can he be regarded as India's representative when he interrogates the very idea of nationhood? Tagore, once a champion of the Swadeshi movement, had, after all, withdrawn from active participation in it when it took a violent, communal turn. In his eyes, the modern nation state, a model adopted from the West, is a limited ideal, for at its core resides a divisive politics, one based on hunger for power and domination rather than cooperation and understanding. In *Nationalism* (1917) he says: 'A political civilization has sprung up from the soil of Europe and is overrunning the world . . . It is carnivorous and cannibalistic in its tendencies, it feeds upon the resources of other peoples and tries to swallow their whole future' (Das 1996: 440). The India Tagore seeks to represent is not so much the geopolitical entity defined in terms of Western paradigms of nationhood, but rather, a cultural construct associated with a cluster of ideas, values and practices. In the novel *Gora*, this is the image of 'Bharatvarsha' that emerges so powerfully. As Ramachandra Guha says, 'Tagore was a patriot who loved his country, without being a nationalist who saw his nation as necessarily superior to other nations' (Guha: 177).

Tagore sees the process of cultural mediation as a mutual exchange. Seeking to generate awareness about Indian culture, he also remains open to influences from abroad. His syncretic genius is eminently suited to the task of combining elements from diverse cultures, and channelizing them in new creative directions. This is no mere aesthetic exercise. At its core lies a profound philosophy of humanism, a conviction that mutual tolerance is the way forward for the world. In the language of literature, art and music, he communicates with the peoples of the world. In this respect, he anticipates recent theories of intercultural communication. According to Kanwar B.

Mathur, 'Intercultural communication is the process of exchange of thoughts and meaning between people of different cultures which despite many inner diversities, nevertheless reveal a substantial unity in basic beliefs and forms of experience – in customs, norms and behaviourial characteristics' (Mathur 37).[3]

But such a goal is not easy to achieve, for power operates at the interfaces between cultures, sometimes in ways that generate hostility rather than cooperation. According to Milton Cummings, cultural diplomacy is 'the exchange of ideas, information, art and other aspects of culture among nations and their peoples in order to foster mutual understanding.' But Cummings warns us that this 'can also be more of a one-way street than a twoway exchange, as when one nation concentrates its efforts on promoting the national language, explaining its policies and point of view, or 'telling its story' to the rest of the world' (Cummings: 1). In Tagore's eyes, imperialism provides an instance of such one-sided cultural dominance. His own ideas of cultural interaction are premised, not on the ideal of an imposed homogeneity, but upon mutual tolerance and acceptance of difference and heterogeneity. In the essay 'Bharatvarsher Itihas' (1902), he says: 'We can see that the aim of Bharatvarsha has always been to establish unity amidst differences . . . If there be genuine differences, it is possible to accommodate in its appropriate place such differences. You cannot legislate unity into existence. Elements which cannot assimilate need to be recognised and put in their appropriate separate places . . . . keeping them separate and at the same time engaged in a common task.'[4] In the dialectic of difference and connectedness, Tagore found his mantra. This is also the mantra of Indian democracy: unity in diversity.

Difference, though, can also lead to hostility between cultures. According to Mathur, 'The extent of intercultural understanding, misunderstanding or non-understanding is determined by the extent of likeness and difference in frames of reference, value systems, or world views of the cultures involved, from their cognitive and affective distance from each other.' Tagore was not unaware of such tensions. While advocating the theoretical goal of unity in diversity, in practice he encountered such failures of understanding at several moments in his career. For instance, the views expressed in the essays compiled in *Nationalism* (1917) met with a hostile response from several sectors in the West. His critique of Japan's attempt to imitate Western models of progress also generated a negative response there. In China, too, he received a mixed reception, for though he had many admirers there, his emphasis on civilizational ties was read by the progressive radicals as a backward-looking stance opposed to their ideology of change.

Tagore remained convinced, all the same, that humanism was a greater ideal than nationalism. He was also sure that a resurgent Asia would lead the world in this direction, in the future. His vision was premised on a worldview that saw tradition and modernity, not as antithetical forces, but at a continuum where change could be premised on connectedness with the past. It was also a worldview that saw no essential contradiction between rootedness in one's own culture, and openness to the influence of others. This is sometimes described as a 'rooted cosmopolitanism', in the sense invoked by K Anthony Appiah – an acknowledgement of multiple affiliations, based primarily on local roots, but branching out into a sense of the interconnectedness of the different parts of the world (Appiah 1996) .

Tagore's experiment with Visva-Bharati exemplifies this approach. The university, premised on the ancient Indian 'Tapovan' model of learning in harmony with nature and the environment, was devoted to the promotion of international learning. Tagore explains this idea in the essay 'Vidya Samavaya' (Aggregation of Knowledge, 1919): 'gone are the days for nurturing exclusive national systems of knowledge in isolation. The age of cooperative learning has arrived. . . . Hence we need an extensive field for educational cooperation in our country, where forms of Indian knowledge must be judged in the context of the evolution of human knowledge as a whole'.[5]

In his travels abroad, while searching for funds to support his dream university, Tagore interacted with large number of people, including major creative, intellectual and political personages across the world. In his interactions, he actively promoted Indian music dance, art and literature. His approach, at times, seems to anticipate the recent idea of 'soft power', the power to attract and persuade through popular outreach, promoted in the field of international relations by Joseph Nye.[6] But this would be a limited understanding of his approach, for during his sojourns in foreign countries, Tagore also absorbed influences from other cultures, and made them part of his own distinctive oeuvre. For instance, he imported batik from Indonesia into the rural reconstruction scheme of Sriniketan, and incorporated elements of Balinese dance and Kandyan dance from Sri Lanka into Rabindra Nritya, the dance form associated with Santiniketan. Rabindrasangeet is a unique blend of indigenous and foreign elements, and Tagore's experiments with Haiku are well known.

Tagore had a vision, but he did not achieve all the goals he had set for the India of his dreams. At the end of his life, in *Crisis of Civilization*, he lamented: 'As I look around I see the crumbling ruins of a proud civilization strewn like a vast heap of futility. . . . From one end of the world to the other the poisonous fumes of hatred darken the atmosphere. . . . The wheels of fate will some day compel the English to give up their Indian empire. But what kind of India will they leave behind, what stark misery?' (Tagore 2011: 215). Tagore ends on a note of affirmation: 'And yet I shall not commit the grievous sin of losing faith in man' (ibid.). All the same, his vision of the present appears bleak.

Tagore did not live to fulfil his goals, but his vision had a profound impact on those who came after. Humayun Kabir describes Tagore as 'the first great Indian in recent times who went out on a cultural mission for restoring contacts and establishing friendships with peoples of other countries without any immediate or specific educational, economic, political or religious aim. It is also remarkable that his cultural journeys were not confined to the western world'.[7] Upon Nehru, the architect of India's foreign policy, Tagore exercised a powerful influence. According to Sunil Khilnani, 'it was from Tagore's civilizational views that Nehru drew intellectual inspiration. The legend of the Indian past conjured by Nehru exemplifies many of Tagore's beliefs about India's distinctive capacity to absorb different influences and to make something new out of them' (Khilnani xxiii). When Nehru was leaving on a journey to China in 1939, Tagore wrote to him, exhorting him to be the torch bearer in the field of India's diplomatic relations with other countries. He said: 'The pathway which led from India to her great neighbours is partly closed through centuries of neglect; we have to remove the weeds, and also the recent barriers erected by fratricidal politics so that once more the traffic of human interchange can continue, linking our country with Japan and China. India's great awakening had crossed deserts and mountains, the overflow of her glorious epoch of culture touched far continents and left permanent deposits in distant shores of Asia. . . . I cannot help hoping that as a messenger from India's youth you would give strength to the historic forces of Asiatic unity, bringing new urge of neighbourly understanding to our eastern peoples' (Tagore 2005: 512-3). According to Sib S. Ray 'What Tagore started continued after him. Jawaharlal Nehru, a great admirer of Tagore, visited China in 1939.. . . After Independence Nehru's foreign policy towards China was largely influenced by Tagore's vision. We can trace it to his message, sent through his daughter, Indira Nehru, on the occasion of the inauguration of Cheena Bhavana: '. . . it will forge a new link which will enable our two countries, India and China, to come closer to each other in due time" (Sib S. Ray: 137). In *The Discovery of India*, Nehru writes: '*Nationalism* is a narrowing creed, and nationalism in conflict with dominating imperialism produces all manner of frustrations and complexes. It was Tagore's immense service to India, as it has been Gandhi's in a different plane, that he forced the people in some measure out of their narrow grooves of thought and made them think of broader issues affecting humanity' (Nehru 1982: 340).

Tagore, however, was more than a representative of India. He saw himself as a representative of mankind, and his influence on other cultures, in some parts of the world, was immense. He was the only poet in the world to have composed songs that were adopted as the national anthems of two nations, India and Bangladesh. He also influenced the anthem of Sri Lanka, and inspired a cultural renaissance there. Bangladesh did not exist as an independent nation during Tagore's lifetime, but the region, then known as East Bengal, had a formative influence on Tagore's life and works; it was while managing the family estates here that he developed a first-hand knowledge of the lives of ordinary people and their problems, and also realized his fascination for the landscape that would inspire some of his best poetry and music. In the evolution of Bangladeshi nationalism and the emergence of Bangladesh as a independent nation, the ideas of Tagore, and the very act of singing Tagore songs, played a crucial political role. Today, Tagore forms a vital cultural link between Bangladesh and India; the joint celebrations of his 150th birth anniversary constitute a tribute to Tagore's own credo of cultural contact and cooperation.

Tagore's ideas also anticipate some of the key principles of the United Nations Educational Scientific and Cultural Organization (UNESCO). The UNESCO's Declaration of the Principles of International Cultural Cooperation (1966) states: 'In their rich variety and diversity, and in the reciprocal influences they exert on one another, all cultures form part of the common heritage belonging to mankind. . . International cultural cooperation shall cover all aspects of intellectual and creative activities relating to education, science and culture' (Mathur 41). Its aims include the following: 'to spread knowledge, to stimulate and to enrich culture . . . to develop peaceful relations and friendship among the peoples and bring about a better understanding of each other's way of life . . . to enable everyone to have access to knowledge, to enjoy arts and literature of all peoples, to share in advances made in science in all parts of the world and in the resulting benefits, and to contribute to the enrichment of cultural life . . . [and] to raise the level of spiritual and moral life of man in all parts of the world.' The Declaration also advocates mutual respect between cultures: 'International cooperation shall respect the distinctive character of each. . . .In intercultural cooperation stress shall be laid on ideas and values conducive to the creation of a climate of friendship and peace' (Mathur 42). This advocacy of cooperation based on mutual respect is also Tagore's credo.

Above all, Tagore valued freedom, a concept also at the heart of Nehru's vision of independent India. 'What does independence consist of?' asks Nehru, adding: 'It consists fundamentally and basically of foreign relations. That is the test of independence. All else is local autonomy' (Nehru 1949: 237). He declares: 'A free India would inevitably play a growing part in international affairs' (Nehru 1978, 55). Such was also Tagore's vision: 'Where the mind is without fear and the head is held high,/ Into that heaven of freedom, my father, let my country awake' (Das 1994: 53).

But then, for Tagore, his country was not just India, but the world at large, and the universe. To his dying day, Tagore never ceased to wonder at his place in the greater scheme of things. Beyond the nation, beyond the human world, he was aware of a vaster reality, of his oneness with it. That was the source of his wonder – best evoked in song:

> Stars fill the sky, the world teems with life,
> And amidst it all I find my place!
> I wonder, and so I sing.'[8]
>
> (Tagore 2011: 347)

Notes

1 Tagore, holograph message dated 3 May 1930, reproduced in *Indo-Asian Culture* 10:1 (Kolkata, July 1961) 115; cited in Hay 331.

2 Speech delivered at Jaffna Central College, 12 June,1934. Reported in *Ceylon Observer*, 13 June 1934, cited in *Rabindranath Tagore in Ceylon*, ed. Kalidas Nag, (a pamphlet published by the Calcutta Art Society in 1944 on the occasion of presenting to Ceylon an oil portrait of Tagore), p.5.

3 Mathur also cites Gerhard Maletzke, 'International Communication in International and Intercultural Communication', in Hans Dietrich Fischer and John Merrill, *Forms and Functions of Supranational Communication in International and Intercultural Communication.* (New York: Hastings House, 1976).

4 Tagore, 'Bharatvarsher Itihas' (The history of India), *Bangadarshan* (1902), rept. *Itihas*, pp. 10-11, trans. Sabyasachi Bhattacharya.

5 Tagore, 'Vidya Samavaya', p. 302. Translation Radha Chakravarty.

6 See Joseph Nye, *Soft Power: The Means to Success in World Politics* (2004).

7 Humayun Kabir, 'Introduction', *Towards Universal Man*, p. 24.

8 Translated by Ratna Prakash.

References

Appiah, Kwame Anthony. 'Cosmopolitan Patriots'. *For the Love of the Country*, ed. Joshua Cohen. (Boston: Beacon Press, 1996), p. 21-30.

Cummings, Milton C. *Cultural Diplomacy and the United States Government: A Survey*. (Washington, D.C.: Centre for Arts and Culture, 2003).

Das, Sisir K. *The English Writings of Rabindranath Tagore. Vols.1, 2 and 3.* 1994-1996.

Das Gupta, Shakti. *Tagore's Asian Outlook.* (Calcutta: Nava Bharati, 1961).

Dutta, Krishna and Andrew Robinson. *Rabindranath Tagore: The Myriad Minded Man. (*London: Bloomsbury, 1995).

Guha, Ramachandra. 'Travelling with Tagore'. *Theorizing the Present: Essays for Partha Chatterjee*, ed. Anjan Ghosh, Tapati-Thakurta, and Janaki Nair. (New Delhi: Oxford University Press, 2011), pp. 152-187.

Hay, Stephen N. *Asian Ideas of East and West: Tagore and His Critics in Japan, China and India.* (Cambridge, Mass.: Harvard University Press, 1970).

Kabir, Humayun. 'Introduction'. Rabindranath Tagore, *Towards Universal Man.* (Bombay: Asia Publishing House, 1961).

Khilnani, Sunil. 'Introduction'. Jawaharlal Nehru, *The Discovery of India*. 1946. (New Delhi: Penguin India, 2010). pp. xv-xxvi.

Mathur, Kanwar B. *Intercultural Communication: Agenda for Developing Countries*. (New Delhi: Allied Publishers, 2001).

Nag, Kalidas, ed. *Rabindranath Tagore in Ceylon*. Pamphlet published by the Calcutta Art Society, 1944.

Nehru, Jawaharlal. *Jawaharlal Nehru's Speeches*, Vol. 1. (New Delhi: Govt. of India, Publications Division, 1949).

____. *Selected Works,* First Series, vol. 11. (New Delhi: Orient Longman, 1978).

____. *The Discovery of India*. (Delhi: Jawaharlal Nehru Memorial Fund/ Oxford University Press, 1982).

Nye, Joseph. *Soft Power: The Means to Success in World Politics.* (New York: Public Affairs, 2004).

Ray, Sib. S. 'Rabindranath Tagore in China.' *Celebrating Tagore*. Ed. Rama Datta and Clinton Seely. (New Delhi: Allied Publishers, 2009), p. 127-138.

Tagore, Rabindranath. *Itihas* (Bengali). (Kolkata: Visva-Bharati, 1955).

____.'Vidya Samavaya' (Bengali). *Rabindra Rachanabali* (Sulabh Sanskaran) Vol. 16. (Kolkata: Visva-Bharati, 2000), p. 301-3.

____. *Selected Letters of Rabindranath Tagore.* Ed. Krishna Dutta and Andrew Robinson. (New Delhi: Foundation Books/Cambridge Univ. Press, 2005).

____. *The Essential Tagore*. Ed. Fakrul Alam and Radha Chakravarty. (Belknap Press/Harvard University Press, 2011).

www.ingramcontent.com/pod-product-compliance
Lightning Source LLC
Chambersburg PA
CBHW081415220926
51613CB00010B/196
9789382652953